MW01204598

A Course *In* Game Theory

A Course *In* Game Theory

Thomas S Ferguson

UCLA

World Scientific

EW JERSEY • LONDON • SINGAPORE • BEIJING • SHANGHAI • HONG KONG • TAIPEI • CHENNAI • TOKYO

Published by

World Scientific Publishing Co. Pte. Ltd.
5 Toh Tuck Link, Singapore 596224
USA office: 27 Warren Street, Suite 401-402, Hackensack, NJ 07601
UK office: 57 Shelton Street, Covent Garden, London WC2H 9HE

Library of Congress Control Number: 2020029665

British Library Cataloguing-in-Publication Data
A catalogue record for this book is available from the British Library.

A COURSE IN GAME THEORY

ISBN 978-981-3227-34-7 (hardcover)
ISBN 978-981-3227-36-1 (ebook for institutions)
ISBN 978-981-3227-37-8 (ebook for individuals)

For any available supplementary material, please visit
https://www.worldscientific.com/worldscibooks/10.1142/10634#t=suppl

Typeset by Stallion Press
Email: enquiries@stallionpress.com

Preface

This book is intended as an introduction to game theory for undergraduate students. It has many exercises with solutions and so is also suitable for self-study by the student working alone. There is enough material in the text for a year course in game theory. However, for a semester or quarter course, some choice must be made. So several courses may be considered. Part I on combinatorial games is not explicitly used in the the rest of the text and so may be omitted. I suggest three possible courses.

1. Mathematical Games: Part I and the first four or five sections on Part II. The prettiest mathematics appears in this course.
2. Classical Game Theory; All of Part II, possibly with some use of Part I as an introduction. A course on two-person zero-sum games.
3. Standard Game Theory: The first four or five chapters of Part II, plus all of Parts III and IV. A course on general games.

The creation of this book is a project that has been evolving for over 40 years. Back in the 1960's the Mathematics Department at UCLA had a course entitled "Game Theory and Linear Programming" that was taught each quarter. I taught this course from time to time, but since I had a love of game theory, I eventually reduced the Linear Programming part to 4 weeks to allow me to devote six weeks to game theory. (The 4 week course has appeared on my web page as an electronic text under the title "Linear Programming — A Concise Introduction".)

In the 1970s, when the Mathematics Department offered a full quarter course in Game Theory, work on the present text was begun. Various preliminary versions were put on the web, but the "first edition" appeared as a pdf file in 1994 under the simple title "Game Theory". There are advantages for placing the text on the web. First, readers may take the trouble to write to you pointing out errors, misprints, and suggestions for clarity. Second, it allows one to add explanations, replace an example with a more interesting one and add new exercises. The only major change to the first edition occurred in 2003 with the addition of the chapter on Coin-Turning Games.

The "second edition" of this book appeared on my web page in 2014. The main changes made were the following. First, material on the Fictitious Play Algorithm to approximate the value and optimal strategies in two-person zero-sum games has been added in Part II, Chapter 4, Section 6. Second, the material in Part II, Chapter 7 on Poker Models has been replaced by a more general treatment of Infinite Games, with Poker Models reduced to a single section. Third, the very simple proof of the Minimax Theorem, due to Guillermo Owen, was placed in the Appendix. All other changes made were for correcting simple errors and misprints and for clarification of certain obscurities.

For this we thank many readers. First, there are the many students who took the course and whose suggestions and complaints have improved the readability of the text greatly. Second, there are the many teachers at diverse universities across the country who have used some part of the text in their classrooms, and sent me opinions and suggestions for improvement. Third, there are individual students around the world who saw the text on the web and asked me for help or clarification. To all these, I am very grateful. Finally I had the good fortune to have had Lloyd Shapely as a friend and colleague for many years; I have learned much from him. His influence has shaped much of the outlook on game theory that appears in this book.

Introduction

Game theory is a fascinating subject. We all know many entertaining games, such as chess, poker, tic-tac-toe, bridge, baseball, computer games — the list is quite varied and almost endless. In addition, there is a vast area of economic games, discussed in Myerson (1991) and Kreps (1990), and the related political games, Ordeshook (1986), Shubik (1982), and Taylor (1995). The competition between firms, the conflict between management and labor, the fight to get bills through congress, the power of the judiciary, war and peace negotiations between countries, and so on, all provide examples of games in action. There are also psychological games played on a personal level, where the weapons are words, and the payoffs are good or bad feelings, Berne (1964). There are biological games, the competition between species, where natural selection can be modeled as a game played between genes, Smith (1982). There is a connection between game theory and the mathematical areas of logic and computer science. One may view theoretical statistics as a two-person game in which nature takes the role of one of the players, as in Blackwell and Girshick (1954) and Ferguson (1968).

Games are characterized by a number of players or decision makers who interact, possibly threaten each other and form coalitions, take actions under uncertain conditions, and finally receive some benefit or reward or possibly some punishment or monetary loss. In this text, we present various mathematical models of games and study the phenomena that arise. In some cases, we will be able to suggest what courses of action should be taken by the players. In others, we hope simply to be able to understand what is happening in order to make better predictions about the future.

As we outline the contents of this text, we introduce some of the keywords and terminology used in game theory. First there is the *number of players* which will be denoted by n. Let us label the players with the integers 1 to n, and denote the *set of players* by $N = \{1, 2, \ldots, n\}$. We study mostly two-person games, $n = 2$, where the concepts are clearer and the conclusions are more definite. When specialized to one-player, the theory is simply called decision theory. Games of solitaire and puzzles are examples of one-person games as are various sequential optimization problems found in operations research, and optimization, (see Papadimitriou and Steiglitz (1982) for example), or linear programming, (see Chvátal (1983)), or gambling (see Dubins and Savage (1965)). There are even things called "zero-person games", such as the "game of life" of Conway (see Berlekamp *et al.* (1982) Chap. 25); once an automaton gets set in motion, it keeps going without any person making decisions. We assume throughout that there are at least two players, that is, $n \geq 2$. In macroeconomic models, the number of players can be very large, ranging into the millions. In such models it is often preferable to assume that there are an infinite number of players. In fact it has been found useful in many situations to assume there are a continuum of players, with each player having an infinitesimal influence on the outcome as in Aumann and Shapley (1974). (Incidentally, both authors were later to win Nobel Prizes in Economics.) In this course, we take n to be finite.

There are three main mathematical models or forms used in the study of games, the *extensive form*, the *strategic form* and the *coalitional form*. These differ in the amount of detail on the play of the game built into the model. The most detail is given in the extensive form, where the structure closely follows the actual rules of the game. In the extensive form of a game, we are able to speak of a *position* in the game, and of a *move* of the game as moving from one position to another. The set of possible moves from a position may depend on the player whose turn it is to move from that position. In the extensive form of a game, some of the moves may be *random moves*, such as the dealing of cards or the rolling of dice. The rules of the game specify the probabilities of the outcomes of the random moves. One may also speak of the *information* players have when they move. Do they know all past moves in the game by the other players? Do they know the outcomes of the random moves?

When the players know all past moves by all the players and the outcomes of all past random moves, the game is said to be of **perfect information**. Two-person games of perfect information with win or lose outcome and no chance moves are known as *combinatorial games.* There is a beautiful and deep mathematical theory of such games. You may find an exposition of it in Conway (1976) and in Berlekamp *et al.* (1982). Such a game is said to be *impartial* if the two players have the same set of legal moves from each position, and it is said to be *partizan* otherwise. Part I of this text contains an introduction to the theory of impartial combinatorial games. For another elementary treatment of impartial games see the book by Guy (1989).

We begin Part II by describing the *strategic form* or normal form of a game. In the strategic form, many of the details of the game such as position and move are lost; the main concepts are those of a strategy and a payoff. In the strategic form, each player chooses a strategy from a set of possible strategies. We denote the *strategy set* or *action space* of player i by A_i, for $i = 1, 2, \dots, n$. Each player considers all the other players and their possible strategies, and then chooses a specific strategy from his strategy set. All players make such a choice simultaneously, the choices are revealed and the game ends with each player receiving some payoff. Each player's choice may influence the final outcome for all the players.

We model the payoffs as taking on numerical values. In general the payoffs may be quite complex entities, such as "you receive a ticket to a baseball game tomorrow when there is a good chance of rain, and your raincoat is torn". The mathematical and philosophical justification behind the assumption that each player can replace such payoffs with numerical values is discussed in the Appendix under the title, *Utility Theory.* This theory is treated in detail in the books of Savage (1954) and of Fishburn (1988). We therefore assume that each player receives a numerical payoff that depends on the actions chosen by all the players. Suppose player 1 chooses $a_1 \in A_1$, player 2 chooses $a_2 \in A_2$, etc. and player n chooses $a_n \in A_n$. Then we denote the payoff to player j, for $j = 1, 2, \dots, n$, by $f_j(a_1, a_2, \dots, a_n)$, and call it the *payoff function for player* j.

The *strategic form of a game* is defined then by the three objects:

(1) the set, $N = \{1, 2, \dots, n\}$, of players,
(2) the sequence, $A_1, \dots, A_n$, of strategy sets of the players, and
(3) the sequence, $f_1(a_1, \dots, a_n), \dots, f_n(a_1, \dots, a_n)$, of real-valued payoff functions of the players.

A game in strategic form is said to be *zero-sum* if the sum of the payoffs to the players is zero no matter what actions are chosen by the players. That is, the game is zero-sum if

$$\sum_{i=1}^{n} f_i(a_1, a_2, \ldots, a_n) = 0$$

for all $a_1 \in A_1$, $a_2 \in A_2, \ldots, a_n \in A_n$. In the first four chapters of Part II, we restrict attention to the strategic form of finite, two-person, zero-sum games. Such a game is said to be *finite* if both the strategy sets are finite sets. Theoretically, such games have clear-cut solutions, thanks to a fundamental mathematical result known as the *minimax theorem.* Each such game has a *value*, and both players have *optimal strategies* that guarantee the value.

In the last three chapters of Part II, we treat two-person zero-sum games in extensive form, and show the connection between the strategic and extensive forms of games. In particular, one of the methods of solving extensive form games is to solve the equivalent strategic form. Here, we give an introduction to Recursive Games and Stochastic Games, an area of intense contemporary development (see Filar and Vrieze (1997), Maitra and Sudderth (1996) and Sorin (2002)). In the last chapter, we investigate the problems that arise when at least one of the strategy sets of the players is an infinite set.

In Part III, the theory is extended to two-person *non-zero-sum* games. Here the situation is more nebulous. In general, such games do not have values and players do not have optimal strategies. The theory breaks naturally into two parts. There is the *noncooperative theory* in which the players, if they may communicate, may not form binding agreements. This is the area of most interest to economists, see Gibbons (1992), and Bierman and Fernandez (1993), for example. In 1994, John Nash, John Harsanyi and Reinhard Selten received the Nobel Prize in Economics for work in this area. Such a theory is natural in negotiations between nations when there is no overseeing body to enforce agreements, and in business dealings where companies are forbidden to enter into agreements by laws concerning constraint of trade. The main concept, replacing value and optimal strategy is the notion of a *strategic equilibrium*, also called a *Nash equilibrium.* This theory is treated in the first three chapters of Part III.

On the other hand, in the *cooperative theory* the players are allowed to form binding agreements, and so there is strong incentive to work together to receive the largest total payoff. The problem then is how to split the total payoff between or among the players. This theory also splits into two parts.

If the players measure utility of the payoff in the same units and there is a means of exchange of utility such as *side payments*, we say the game has *transferable utility*; otherwise *non-transferable utility*. The last chapter of Part III treat these topics.

When the number of players grows large, even the strategic form of a game, though less detailed than the extensive form, becomes too complex for analysis. In the *coalitional form* of a game, the notion of a strategy disappears; the main features are those of a *coalition* and the *value* or *worth* of the coalition. In many-player games, there is a tendency for the players to form coalitions to favor common interests. It is assumed each coalition can guarantee its members a certain amount, called the value of the coalition. The coalitional form of a game is a part of cooperative game theory with transferable utility, so it is natural to assume that the *grand coalition*, consisting of all the players, will form, and it is a question of how the payoff received by the grand coalition should be shared among the players. We will treat the coalitional form of games in Part IV. There we introduce the important concepts of the *core* of an economy. The core is a set of payoffs to the players where each coalition receives at least its value. An important example is two-sided matching treated in Roth and Sotomayor (1990). We will also look for principles that lead to a unique way to split the payoff from the grand coalition, such as the *Shapley value* and the *nucleolus*. This will allow us to speak of the power of various members of legislatures. We will also examine cost allocation problems (how should the cost of a project be shared by persons who benefit unequally from it).

Related Texts. There are many texts at the undergraduate level that treat various aspects of game theory. Accessible texts that cover certain of the topics treated in this text are the books of Straffin (1993), Morris (1994) and Tijs (2003). The book of Owen (1982) is another undergraduate text, at a slightly more advanced mathematical level. The economics perspective is presented in the entertaining book of Binmore (1992). The New Palmgrave book on game theory, Eatwell *et al.* (1987), contains a collection of historical sketches, essays and expositions on a wide variety of topics. Older texts by Luce and Raiffa (1957) and Karlin (1959) were of such high quality and success that they have been reprinted in inexpensive Dover Publications editions. The elementary and enjoyable book by Williams (1966) treats the two-person zero-sum part of the theory. Also recommended are the lectures on game theory by Robert Aumann (1989), one of the leading scholars of the field. And last, but actually first, there is the book by von Neumann and Morgenstern (1944), that started the whole field of game theory.

Contents

Preface v

Introduction vii

Part I. Impartial Combinatorial Games 1

1. Take-Away Games 3
   - 1.1 A Simple Take-Away Game 4
   - 1.2 What is a Combinatorial Game? 4
   - 1.3 P-positions, N-positions 5
   - 1.4 Subtraction Games 6
   - 1.5 Exercises 7

2. The Game of Nim 11
   - 2.1 Preliminary Analysis 11
   - 2.2 Nim-Sum 12
   - 2.3 Nim with a Larger Number of Piles 13
   - 2.4 Proof of Bouton's Theorem 13
   - 2.5 Misère Nim 14
   - 2.6 Exercises 15

3. Graph Games 18
   - 3.1 Games Played on Directed Graphs 18
   - 3.2 The Sprague-Grundy Function 19
   - 3.3 Examples 20
   - 3.4 The Sprague-Grundy Function on More General Graphs 22
   - 3.5 Exercises 24

4. Sums of Combinatorial Games 27
4.1 The Sum of n Graph Games 27
4.2 The Sprague-Grundy Theorem 28
4.3 Applications 29
4.4 Take-and-Break Games 30
4.5 Exercises 34

5. Coin Turning Games 37
5.1 Examples 38
5.2 Two-dimensional Coin-Turning Games 41
5.3 Nim Multiplication 44
5.4 Tartan Games 45
5.5 Exercises 49

6. Green Hackenbush 52
6.1 Bamboo Stalks 52
6.2 Green Hackenbush on Trees 53
6.3 Green Hackenbush on General Rooted Graphs 56
6.4 Exercise 58

Part II. Two-Person Zero-Sum Games 59

7. The Strategic Form of a Game 61
7.1 Strategic Form 61
7.2 Example: Odd or Even 62
7.3 Pure Strategies and Mixed Strategies 64
7.4 The Minimax Theorem 65
7.5 Exercises 66

8. Matrix Games — Domination 68
8.1 Saddle Points 69
8.2 Solution of All 2×2 Matrix Games 69
8.3 Removing Dominated Strategies 71
8.4 Solving $2 \times n$ and $m \times 2$ Games 73
8.5 Latin Square Games 74
8.6 Exercises 75

9. The Principle of Indifference 79
9.1 The Equilibrium Theorem 80
9.2 Nonsingular Game Matrices 82
9.3 Diagonal Games 84

9.4 Triangular Games 85
9.5 Symmetric Games 86
9.6 Invariance 88
9.7 Exercises 94

10. Solving Finite Games 100
10.1 Best Responses 101
10.2 Upper and Lower Values of a Game 102
10.3 Invariance under Change of Location and Scale 104
10.4 Reduction to a Linear Programming Problem 105
10.5 Description of the Pivot Method for Solving Games . . . 108
10.6 A Numerical Example 109
10.7 Approximating the Solution: Fictitious Play 112
10.8 Exercises 115

11. The Extensive Form of a Game 118
11.1 The Game Tree 118
11.2 Basic Endgame in Poker 119
11.3 The Kuhn Tree 121
11.4 The Representation of a Strategic Form Game in Extensive Form 124
11.5 Reduction of a Game in Extensive Form to Strategic Form 125
11.6 Example 126
11.7 Games of Perfect Information 128
11.8 Behavioral Strategies 129
11.9 Exercises 130

12. Recursive and Stochastic Games 134
12.1 Matrix Games with Games as Components 134
12.2 Multistage Games 136
12.3 Recursive Games. ϵ-Optimal Strategies 140
12.4 Stochastic Movement Among Games 144
12.5 Stochastic Games 146
12.6 Approximating the Solution 149
12.7 Exercises 150

13. Infinite Games 153
13.1 The Minimax Theorem for Semi-Finite Games 153
13.2 Continuous Games 158

13.3 Concave Games and Convex Games 160
13.4 Solving Games 162
13.5 Uniform [0,1] Poker Models 166
13.6 Exercises 171

Part III. Two-Person General-Sum Games 177

14. Bimatrix Games — Safety Levels 179
14.1 General-Sum Strategic Form Games 179
14.2 General-Sum Extensive Form Games 180
14.3 Reducing Extensive Form to Strategic Form 181
14.4 Overview 182
14.5 Safety Levels 182
14.6 Exercises 184

15. Noncooperative Games 186
15.1 Strategic Equilibria 186
15.2 Examples 189
15.3 Finding All PSE's 192
15.4 Iterated Elimination of Strictly Dominated Strategies 192
15.5 Exercises 194

16. Models of Duopoly 197
16.1 The Cournot Model of Duopoly 197
16.2 The Bertrand Model of Duopoly 200
16.3 The Stackelberg Model of Duopoly 202
16.4 Entry Deterrence 204
16.5 Exercises 206

17. Cooperative Games 208
17.1 Feasible Sets of Payoff Vectors 208
17.2 Cooperative Games with Transferable Utility 210
17.3 Cooperative Games with Non-Transferable Utility 215
17.4 End-Game with an All-In Player 222
17.5 Exercises 225

Part IV. Games in Coalitional Form 229

18. Many-Person TU Games 231
18.1 Coalitional Form. Characteristic Functions 231
18.2 Relation to Strategic Form 232

18.3 Constant-Sum Games 234
18.4 Example 234
18.5 Exercises 235

19. Imputations and the Core 237
19.1 Imputations 237
19.2 Essential Games 238
19.3 The Core 239
19.4 Examples 240
19.5 Exercises 241

20. The Shapley Value 243
20.1 Value Functions — The Shapley Axioms 243
20.2 Computation of the Shapley Value 246
20.3 An Alternative Form of the Shapley Value 246
20.4 Simple Games. The Shapley–Shubik Power Index 248
20.5 Exercises 250

21. The Nucleolus 255
21.1 Definition of the Nucleolus 255
21.2 Properties of the Nucleolus 258
21.3 Computation of the Nucleolus 259
21.4 Exercises 260

Appendix 1 Utility Theory 263

Appendix 2 Owen's Proof of the Minimax Theorem 269

Appendix 3 Contraction Maps and Fixed Points 271

Appendix 4 Existence of Equilibria in Finite Games 275

Solutions to Exercises of Part I 279
Solutions to Chap. 1 280
Solutions to Chap. 2 283
Solutions to Chap. 3 286
Solutions to Chap. 4 291
Solutions to Chap. 5 295
Solution to Chap. 6 299

Solutions to Exercises of Part II 301
Solutions to Chap. 7 302
Solutions to Chap. 8 304
Solutions to Chap. 9 308
Solutions to Chap. 10 317
Solutions to Chap. 11 321
Solutions to Chap. 12 327
Solutions to Chap. 13 334

Solutions to Exercises of Part III 341
Solutions to Chap. 14 342
Solutions to Chap. 15 344
Solutions to Chap. 16 349
Solutions to Chap. 17 354

Solutions to Exercises of Part IV 359
Solutions to Chap. 18 360
Solutions to Chap. 19 363
Solutions to Chap. 20 366
Solutions to Chap. 21 374

Solutions to Exercises of Appendix 1 381

References 383

Index 389

Part I

Impartial Combinatorial Games

1 Take-Away Games

Combinatorial games are two-person games with perfect information and no chance moves, and with a win-or-lose outcome. Such a game is determined by a set of positions, including an initial position, and the player whose turn it is to move. Play moves from one position to another, with the players usually alternating moves, until a terminal position is reached. A terminal position is one from which no moves are possible. Then one of the players is declared the winner and the other the loser.

There are two main references for the material on combinatorial games. One is the research book, *On Numbers and Games* by J. H. Conway, Academic Press, 1976. This book introduced many of the basic ideas of the subject and led to a rapid growth of the area that continues today. The other reference, more appropriate for this text, is the two-volume book, *Winning Ways for Your Mathematical Plays* by Berlekamp, Conway and Guy, Academic Press, 1982, in paperback. There are many interesting games described in this book and much of it is accessible to the undergraduate mathematics student. This theory may be divided into two parts, ***impartial games*** in which the set of moves available from any given position is the same for both players, and ***partizan games*** in which each player has a different set of possible moves from a given position. Games like chess or checkers in which one player moves the white pieces and the other moves the black pieces are partizan. In Part I, we treat only the theory of impartial games. An elementary introduction to impartial combinatorial games is given in the book *Fair Game* by Richard K. Guy, published in the COMAP Mathematical Exploration Series, 1989. We start with a simple example.

1.1 A Simple Take-Away Game

Here are the rules of a very simple impartial combinatorial game of removing chips from a pile of chips.

(1) *There are two players. We label them I and II.*
(2) *There is a pile of 21 chips in the center of a table.*
(3) *A move consists of removing one, two, or three chips from the pile. At least one chip must be removed, but no more than three may be removed.*
(4) *Players alternate moves with Player I starting.*
(5) *The player that removes the last chip wins. (The last player to move wins. If you can't move, you lose.)*

How can we analyze this game? Can one of the players force a win in this game? Which player would you rather be, the player who starts or the player who goes second? What is a good strategy?

We analyze this game from the end back to the beginning. This method is sometimes called ***backward induction***.

If there are just one, two, or three chips left, the player who moves next wins simply by taking all the chips.

Suppose there are four chips left. Then the player who moves next must leave either one, two or three chips in the pile and his opponent will be able to win. So four chips left is a loss for the next player to move and a win for the previous player, i.e. the one who just moved.

With 5, 6, or 7 chips left, the player who moves next can win by moving to the position with four chips left.

With 8 chips left, the next player to move must leave 5, 6, or 7 chips, and so the previous player can win.

We see that positions with $0, 4, 8, 12, 16, \ldots$ chips are target positions; we would like to move into them. We may now analyze the game with 21 chips. Since 21 is not divisible by 4, the first player to move can win. The unique optimal move is to take one chip and leave 20 chips which is a target position.

1.2 What is a Combinatorial Game?

We now define the notion of a combinatorial game more precisely. It is a game that satisfies the following conditions.

(1) *There are two players.*
(2) *There is a set, usually finite, of possible positions of the game.*
(3) *The rules of the game specify for both players and each position which moves to other positions are legal moves. If the rules make no*

distinction between the players, that is if both players have the same options of moving from each position, the game is called ***impartial;*** *otherwise, the game is called* ***partizan.***

(4) *The players alternate moving.*

(5) *The game ends when a position is reached from which no moves are possible for the player whose turn it is to move. Under the* ***normal play rule****, the last player to move wins. Under the* ***misère play rule*** *the last player to move loses.*

If the game never ends, it is declared a draw. However, we shall nearly always add the following condition, called ***the Ending Condition***. This eliminates the possibility of a draw.

(6) *The game ends in a finite number of moves no matter how it is played.*

It is important to note what is omitted in this definition. No random moves such as the rolling of dice or the dealing of cards are allowed. This rules out games like backgammon and poker. A combinatorial game is a game of perfect information: simultaneous moves and hidden moves are not allowed. This rules out battleship and scissors-paper-rock. No draws in a finite number of moves are possible. This rules out tic-tac-toe. Unless stated otherwise, we restrict attention to impartial games under the normal play rule.

1.3 P-positions, N-positions

Returning to the take-away game of Sec. 1.1, we see that $0, 4, 8, 12, 16, \ldots$ are positions that are winning for the Previous player (the player who just moved) and that $1, 2, 3, 5, 6, 7, 9, 10, 11, \ldots$ are winning for the Next player to move. The former are called P-positions, and the latter are called N-positions. The P-positions are just those with a number of chips divisible by 4, called target positions in Sec. 1.1.

In impartial combinatorial games, one can find in principle which positions are P-positions and which are N-positions by (possibly transfinite) induction using the following labeling procedure starting at the terminal positions. We say a position in a game is a ***terminal position***, if no moves from it are possible. This algorithm is just the method we used to solve the take-away game of Sec. 1.1.

Step 1: Label every terminal position as a P-position.

Step 2: Label every position that can reach a labelled P-position in one move as an N-position.

Step 3: Find those positions whose only moves are to labelled N-positions; label such positions as P-positions.

Step 4: If no new P-positions were found in Step 3, stop; otherwise return to Step 2.

It is easy to see that the strategy of moving to P-positions wins. From a P-position, your opponent can move only to an N-position (3). Then you may move back to a P-position (2). Eventually the game ends at a terminal position and since this is a P-position, you win (1).

Here is a characterization of P-positions and N-positions that is valid for impartial combinatorial games satisfying the ending condition, under the normal play rule.

Characteristic Property. *P-positions and N-positions are defined recursively by the following three statements:*

(1) *All terminal positions are P-positions.*
(2) *From every N-position, there is at least one move to a P-position.*
(3) *From every P-position, every move is to an N-position.*

For games using the misère play rule, the characteristic property still holds provided condition (1) is replaced by the condition that all terminal positions are N-positions.

1.4 Subtraction Games

Let us now consider a class of combinatorial games that contains the take-away game of Sec. 1.1 as a special case. Let S be a set of positive integers. The subtraction game with subtraction set S is played as follows. From a pile with a large number, say n, of chips, two players alternate moves. A move consists of removing s chips from the pile where $s \in S$. Last player to move wins.

The take-away game of Sec. 1.1 is the subtraction game with subtraction set $S = \{1, 2, 3\}$. In Exercise 1.5.2, you are asked to analyze the subtraction game with subtraction set $S = \{1, 2, 3, 4, 5, 6\}$.

For illustration, let us analyze the subtraction game with subtraction set $S = \{1, 3, 4\}$ by finding its P-positions. There is exactly one terminal position, namely 0. Then 1, 3, and 4 are N-positions, since they can be moved to 0. But 2 then must be a P-position since the only legal move from 2 is to 1, which is an N-position. Then 5 and 6 must be N-positions since they can be moved to 2. Now we see that 7 must be a P-position since the only moves from 7 are to 6, 4, or 3, all of which are N-positions.

Now we continue similarly: we see that 8, 10 and 11 are N-positions, 9 is a P-position, 12 and 13 are N-positions and 14 is a P-position. This extends by induction. We find that the set of P-positions is $P = \{0, 2, 7, 9, 14, 16, \ldots\}$, the set of nonnegative integers leaving remainder 0 or 2 when divided by 7. The set of N-positions is the complement, $N = \{1, 3, 4, 5, 6, 8, 10, 11, 12, 13, 15, \ldots\}$.

Table 1. The pattern $PNPNNNN$ of length 7 repeats forever.

x	0	1	2	3	4	5	6	7	8	9	10	11	12	13	14	...
Position	P	N	P	N	N	N	N	P	N	P	N	N	N	N	P	...

Who wins the game with 100 chips, the first player or the second? The P-positions are the numbers equal to 0 or 2 modulus 7. Since 100 has remainder 2 when divided by 7, 100 is a P-position; the second player to move can win with optimal play.

1.5 Exercises

1. Consider the misère version of the take-away game of Sec. 1.1, where the last player to move loses. The objective is to force your opponent to take the last chip. Analyze this game. What are the target positions (P-positions)?
2. Generalize the Take-Away Game: (a) Suppose in a game with a pile containing a large number of chips, you can remove any number from 1 to 6 chips at each turn. What is the winning strategy? What are the P-positions? (b) If there are initially 31 chips in the pile, what is your winning move, if any?
3. **The Thirty-one Game.** (Mott-Smith (1954)) From a deck of cards, take the Ace, 2, 3, 4, 5, and 6 of each suit. These 24 cards are laid out face up on a table. The players alternate turning over cards and the sum of the turned over cards is computed as play progresses. Each Ace counts as one. The player who first makes the sum go above 31 loses. It would seem that this is equivalent to the game of the previous exercise played on a pile of 31 chips. But there is a catch. No integer may be chosen more than four times.

   (a) If you are the first to move, and if you use the strategy found in the previous exercise, what happens if the opponent keeps choosing 4?
   (b) Nevertheless, the first player can win with optimal play. How?

4. Find the set of P-positions for the subtraction games with subtraction sets
   (a) $S = \{1, 3, 5, 7\}$.
   (b) $S = \{1, 3, 6\}$.
   (c) $S = \{1, 2, 4, 8, 16, \ldots\}$ = all powers of 2.
   (d) Who wins each of these games if play starts at 100 chips, the first player or the second?
5. **Empty and Divide**. (Ferguson (1998)) There are two boxes. Initially, one box contains m chips and the other contains n chips. Such a position is denoted by (m, n), where $m > 0$ and $n > 0$. The two players alternate moving. A move consists of emptying one of the boxes, and dividing the contents of the other between the two boxes with at least one chip in each box. There is a unique terminal position, namely $(1, 1)$. Last player to move wins. Find all P-positions.
6. **Chomp!** A game invented by Schuh (1952) in an arithmetical form was discovered independently in a completely different form by Gale (1974). Gale's version of the game involves removing squares from a rectangular board, say an $m \times n$ board. A move consists in taking a square and removing it and all squares to the right and above. Players alternate moves, and the person to take square $(1, 1)$ loses. The name "Chomp" comes from imagining the board as a chocolate bar, and moves involving breaking off some corner squares to eat. The square $(1, 1)$ is poisoned though; the player who chomps it loses. You can play this game on the web at http://www.tomsferguson.com/Games/chomp.html.

   For example, starting with an 8×3 board, suppose the first player chomps at $(6, 2)$ gobbling 6 pieces, and then second player chomps at $(2, 3)$ gobbling 4 pieces, leaving the following board, where $\otimes$ denotes the poisoned piece.

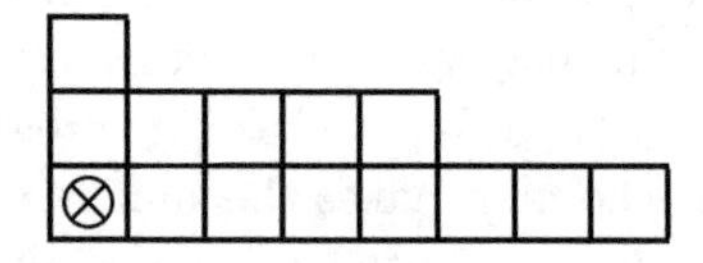

   (a) Show that this position is an N-position by finding a winning move for the first player. (It is unique.)
   (b) It is known that the first player can win all *rectangular* starting positions. The proof, though ingenious, is not hard. However, it is an "existence" proof. It shows that there is a winning strategy for

the first player, but gives no hint on how to find the first move! See if you can find the proof. Here is a hint: Does removing the upper right corner constitute a winning move?

7. **Dynamic subtraction.** One can enlarge the class of subtraction games by letting the subtraction set depend on the last move of the opponent. Many early examples appear in Chap. 12 of Schuh (1968). Here are two other examples. (For a generalization, see Schwenk (1970).)

   (a) There is one pile of n chips. The first player to move may remove as many chips as desired, at least one chip but not the whole pile. Thereafter, the players alternate moving, each player not being allowed to remove more chips than his opponent took on the previous move. What is an optimal move for the first player if $n = 44$? For what values of n does the second player have a win?

   (b) **Fibonacci Nim.** (Whinihan (1963)) The same rules as in (a), except that a player may take at most twice the number of chips his opponent took on the previous move. The analysis of this game is more difficult than the game of part (a) and depends on the sequence of numbers named after Leonardo Pisano Fibonacci, which may be defined as $F_1 = 1$, $F_2 = 2$, and $F_{n+1} = F_n + F_{n-1}$ for $n \geq 2$. The Fibonacci sequence is thus: $1, 2, 3, 5, 8, 13, 21, 34, 55, \ldots$. The solution is facilitated by

   Zeckendorf's Theorem. Every positive integer can be written uniquely as a sum of distinct non-neighboring Fibonacci numbers.

   There may be many ways of writing a number as a sum of Fibonacci numbers, but there is only one way of writing it as a sum of non-neighboring Fibonacci numbers. Thus, $43 = 34 + 8 + 1$ is the unique way of writing 43, since although $43 = 34 + 5 + 3 + 1$, 5 and 3 are neighbors. What is an optimal move for the first player if $n = 43$? For what values of n does the second player have a win? Try out your solution on http://www.tomsferguson.com/Games/fibonim.html.

8. **The *SOS* Game.** (From the 28th Annual USA Mathematical Olympiad, 1999) The board consists of a row of n squares, initially empty. Players take turns selecting an empty square and writing either an S or an O in it. The player who first succeeds in completing SOS in consecutive squares wins the game. If the whole board gets filled up without an SOS appearing consecutively anywhere, the game is a draw.

(a) Suppose $n = 4$ and the first player puts an S in the first square. Show the second player can win.
(b) Show that if $n = 7$, the first player can win the game.
(c) Show that if $n = 2000$, the second player can win the game.
(d) Who, if anyone, wins the game if $n = 14$?

2 The Game of Nim

The most famous take-away game is the game of Nim, played as follows. There are three piles of chips containing x_1, x_2, and x_3 chips respectively. (Piles of sizes 5, 7, and 9 make a good game.) Two players take turns moving. Each move consists of selecting one of the piles and removing chips from it. You may not remove chips from more than one pile in one turn, but from the pile you selected you may remove as many chips as desired, from one chip to the whole pile. The winner is the player who removes the last chip. You can play this game on the web at http://www.tomsferguson.com/Games/Nim.html.

2.1 Preliminary Analysis

There is exactly one terminal position, namely $(0,0,0)$, which is therefore a P-position. The solution to one-pile Nim is trivial: you simply remove the whole pile. Any position with exactly one non-empty pile, say $(0,0,x)$ with $x > 0$ is therefore an N-position. Consider two-pile Nim. It is easy to see that the P-positions are those for which the two piles have an equal number of chips, $(0,1,1)$, $(0,2,2)$, etc. This is because if it is the opponent's turn to move from such a position, he must change to a position in which the two piles have an unequal number of chips, and then you can immediately return to a position with an equal number of chips (perhaps the terminal position).

If all three piles are non-empty, the situation is more complicated. Clearly, $(1,1,1)$, $(1,1,2)$, $(1,1,3)$ and $(1,2,2)$ are all N-positions because they can be moved to $(1,1,0)$ or $(0,2,2)$. The next simplest position is $(1,2,3)$ and it must be a P-position because it can only be moved to one of the previously discovered N-positions. We may go on and discover that the next most simple P-positions are $(1,4,5)$, and $(2,4,6)$, but it is difficult to see how to generalize this. Is $(5,7,9)$ a P-position? Is $(15,23,30)$ a P-position?

If you go on with the above analysis, you may discover a pattern. But to save us some time, I will describe the solution to you. Since the solution is somewhat fanciful and involves something called nim-sum, the validity of the solution is not obvious. Later, we prove it to be valid using the elementary notions of P-position and N-position.

2.2 Nim-Sum

The nim-sum of two nonnegative integers is their addition without carry in base 2. Let us make this notion precise.

Every nonnegative integer x has a unique base 2 representation of the form $x = x_m 2^m + x_{m-1} 2^{m-1} + \cdots + x_1 2 + x_0$ for some m, where each x_i is either zero or one. We use the notation $(x_m x_{m-1} \cdots x_1 x_0)_2$ to denote this representation of x to base two. Thus, $22 = 1 \cdot 16 + 0 \cdot 8 + 1 \cdot 4 + 1 \cdot 2 + 0 \cdot 1 = (10110)_2$. The nim-sum of two integers is found by expressing the integers to base two and using addition modulo 2 on the corresponding individual components:

Definition. The nim-sum of $(x_m \cdots x_0)_2$ and $(y_m \cdots y_0)_2$ is $(z_m \cdots z_0)_2$, and we write $(x_m \cdots x_0)_2 \oplus (y_m \cdots y_0)_2 = (z_m \cdots z_0)_2$, where for all k, $z_k = x_k + y_k \pmod 2$, that is, $z_k = 1$ if $x_k + y_k = 1$ and $z_k = 0$ otherwise.

For example, $(10110)_2 \oplus (110011)_2 = (100101)_2$. This says that $22 \oplus 51 = 37$. This is easier to see if the numbers are written vertically (we also omit the parentheses for clarity):

$$\begin{aligned} 22 &= 10110_2 \\ 51 &= 110011_2 \\ \hline \text{nim-sum} &= 100101_2 = 37 \end{aligned}$$

Nim-sum is associative (i.e. $x \oplus (y \oplus z) = (x \oplus y) \oplus z$) and commutative (i.e. $x \oplus y = y \oplus x$), since addition modulo 2 is. Thus we may write $x \oplus y \oplus z$ without specifying the order of addition. Furthermore, 0 is an identity for addition ($0 \oplus x = x$), and every number is its own negative ($x \oplus x = 0$), so that the cancellation law holds: $x \oplus y = x \oplus z$ implies $y = z$. (If $x \oplus y = x \oplus z$, then $x \oplus x \oplus y = x \oplus x \oplus z$, and so $y = z$.)

Thus, nim-sum has a lot in common with ordinary addition, but what does it have to do with playing the game of Nim? The answer is contained in the following theorem of Bouton (1902).

Theorem 1. *A position, (x_1, x_2, x_3), in Nim is a P-position if and only if the nim-sum of its components is zero, $x_1 \oplus x_2 \oplus x_3 = 0$.*

As an example, take the position $(x_1, x_2, x_3) = (13, 12, 8)$. Is this a P-position? If not, what is a winning move? We compute the nim-sum of 13, 12 and 8:

$$\begin{array}{rl} 13 = & 1101_2 \\ 12 = & 1100_2 \\ 8 = & 1000_2 \\ \hline \text{nim-sum} = & 1001_2 = 9 \end{array}$$

Since the nim-sum is not zero, this is an N-position according to Theorem 1. Can you find a winning move? You must find a move to a P-position, that is, to a position with an even number of 1's in each column. One such move is to take away 9 chips from the pile of 13, leaving 4 there. The resulting position has nim-sum zero:

$$\begin{array}{rl} 4 = & 100_2 \\ 12 = & 1100_2 \\ 8 = & 1000_2 \\ \hline \text{nim-sum} = & 0000_2 = 0 \end{array}$$

Another winning move is to subtract 7 chips from the pile of 12, leaving 5. Check it out. There is also a third winning move. Can you find it?

2.3 Nim with a Larger Number of Piles

We saw that 1-pile nim is trivial, and that 2-pile nim is easy. Since 3-pile nim is much more complex, we might expect 4-pile nim to be much harder still. But that is not the case. Theorem 1 also holds for a larger number of piles! A nim position with four piles, (x_1, x_2, x_3, x_4), is a P-position if and only if $x_1 \oplus x_2 \oplus x_3 \oplus x_4 = 0$. The proof below works for an arbitrary finite number of piles.

2.4 Proof of Bouton's Theorem

Let $\mathcal{P}$ denote the set of nim positions with nim-sum zero, and let $\mathcal{N}$ denote the complement set, the set of positions of positive nim-sum. We check the three conditions of the definition in Sec. 1.3.

(1) *All terminal positions are in $\mathcal{P}$.* That's easy. The only terminal position is the position with no chips in any pile, and $0 \oplus 0 \oplus \cdots = 0$.

(2) *From each position in $\mathcal{N}$, there is a move to a position in $\mathcal{P}$.* Here's how we construct such a move. Form the nim-sum as a column addition, and look at the leftmost (most significant) column with an odd number of 1's. Change any of the numbers that have a 1 in that column to a number such that there are an even number of 1's in each column. This makes that number smaller because the 1 in the most significant position changes to a 0. Thus this is a legal move to a position in $\mathcal{P}$.

(3) *Every move from a position in $\mathcal{P}$ is to a position in $\mathcal{N}$.* If $(x_1, x_2, \ldots)$ is in $\mathcal{P}$ and x_1 is changed to $x_1' < x_1$, then we cannot have $x_1 \oplus x_2 \oplus \cdots = 0 = x_1' \oplus x_2 \oplus \cdots$, because the cancellation law would imply that $x_1 = x_1'$. So $x_1' \oplus x_2 \oplus \cdots \neq 0$, implying that $(x_1', x_2, \ldots)$ is in $\mathcal{N}$.

These three properties show that $\mathcal{P}$ is the set of P-positions. □

It is interesting to note from (2) that in the game of nim the number of winning moves from an N-position is equal to the number of 1's in the leftmost column with an odd number of 1's. In particular, there is always an odd number of winning moves.

2.5 Misère Nim

What happens when we play nim under the misère play rule? Can we still find who wins from an arbitrary position, and give a simple winning strategy? This is one of those questions that at first looks hard, but after a little thought turns out to be easy.

Here is Bouton's method for playing misère nim optimally. Play it as you would play nim under the normal play rule as long as there are at least two heaps of size greater than one. When your opponent finally moves so that there is exactly one pile of size greater than one, reduce that pile to zero or one, whichever leaves an *odd* number of piles of size one remaining.

This works because your optimal play in nim never requires you to leave exactly one pile of size greater than one (the nim sum must be zero), and your opponent cannot move from two piles of size greater than one to no piles greater than one. So eventually the game drops into a position with exactly one pile greater than one and it must be your turn to move. Try it out at http://www.tomsferguson.com/Games/Nim.html.

A similar analysis works in many other games. But in general the misère play theory is much more difficult than the normal play theory. Some games have a fairly simple normal play theory but an extraordinarily difficult misère theory, such as the games of Kayles and Dawson's chess, presented in Sec. 4.4.

2.6 Exercises

1. (a) What is the nim-sum of 27 and 17?
   (b) The nim-sum of 38 and x is 25. Find x.
   (c) If $x \oplus y = 17$ and $x \oplus z = 13$, find $y \oplus z$.
2. Find all winning moves in the game of nim,
   (a) with three piles of 12, 19, and 27 chips.
   (b) with four piles of 13, 17, 19, and 23 chips.
   (c) What is the answer to (a) and (b) if the misére version of nim is being played?
3. Nimble. Nimble is played on a game board consisting of a line of squares labelled: $0, 1, 2, 3, \ldots$. A finite number of coins is placed on the squares with possibly more than one coin on a single square. A move consists in taking one of the coins and moving it to any square to the left, possibly moving over some of the coins, and possibly onto a square already containing one or more coins. The players alternate moves and the game ends when all coins are on the square labelled 0. The last player to move wins. Show that this game is just nim in disguise. In the following position with 6 coins, who wins, the next player or the previous player? If the next player wins, find a winning move.

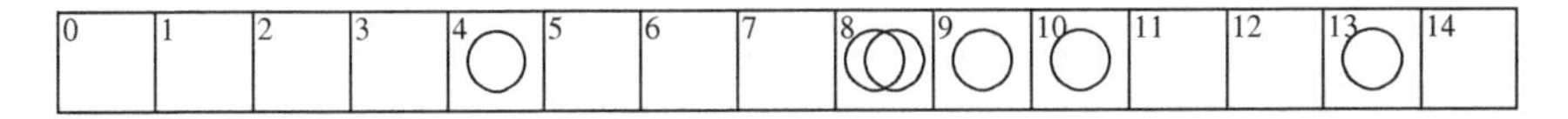

4. Turning Turtles. Another class of games, due to Lenstra, is played with a long line of coins, with moves involving turning over some coins from heads to tails or from tails to heads. See Chap. 5 for some of the remarkable theory. Here is a simple example called Turning Turtles.

   A horizontal line of n coins is laid out randomly with some coins showing heads and some tails. A move consists of turning over one of the coins from heads to tails, and in addition, if desired, turning over one other coin to the left of it (from heads to tails or tails to heads). For example consider the sequence of $n = 13$ coins:

T	H	T	T	H	T	T	T	H	H	T	H	T
1	2	3	4	5	6	7	8	9	10	11	12	13

   One possible move in this position is to turn the coin in place 9 from heads to tails, and also the coin in place 4 from tails to heads.

   (a) Show that this game is just nim in disguise if an H in place n is taken to represent a nim pile of n chips.

(b) Assuming (a) to be true, find a winning move in the above position.
(c) Try this game and some other related games at http://www.cut-the-knot.org/.

5. Northcott's Game. A position in Northcott's game is a checkerboard with one black and one white checker on each row. "White" moves the white checkers and "Black" moves the black checkers. A checker may move any number of squares along its row, but may not jump over or onto the other checker. Players move alternately and the last to move wins.
Note two points:

 1. This is a partizan game, because Black and White have different moves from a given position.
 2. This game does not satisfy the Ending Condition, (6) of Sec. 1.2. The players could move around endlessly.

 Nevertheless, knowing how to play nim is a great advantage in this game. In the position below, who wins, Black or White? or does it depend on who moves first?

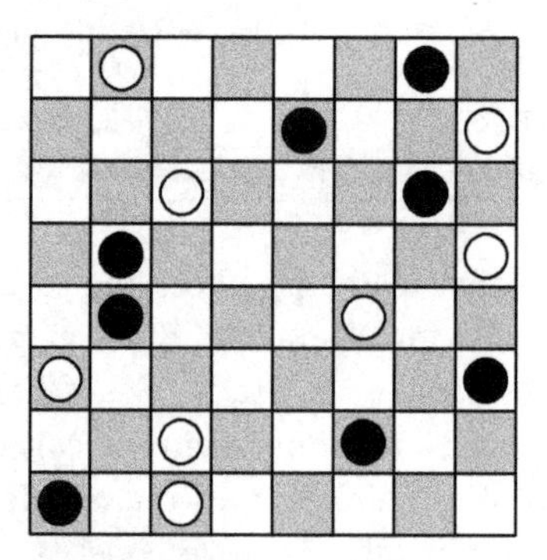

6. Staircase Nim. (Sprague (1937)) A staircase of n steps contains coins on some of the steps. Let $(x_1, x_2, \ldots, x_n)$ denote the position with x_j coins on step j, $j = 1, \ldots, n$. A move of staircase nim consists of moving any positive number of coins from any step, j, to the next lower step, $j-1$. Coins reaching the ground (step 0) are removed from play. A move taking, say, x chips from step j, where $1 \le x \le x_j$, and putting them on step $j-1$, leaves $x_j - x$ coins on step j and results in $x_{j-1} + x$ coins on step $j-1$. The game ends when all coins are on the ground. Players alternate moves and the last to move wins.

Show that $(x_1, x_2, \ldots, x_n)$ is a P-position if and only if the numbers of coins on the odd numbered steps, $(x_1, x_3, \ldots, x_k)$ where $k = n$ if n is odd and $k = n - 1$ if n is even, forms a P-position in ordinary nim.

7. Moore's Nim$_k$. A generalization of nim with a similar elegant theory was proposed by E. H. Moore (1910), called Nim$_k$. There are n piles of chips and play proceeds as in nim except that in each move a player may remove as many chips as desired from any k piles, where k is fixed. At least one chip must be taken from some pile. For $k = 1$ this reduces to ordinary nim, so ordinary nim is Nim$_1$. Try playing Nim$_2$ at http://www.tomsferguson.com/Games/Moore.html.

   Moore's Theorem states that a position $(x_1, x_2, \ldots, x_n)$, is a P-position in Nim$_k$ if and only if when x_1 to x_n are expanded in base 2 and added in base $k+1$ without carry, the result is zero. (In other words, the number of 1's in each column must be divisible by $k + 1$.)

   (a) Consider the game of Nimble of Exercise 3 but suppose that at each turn a player may move one or two coins to the left as many spaces as desired. Note that this is really Moore's Nim$_k$ with $k = 2$. Using Moore's Theorem, show that the Nimble position of Exercise 3 is an N-position, and find a move to a P-position.
   (b) Prove Moore's Theorem.
   (c) What constitutes optimal play in the misère version of Moore's Nim$_k$?

3 Graph Games

We now give an equivalent description of a combinatorial game as a game played on a directed graph. This will contain the games described in Chaps. 1 and 2. This is done by identifying positions in the game with vertices of the graph and moves of the game with edges of the graph. Then we will define a function known as the Sprague-Grundy function that contains more information than just knowing whether a position is a P-position or an N-position.

3.1 Games Played on Directed Graphs

We first give the mathematical definition of a directed graph.

Definition. A *directed graph*, G, is a pair (X, F) where X is a nonempty set of vertices (*positions*) and F is a function that gives for each $x \in X$ a subset of X, $F(x) \subset X$. For a given $x \in X$, $F(x)$ represents the positions to which a player may move from x (called the *followers* of x). If $F(x)$ is empty, x is called a terminal position.

A two-person win-lose game may be played on such a graph $G = (X, F)$ by stipulating a starting position $x_0 \in X$ and using the following rules:

(1) Player I moves first, starting at x_0.
(2) Players alternate moves.
(3) At position x, the player whose turn it is to move chooses a position $y \in F(x)$.
(4) The player who is confronted with a terminal position at his turn, and thus cannot move, loses.

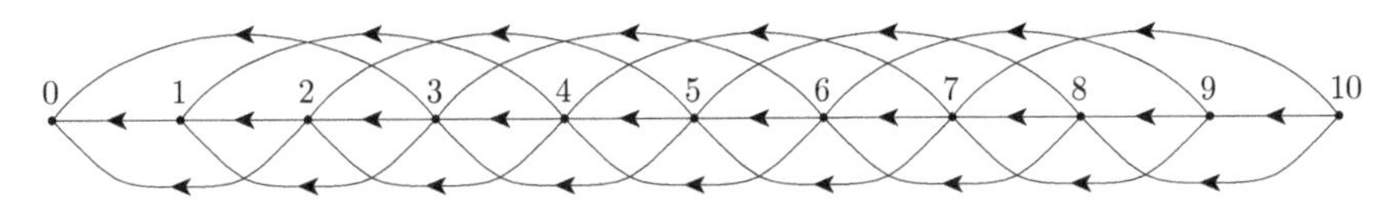

Fig. 3.1 The subtraction game with $S = \{1, 2, 3\}$.

As defined, graph games could continue for an infinite number of moves. To avoid this possibility and a few other problems, we first restrict attention to graphs that have the property that no matter what starting point x_0 is used, there is a number n, possibly depending on x_0, such that every path from x_0 has length less than or equal to n. (A *path* of length m is a sequence $x_0, x_1, x_2, \ldots, x_m$ such that $x_i \in F(x_{i-1})$ for all $i = 1, \ldots, m$.) Such graphs are called ***progressively bounded***. If X itself is finite, this merely means that there are no ***cycles***. (A cycle is a path, $x_0, x_1, \ldots, x_m$, with $x_0 = x_m$ and distinct vertices $x_0, x_1, \ldots, x_{m-1}$, for some $m \geq 1$.)

As an example, the subtraction game with subtraction set $S = \{1, 2, 3\}$, analyzed in Sec. 1.1, that starts with a pile of n chips has a representation as a graph game. Here $X = \{0, 1, \ldots, n\}$ is the set of vertices. The empty pile is terminal, so $F(0) = \emptyset$, the empty set. We also have $F(1) = \{0\}$, $F(2) = \{0, 1\}$, and for $2 \leq k \leq n$, $F(k) = \{k-3, k-2, k-1\}$. This completely defines the game.

It is useful to draw a representation of the graph. This is done using dots to represent vertices and lines to represent the possible moves. An arrow is placed on each line to indicate which direction the move goes. The graphic representation of this subtraction game played on a pile of 10 chips is given in Fig. 3.1.

3.2 The Sprague-Grundy Function

Graph games may be analyzed by considering P-positions and N-positions. It may also be analyzed through the Sprague-Grundy function.

Definition. The *Sprague-Grundy function* of a graph, (X, F), is a function, g, defined on X and taking nonnegative integer values, such that

$$g(x) = \min\{n \geq 0 : \ n \neq g(y) \quad \text{for} \quad y \in F(x)\}. \tag{1}$$

In words, $g(x)$ the smallest nonnegative integer not found among the Sprague-Grundy values of the followers of x. If we define the ***minimal excludant***, or ***mex***, of a set of nonnegative integers as the smallest nonnegative integer not in the set, then we may write simply

$$g(x) = \text{mex}\{g(y) : y \in F(x)\}. \quad (2)$$

Note that $g(x)$ is defined recursively. That is, $g(x)$ is defined in terms of $g(y)$ for all followers y of x. Moreover, the recursion is self-starting. For terminal vertices, x, the definition implies that $g(x) = 0$, since $F(x)$ is the empty set for terminal x. For non-terminal x, all of whose followers are terminal, $g(x) = 1$. In the examples in the next section, we find $g(x)$ using this inductive technique. This works for all progressively bounded graphs, and shows that for such graphs, the Sprague-Grundy function exists, is unique and is finite. However, other graphs require more subtle techniques and are treated in Sec. 3.4.

Given the Sprague-Grundy function g of a graph, it is easy to analyze the corresponding graph game. Positions x for which $g(x) = 0$ are P-positions and all other positions are N-positions. The winning procedure is to choose at each move to move to a vertex with Sprague-Grundy value zero. This is easily seen by checking the conditions of Sec. 1.3:

(1) If x is a terminal position, $g(x) = 0$.
(2) At positions x for which $g(x) = 0$, every follower y of x is such that $g(y) \neq 0$, and
(3) At positions x for which $g(x) \neq 0$, there is at least one follower y such that $g(y) = 0$.

The Sprague-Grundy function thus contains a lot more information about a game than just the P- and N-positions. What is this extra information used for? As we will see in Chap. 4, the Sprague-Grundy function allows us to analyze *sums* of graph games.

3.3 Examples

1. We use Fig. 3.2 to describe the inductive method of finding the SG-values, i.e. the values that the Sprague-Grundy function assigns to the vertices. The method is simply to search for a vertex all of whose followers have SG-values assigned. Then apply (1) or (2) to find its SG-value, and repeat until all vertices have been assigned values.

To start, all terminal positions are assigned SG-value 0. There are exactly four terminal positions, to the left and below the graph. Next, there is only one vertex all of whose followers have been assigned values, namely vertex a. This is assigned value 1, the smallest number not among

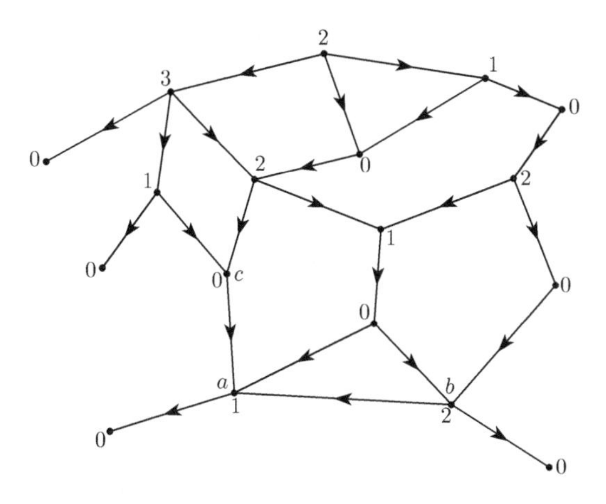

Fig. 3.2

the followers. Now there are two more vertices, b and c, all of whose followers have been assigned SG-values. Vertex b has followers with values 0 and 1 and so is assigned value 2. Vertex c has only one follower with SG-value 1. The smallest nonnegative integer not equal to 1 is 0, so its SG-value is 0. Now we have three more vertices whose followers have been assigned SG-values. Check that the rest of the SG-values have been assigned correctly.

2. What is the Sprague-Grundy function of the subtraction game with subtraction set $S = \{1, 2, 3\}$? The terminal vertex, 0, has SG-value 0. The vertex 1 can only be moved to 0 and $g(0) = 0$, so $g(1) = 1$. Similarly, 2 can move to 0 and 1 with $g(0) = 0$ and $g(1) = 1$, so $g(2) = 2$, and 3 can move to 0, 1 and 2, with $g(0) = 0$, $g(1) = 1$ and $g(2) = 2$, so $g(3) = 3$. But 4 can only move to 1, 2 and 3 with SG-values 1, 2 and 3, so $g(4) = 0$. Continuing in this way we see

x	0	1	2	3	4	5	6	7	8	9	10	11	12	13	14	...
$g(x)$	0	1	2	3	0	1	2	3	0	1	2	3	0	1	2	...

In general $g(x) = x \pmod 4$, i.e. $g(x)$ is the remainder when x is divided by 4.

3. *At-Least-Half.* Consider the one-pile game with the rule that you must remove at least half of the counters. The only terminal position is zero.

We may compute the Sprague-Grundy function inductively as

x	0	1	2	3	4	5	6	7	8	9	10	11	12	...
$g(x)$	0	1	2	2	3	3	3	3	4	4	4	4	4	...

We see that $g(x)$ may be expressed as the exponent in the smallest power of 2 greater than x: $g(x) = \min\{k : 2^k > x\}$.

In reality, this is a rather silly game. One can win it at the first move by taking all the counters! What good is it to do all this computation of the Sprague-Grundy function if one sees easily how to win the game anyway?

The answer is given in the next chapter. If the game is played with several piles instead of just one, it is no longer so easy to see how to play the game well. The theory of the next chapter tells us how to use the Sprague-Grundy function together with nim-addition to find optimal play with many piles.

3.4 The Sprague-Grundy Function on More General Graphs

Let us look briefly at the problems that arise when the graph may not be progressively bounded, or when it may even have cycles.

First, suppose the hypothesis that the graph be progressively bounded is weakened to requiring only that the graph be progressively finite. A graph is *progressively finite* if *for every vertex* x_0, *every path starting at* x_0 *has a finite length.* This condition is essentially equivalent to the Ending Condition (6) of Sec. 1.2. Cycles would still be ruled out in such graphs.

As an example of a graph that is progressively finite but not progressively bounded, consider the graph of the game in Fig. 3.3 in which the first move is to choose the number of chips in a pile, and from then on to treat the pile as a nim pile. From the initial position each path has a finite length so the graph is progressively finite. But the graph is not

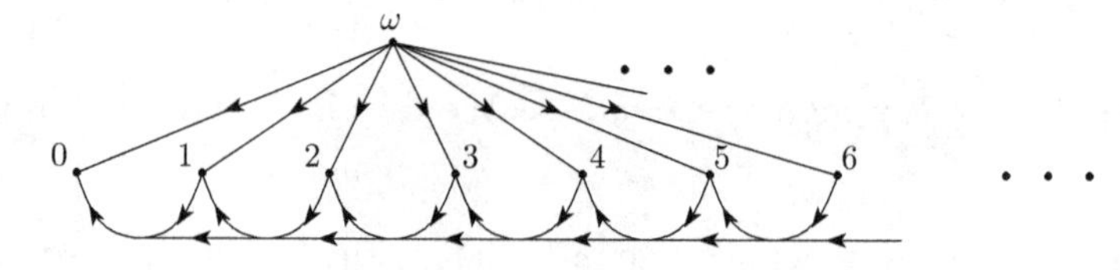

Fig. 3.3 A progressively finite graph that is not progressively bounded.

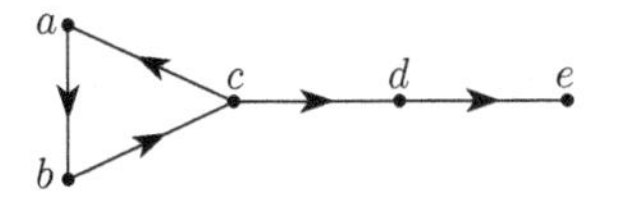

Fig. 3.4 A cyclic graph.

progressively bounded since there is no upper limit to the length of a path from the initial position.

The Sprague-Grundy theory can be extended to progressively finite graphs, but transfinite induction must be used. The SG-value of the initial position in Fig. 3.3 above would be the smallest ordinal greater than all integers, usually denoted by ω. We may also define nim positions with SG-values $\omega+1, \omega+2, \ldots, 2\omega, \ldots, \omega^2, \ldots, \omega^\omega$, etc. In Exercise 6, you are asked to find several of these transfinite SG-values.

In summary, the set of impartial combinatorial games as defined in Sec. 1.2 is equivalent to the set of graph games with progressively finite graphs. The Sprague-Grundy function on such a graph exists if transfinite values are allowed. For progressively bounded graph games, the Sprague-Grundy function exists and is finite.

If the graph is allowed to have cycles, new problems arise. The SG-function satisfying (1) may not exist. Even if it does, the simple inductive procedure of the previous sections may not suffice to find it. Even if the Sprague-Grundy function exists and is known, it may not be easy to find a winning strategy.

Graphs with cycles do not satisfy the Ending Condition (6) of Sec. 1.2. Play may last forever. In such a case we say the game ends in a tie; neither player wins. Figure 3.4 gives an example where there are tied positions.

The node e is terminal and so has Sprague-Grundy value 0. Since e is the only follower of d, d has Sprague-Grundy value 1. So a player at c will not move to d since such a move obviously loses. Therefore the only reasonable move is to a. After two more moves, the game moves to node c again with the opponent to move. The same analysis shows that the game will go around the cycle $abcabc\ldots$ forever. So positions a, b and c are all tied positions. In this example the Sprague-Grundy function does not exist.

When the Sprague-Grundy function exists in a graph with cycles, more subtle techniques are often required to find it. Some examples for the reader to try his/her skill are found in Exercise 9. But there is another problem. Suppose the Sprague-Grundy function is known and the graph has cycles. If you are at a position with non-zero SG-value, you know you can win by moving to a position with SG-value 0. But which one? You may choose one that takes you on a long trip and after many moves you find yourself back to where you started. An example of this is in Northcott's Game in Exercise 5 of Chap. 2. There it is easy to see how to proceed to end the game, but in general it may be difficult to see what to do. For an efficient algorithm that computes the Sprague-Grundy function along with a counter that gives information on how to play, see Fraenkel (2002).

3.5 Exercises

1.

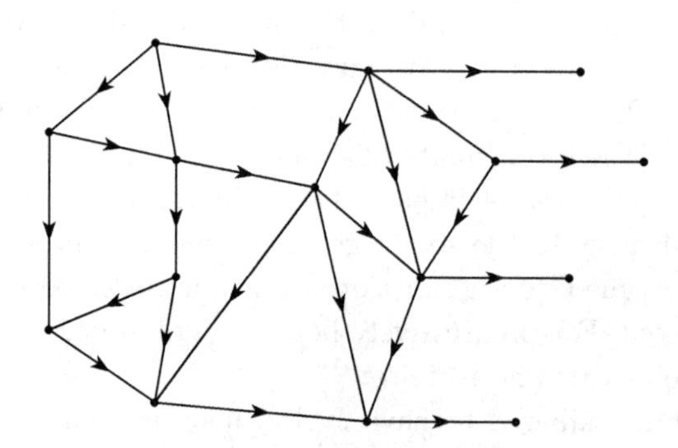

Fig. 3.5 Find the Sprague-Grundy function.

2. Find the Sprague-Grundy function of the subtraction game with subtraction set $S = \{1, 3, 4\}$.
3. Consider the one-pile game with the rule that you may remove at most half the chips. Of course, you must remove at least one, so the terminal positions are 0 and 1. Find the Sprague-Grundy function.
4. (a) Consider the one-pile game with the rule that you may remove c chips from a pile of n chips if and only if c is a divisor of n, including 1 and n. For example, from a pile of 12 chips, you may remove 1, 2, 3, 4, 6, or 12 chips. The only terminal position is 0. This game is called **Dim**$^+$ in *Winning Ways*. Find the Sprague-Grundy function.
   (b) Suppose the above rules are in force with the exception that it is not allowed to remove the whole pile. This is called the Aliquot game by Silverman (1971). Thus, if there are 12 chips, you may remove 1, 2, 3, 4, or 6 chips. The only terminal position is 1. Find the Sprague-Grundy function.
5. **Wythoff's Game.** (Wythoff (1907)) The positions of the Wythoff's game are given by a queen on a chessboard. Players, sitting on the same

side of the board, take turns moving the queen. But the queen may only be moved vertically down, or horizontally to the left or diagonally down to the left. When the queen reaches the lower left corner, the game is over and the player to move last wins. Thinking of the squares of the board as vertices and the allowed moves of the queen as edges of a graph, this becomes a graph game. Find the Sprague-Grundy function of the graph by writing in each square of the 8×8 chessboard its Sprague-Grundy value.

6. **Two-Dimensional Nim** is played on a quarter-infinite board with a finite number of counters on the squares. A move consists in taking a counter and moving it any number of squares to the left on the same row, or moving it to any square whatever on any lower row. A square is allowed to contain any number of counters. If all the counters are on the lowest row, this is just the game Nimble of Exercise 3 of Chap. 2.

   (a) Find the Sprague-Grundy values of the squares.
   (b) After you learn the theory contained in the next section, come back and see if you can solve the position represented by the figure below. Is the position below a P-position or an N-position? If it is an N-position, what is a winning move? How many moves will this game last? Can it last an infinite number of moves?

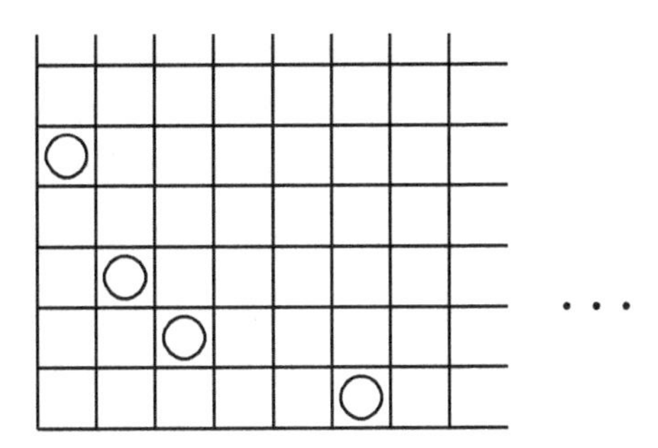

   (c) Suppose we allow adding any finite number of counters to the left of, or to any row below, the counter removed. Does the game still end in a finite number of moves?

7. Show that subtraction games with finite subtraction sets have Sprague-Grundy functions that are eventually periodic.
8. **Impatient subtraction games.** Suppose we allow an extra move for impatient players in subtraction games. In addition to removing

s chips from the pile where s is in the subtraction set, S, we allow the whole pile to be taken at all times. Let $g(x)$ represent the Sprague-Grundy function of the subtraction game with subtraction set S, and let $g^+(x)$ represent the Sprague-Grundy function of impatient subtraction game with subtraction set S. Show that $g^+(x) = g(x-1)+1$ for all $x \geq 1$.

9. The following directed graphs have cycles and so are not progressively finite. See if you can find the P- and N-positions and the Sprague-Grundy function.

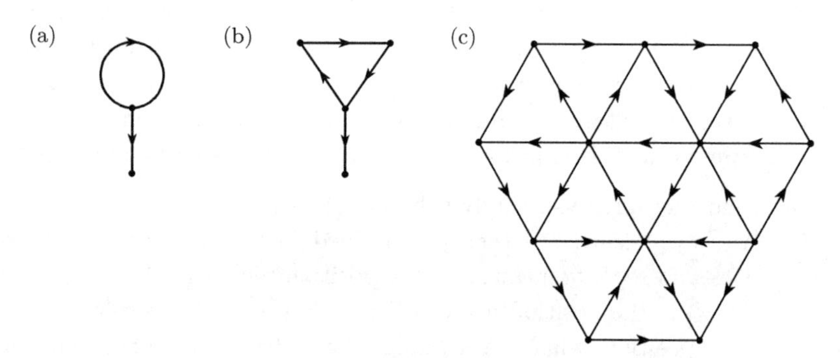

10. Consider the following game with n piles of chips. We may remove any number of chips from one pile as in Nim, or we may remove the same number of chips from any three piles. Analyze this game.

    (For $n = 3$, this may be considered as an extension to 3 dimensions of Wythoff's Game of Exercise 5, where the diagonal moves of the queen are restricted to the main diagonal, that is, $(x, y, z) \to (x-d, y-d, z-d)$, with $0 < d \leq \min(x, y, z)$.)

4 Sums of Combinatorial Games

Given several combinatorial games, one can form a new game played according to the following rules. A given initial position is set up in each of the games. Players alternate moves. A move for a player consists in selecting any one of the games and making a legal move in that game, leaving all other games untouched. Play continues until all of the games have reached a terminal position, when no more moves are possible. The player who made the last move is the winner.

The game formed by combining games in this manner is called the (***disjunctive***) ***sum*** of the given games. We first give the formal definition of a sum of games and then show how the Sprague-Grundy functions for the component games may be used to find the Sprague-Grundy function of the sum. This theory is due independently to R. P. Sprague (1936–7) and P. M. Grundy (1939).

4.1 The Sum of n Graph Games

Suppose we are given n progressively bounded graphs, $G_1 = (X_1, F_1), G_2 = (X_2, F_2), \ldots, G_n = (X_n, F_n)$. One can combine them into a new graph, $G = (X, F)$, called the ***sum*** of $G_1, G_2, \ldots, G_n$ and denoted by $G = G_1 + \cdots + G_n$ as follows. The set X of vertices is the Cartesian product, $X = X_1 \times \cdots \times X_n$. This is the set of all n-tuples $(x_1, \ldots, x_n)$ such that $x_i \in X_i$ for all i. For a vertex $x = (x_1, \ldots, x_n) \in X$, the set of followers of x is defined as

$$\begin{aligned} F(x) = F(x_1, \ldots, x_n) = & \; F_1(x_1) \times \{x_2\} \times \cdots \times \{x_n\} \\ & \cup \{x_1\} \times F_2(x_2) \times \cdots \times \{x_n\} \\ & \cup \cdots \\ & \cup \{x_1\} \times \{x_2\} \times \cdots \times F_n(x_n). \end{aligned}$$

Thus, a move from $x = (x_1, \ldots, x_n)$ consists in moving exactly one of the x_i to one of its followers (i.e. a point in $F_i(x_i)$). The graph game played on G is called the ***sum*** of the graph games $G_1, \ldots, G_n$.

If each of the graphs G_i is progressively bounded, then the sum G is progressively bounded as well. The maximum number of moves from a vertex $x = (x_1, \ldots, x_n)$ is the sum of the maximum numbers of moves in each of the component graphs.

As an example, the 3-pile game of nim may be considered as the sum of three one-pile games of nim. This shows that even if each component game is trivial, the sum may be complex.

4.2 The Sprague-Grundy Theorem

The following theorem gives a method for obtaining the Sprague-Grundy function for a sum of graph games when the Sprague-Grundy functions are known for the component games. This involves the notion of nim-sum defined earlier. The basic theorem for sums of graph games says that the Sprague-Grundy function of a sum of graph games is the nim-sum of the Sprague-Grundy functions of its component games. It may be considered a rather dramatic generalization of Theorem 1 of Bouton.

The proof is similar to the proof of Theorem 1.

Theorem 2. *If g_i is the Sprague-Grundy function of G_i, $i = 1, \ldots, n$, then $G = G_1 + \cdots + G_n$ has Sprague-Grundy function $g(x_1, \ldots, x_n) = g_1(x_1) \oplus \cdots \oplus g_n(x_n)$.*

Proof. Let $x = (x_1, \ldots, x_n)$ be an arbitrary point of X. Let $b = g_1(x_1) \oplus \cdots \oplus g_n(x_n)$. We are to show two things for the function $g(x_1, \ldots, x_n)$:

(1) For every nonnegative integer $a < b$, there is a follower of $(x_1, \ldots, x_n)$ that has g-value a.
(2) No follower of $(x_1, \ldots, x_n)$ has g-value b.

Then the SG-value of x, being the smallest SG-value not assumed by one of its followers, must be b.

To show (1), let $d = a \oplus b$, and k be the number of digits in the binary expansion of d, so that $2^{k-1} \leq d < 2^k$ and d has a 1 in the kth position (from the right). Since $a < b$, b has a 1 in the kth position and a has a 0 there. Since $b = g_1(x_1) \oplus \cdots \oplus g_n(x_n)$, there is at least one x_i such that the binary expansion of $g_i(x_i)$ has a 1 in the kth position. Suppose for simplicity that $i = 1$. Then $d \oplus g_1(x_1) < g_1(x_1)$ so that there is a move from x_1 to some x_1' with $g_1(x_1') = d \oplus g_1(x_1)$. Then the move from $(x_1, x_2, \ldots, x_n)$ to $(x_1', x_2, \ldots, x_n)$ is a legal move in the sum, G, and

$$\begin{aligned} g_1(x_1') &\oplus g_2(x_2) \oplus \cdots \oplus g_n(x_n) \\ &= d \oplus g_1(x_1) \oplus g_2(x_2) \oplus \cdots \oplus g_n(x_n) \\ &= d \oplus b = a. \end{aligned}$$

Finally, to show (2), suppose to the contrary that $(x_1, \ldots, x_n)$ has a follower with the same g-value, and suppose without loss of generality that this involves a move in the first game. That is, we suppose that $(x_1', x_2, \ldots, x_n)$ is a follower of $(x_1, x_2, \ldots, x_n)$ and that $g_1(x_1') \oplus g_2(x_2) \oplus \cdots \oplus g_n(x_n) = g_1(x_1) \oplus g_2(x_2) \oplus \cdots \oplus g_n(x_n)$. By the cancellation law, $g_1(x_1') = g_1(x_1)$. But this is a contradiction since no position can have a follower of the same SG-value. □

One remarkable implication of this theorem is that every progressively bounded impartial game, when considered as a component in a sum of such games, behaves as if it were a nim pile. That is, it may be replaced by a nim pile of appropriate size (its Sprague-Grundy value) without changing the outcome, no matter what the other components of the sum may be. We express this observation by saying that *every (progressively bounded) impartial game is equivalent to some nim pile.*

4.3 Applications

1. *Sums of Subtraction Games.* Let us denote by $G(m)$ the one-pile subtraction game with subtraction set $S_m = \{1, 2, \ldots, m\}$, in which from 1 to m chips may be removed from the pile, has Sprague-Grundy function $g_m(x) = x \pmod{m+1}$, and $0 \le g_m(x) \le m$.

Consider the sum of three subtraction games. In the first game, $m = 3$ and the pile has 9 chips. In the second, $m = 5$ and the pile has 10 chips. And in the third, $m = 7$ and the pile has 14 chips. Thus, we are playing the game $G(3) + G(5) + G(7)$ and the initial position is $(9, 10, 14)$. The Sprague-Grundy value of this position is $g(9, 10, 14) = g_3(9) \oplus g_5(10) \oplus g_7(14) = 1 \oplus 4 \oplus 6 = 3$. One optimal move is to change the position in game $G(7)$ to have Sprague-Grundy value 5. This can be done by removing one chip from the pile of 14, leaving 13. There is another optimal move. Can you find it?

This shows the importance of knowing the Sprague-Grundy function. We present further examples of computing the Sprague-Grundy function for various one-pile games. Note that although many of these one-pile games are trivial, as is one-pile nim, the Sprague-Grundy function has its main use in playing the sum of several such games.

2. *Even if Not All — All if Odd.* Consider the one-pile game with the rule that you can remove (1) any even number of chips provided it is not the whole pile, or (2) the whole pile provided it has an odd number of chips. There are two terminal positions, zero and two. We compute inductively,

x	0	1	2	3	4	5	6	7	8	9	10	11	12	...
$g(x)$	0	1	0	2	1	3	2	4	3	5	4	6	5	...

and we see that $g(2k) = k - 1$ and $g(2k - 1) = k$ for $k \geq 1$.

Suppose this game is played with three piles of sizes 10, 13 and 20. The SG-values are $g(10) = 4$, $g(13) = 7$ and $g(20) = 9$. Since $4 \oplus 7 \oplus 9 = 10$ is not zero, this is an N-position. A winning move is to change the SG-value 9 to a 3. For this we may remove 12 chips from the pile of 20 leaving 8, since $g(8) = 3$.

3. *A Sum of Three Different Games.* Suppose you are playing a three-pile take-away game. For the first pile of 18 chips, the rules of the previous game, Even if Not All — All if Odd, apply. For the second pile of 17 chips, the rules of At-Least-Half apply (Example 3.3.3). For the third pile of 7 chips, the rules of nim apply. First, we find the SG-values of the three piles to be 8, 5, and 7 respectively. This has nim-sum 10 and so is an N-position. It can be changed to a P-position by changing the SG-value of the first pile to 2. From the above table, this occurs for piles of 3 and 6 chips. We cannot move from 18 to 3, but we can move from 18 to 6. Thus an optimal move is to subtract 12 chips from the pile of 18 chips leaving 6 chips.

4.4 Take-and-Break Games

There are many other impartial combinatorial games that may be solved using the methods of this chapter. We describe Take-and-Break Games here, and in Chaps. 5 and 6, we look at coin-turning games and at Green Hackenbush. Take-and-Break Games are games where the rules allow taking and/or splitting one pile into two or more parts under certain conditions, thus increasing the number of piles.

1. *Lasker's Nim.* A generalization of Nim into a Take-and-Break Game is due to Emanuel Lasker, world chess champion from 1894 to 1921, and found in his book, *Brettspiele der Völker* (1931), 183–196.

Suppose that each player at his turn is allowed (1) to remove any number of chips from one pile as in nim, or (2) to split one pile containing at least two chips into two non-empty piles (no chips are removed).

Clearly the Sprague-Grundy function for the one-pile game satisfies $g(0) = 0$ and $g(1) = 1$. The followers of 2 are 0, 1 and $(1, 1)$, with respective Sprague-Grundy values of 0, 1, and $1 \oplus 1 = 0$. Hence, $g(2) = 2$. The followers of 3 are 0, 1, 2, and $(1, 2)$, with Sprague-Grundy values 0, 1, 2, and $1 \oplus 2 = 3$. Hence, $g(3) = 4$. Continuing in this manner, we see

x	0	1	2	3	4	5	6	7	8	9	10	11	12	...
$g(x)$	0	1	2	4	3	5	6	8	7	9	10	12	11	...

We therefore conjecture that $g(4k + 1) = 4k + 1$, $g(4k + 2) = 4k + 2$, $g(4k + 3) = 4k + 4$ and $g(4k + 4) = 4k + 3$, for all $k \geq 0$.

The validity of this conjecture may easily be verified by induction as follows.

(a) The followers of $4k+1$ that consist of a single pile have Sprague-Grundy values from 0 to $4k$. Those that consist of two piles, $(4k, 1), (4k - 1, 2), \ldots, (2k + 1, 2k)$, have even Sprague-Grundy values, and therefore $g(4k + 1) = 4k + 1$.

(b) The followers of $4k + 2$ that consist of a single pile have Sprague-Grundy values from 0 to $4k + 1$. Those that consist of two piles, $(4k + 1, 1), (4k, 2), \ldots, (2k + 1, 2k + 1)$, have Sprague-Grundy values alternately divisible by 4 and odd, so that $g(4k + 2) = 4k + 2$.

(c) The followers of $4k + 3$ that consist of a single pile have Sprague-Grundy values from 0 to $4k + 2$. Those that consist of two piles, $(4k + 2, 1), (4k + 1, 2), \ldots, (2k + 2, 2k + 1)$, have odd Sprague-Grundy values, and in particular $g(4k + 2, 1) = 4k + 3$. Hence $g(4k + 3) = 4k + 4$.

(d) Finally, the followers of $4k+4$ that consist of a single pile have Sprague-Grundy values from 0 to $4k + 2$, and $4k + 4$. Those that consist of two piles, $(4k + 3, 1)(4k + 2, 2), \ldots, (2k + 2, 2k + 2)$, have Sprague-Grundy values alternately equal to 1 (mod 4) and even. Hence, $g(4k + 4) = 4k + 3$.

Suppose you are playing Lasker's nim with three piles of 2, 5, and 7 chips. What is your move? First, find the Sprague-Grundy value of the component positions to be 2, 5, and 8 respectively. The nim-sum of these three numbers is 15. You must change the position of Sprague-Grundy value 8 to a position of Sprague-Grundy value 7. This may be done by splitting the pile of 7 chips into two piles of say 1 and 6 chips. At the next move, your opponent will be faced with a four-pile game of Lasker's nim with 1, 2, 5 and 6 chips. This has Sprague-Grundy value zero and so is a P-position.

2. *The Game of Kayles.* This game was introduced about 1900 by Sam Loyd (see *Mathematical Puzzles of Sam Loyd, Vol. 2*, 1960, Dover), and by H. E. Dudeney (see *The Canterbury Puzzles and Other Curious Problems*, 1958, Dover). Two bowlers face a line of 13 bowling pins in a row with the second pin already knocked down. "It is assumed that the ancient players had become so expert that they could always knock down any single kayle-pin, or any two kayle-pins that stood close together (i.e. adjacent pins). They therefore altered the game, and it was agreed that the player who knocked down the last pin was the winner."

This is one of our graph games played with piles of chips whose moves can be described as follows. You may remove one or two chips from any pile after which, if desired, you may split that pile into two nonempty piles.

Removing one chip from a pile without splitting the pile corresponds to knocking down the end chip of a line. Removing one chip with splitting the pile into two parts corresponds to knocking down a pin in the interior of the line. Similarly for removing two chips.

Let us find the Sprague-Grundy function, $g(x)$, for this game. The only terminal position is a pile with no chips, so $g(0) = 0$. A pile on one chip can be moved only to an empty pile, so $g(1) = 1$. A pile of two chips can be moved either to a pile of one or zero chips, so $g(2) = 2$. A pile of three chips can be moved to a pile of two or one chips, (SG-value 1 and 2) or to two piles of one chip each (SG-value 0), so $g(3) = 3$. Continuing in this way, we find more of the Sprague-Grundy values in Table 4.1.

Table 4.1. The SG-values for Kayles. Entries for the table are for $g(y+z)$ where y is on the left side and z is at the top.

$y\backslash z$	0	1	2	3	4	5	6	7	8	9	10	11
0	**0**	1	2	**3**	1	4	**3**	2	1	**4**	2	**6**
12	4	1	2	**7**	1	4	**3**	2	1	**4**	**6**	7
24	4	1	2	8	**5**	4	**7**	2	1	8	6	7
36	4	1	2	**3**	1	4	7	2	1	8	2	7
48	4	1	2	8	1	4	7	2	1	**4**	2	7
60	4	1	2	8	1	4	7	2	1	8	**6**	7
72	4	1	2	8	1	4	7	2	1	8	2	7

From $x = 72$ onwards, the SG-values are periodic with period 12, with the values in the last line repeating forever. The 14 exceptions to the sequence of the last line are displayed in bold face type. The last exception is at $x = 70$. Since we know all the SG-values, we may consider the game as solved. This solution is due to Guy and Smith (1956).

3. *Dawson's Chess.* One of T. R. Dawson's fanciful problems in *Caissa's Wild Roses* (1935), republished in *Five Classics of Fairy Chess* by Dover (1973), is give-away chess played with pawns. "Given two equal lines of opposing Pawns, White on 3rd rank, Black on 5th, n adjacent files, White to play at losing game, what is the result?" (Captures must be made, last player to move loses.) We treat this game here under the normal play rule, that the last player to move wins.

Those unfamiliar with the movement of the pawn in chess might prefer a different way of describing the game as a kind of misére tic-tac-toe on a line of n squares, with both players using X as the mark. A player may place an X on any empty square provided it is not next to an X already placed. (The player who is forced to move next to another X loses.)

This game may also be described as a game of removing chips from a pile and possibly splitting a pile into two piles. If $n = 1$ there is only one move to $n = 0$, ending the game. For $n > 1$, a move of placing an X at the end of a line removes the two squares at that end of the line from the game. This corresponds to removing two chips from the pile. Similarly, placing an X one square from the end corresponds to removing three chips from the pile. Placing an X in one of the squares not at the end or next to it corresponds to removing three chips from the pile and splitting the pile into two parts. Thus the rules of the game are: (1) You may remove one chip if it is the whole pile, or (2) you may remove two chips from any pile, or (3) you may remove three chips from any pile and if desired split that pile into two parts.

The Sprague-Grundy sequence for Dawson's chess begins

x	0	1	2	3	4	5	6	7	8	9	10	11	12	13	14	15	16	17	18
$g(x)$	0	1	1	2	0	3	1	1	0	3	3	2	2	4	0	5	2	2	3

It is eventually periodic with period 34. There are only 7 exceptions and the last exception occurs at $n = 51$. Can you find the winning move on a line of 18 squares? Armed with the above Sprague-Grundy sequence, you may hone your skill at the game by playing against the computer at http://www.math.ucla.edu/~tom/Games/dawson.html.

4. *Grundy's Game.* In Grundy's Game, the only legal move is to split a single pile into two nonempty piles of different sizes. Thus the only terminal positions are piles of size one or two. The Sprague-Grundy sequence is easy to compute for small pile sizes (see Exercise 10) and the values increase very slowly. Is the Sprague-Grundy sequence eventually periodic? This is unknown though more than 20,000,000,000 values have been computed as of 2002.

4.5 Exercises

1. Consider the take-away game with the rule that (1) you may remove any even number of chips from any pile, or (2) you may remove any pile consisting of one chip. The only terminal position is 0. Find the Sprague-Grundy function.
2. Consider the one-pile game with the rule that you may remove (1) any number of chips divisible by three provided it is not the whole pile, or (2) the whole pile, but only if it has 2 (mod 3) chips (that is, only if it has 2, or 5, or 8, ... chips). The terminal positions are zero, one and three. Find the Sprague-Grundy function.
3. Suppose you are playing a three-pile subtraction game. For the first pile of 18 chips, the rules of Exercise 1 hold. For the second pile of 17 chips, the rules of Exercise 2 apply. For the third pile of 7 chips, the rules of nim apply. What is the Sprague-Grundy value of this position? Find an optimal move.
4. Solve the Kayles problem of Dudeney and Loyd. Of the 13 bowling pins in a row, the second has been knocked down, leaving:

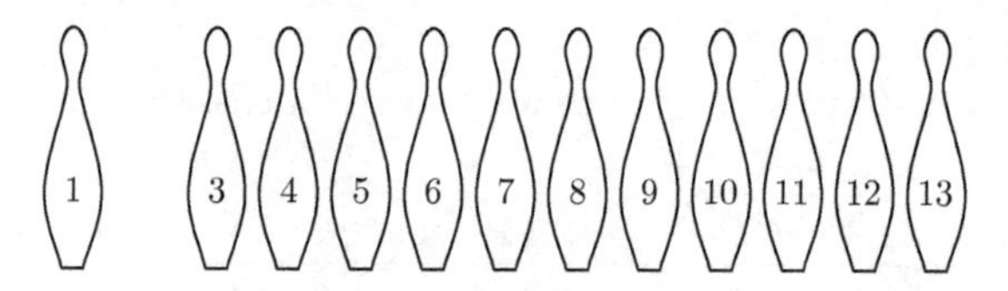

Fig. 4.1 The Kayles problem of Dudeney and Loyd.

   (a) Show this is an N-position. You may use Table 4.1.
   (b) Find a winning move. Which pin(s) should be knocked down?

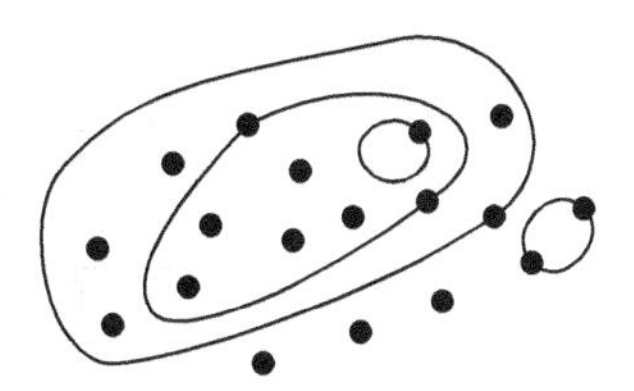

Fig. 4.2 A rims position

(c) Now that you know the theory and have Table 4.1 at hand, you can go to http://www.cut-the-knot.org and beat the computer.

5. Suppose at each turn a player may (1) remove one or two chips, or (2) remove one chip and split the remaining chips into two piles.
   (a) Find the Sprague-Grundy function.
   (b) Suppose the starting position consists of one pile with 15 chips. Find an optimal first move.
6. Suppose that at each turn a player may (1) remove one chip if it is a whole pile, or (2) remove two or more chips and, if desired, split the remaining chips into two piles. Find the Sprague-Grundy function.
7. Suppose that at each turn a player may select one pile and remove c chips if $c = 1 \pmod 3$ and, if desired, split the remaining chips into two piles. Find the Sprague-Grundy function.
8. **Rims.** A position in the game of Rims is a finite set of dots in the plane, possibly separated by some nonintersecting closed loops. A move consists of drawing a closed loop passing through any positive number of dots (at least one) but not touching any other loop. Players alternate moves and the last to move wins.
   (a) Show that this game is a disguised form of nim.
   (b) In the position given in Fig. 4.2, find a winning move, if any.
9. **Rayles.** There are many geometric games like Rims treated in Chap. 17 of *Winning Ways.* In one of them, called Rayles, the positions are those of Rims, but in Rayles, each closed loop must pass through exactly one or two points.
   (a) Show that this game is a disguised form of Kayles.
   (b) Assuming the position given in Fig. 4.2 is a Rayles position, find a winning move, if any.

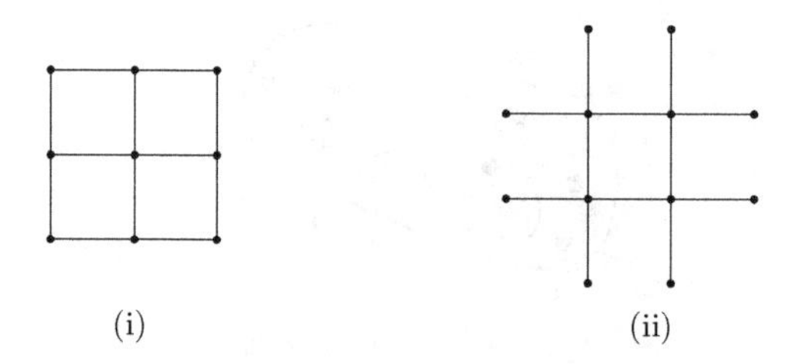

Fig. 4.3

10. **Grundy's Game.**
    (a) Compute the Sprague-Grundy function for Grundy's game, Example 4 Sec. 4.4, for a pile of n chips for $n = 1, 2, \ldots, 13$.
    (b) In Grundy's game with three piles of sizes 5, 8, and 13, find all winning first moves, if any.
11. A game is played on a finite (undirected) graph as follows. Players alternate moves. A move consists of removing a vertex and all edges incident to that vertex, with the exception that a vertex without any incident edges may not be removed. That is, at least one edge must be removed. Last player to move wins. Investigate this game. For example,
    (a) Find the Sprague-Grundy value of S_n, the star with n points. (The star with n points is the graph with $n+1$ vertices and n edges that share a common vertex.)
    (b) Find the Sprague-Grundy value of L_n, the line of n edges and $n+1$ vertices (at least for small values of n).
    (c) Find the Sprague-Grundy value of C_n, the circular graph of n vertices and n edges.
    (d) Find the Sprague-Grundy value of a double star. (A double star is a graph of two stars joined by a single edge.)
    (e) In Fig. 4.3(i) who wins the 3×3 square lattice? In Fig. 4.3(ii) who wins the tic-tac-toe board? Generalize.

5 Coin Turning Games

We have mentioned the game called Turning Turtles in Exercise 2.6.4 as a disguised version of nim. This form of the game of nim is a prototype of a class of other games that have similar descriptions. These ideas are due to H. W. Lenstra. In particular, the extension of these games to two dimensions leads to a beautiful theory involving nim multiplication which, with the definition of nim addition, makes the nonnegative integers into a field.

In this class of games, called Coin Turning Games, we are given a finite number of coins in a row, each showing either heads or tails. A move consists of turning over, from heads to tails or from tails to heads, all coins in a set of coins allowed by the rules of the game. The rules always specify that the rightmost coin turned over must go from heads to tails. The purpose of this is to guarantee that the game will end in a finite number of moves no matter how it is played (the Ending Condition).

The rules also specify that the set of coins that may be turned over depends only on the position of the rightmost coin turned over and not otherwise on the position of the other heads and tails, or on the previous moves of the game, or on time, etc. Moreover, the games are impartial and the last player to move wins.

Under these conditions, the same decomposition that worked for Turning Turtles also works for all these games, namely: A position with k heads in positions $x_1, \ldots, x_k$ is the (disjunctive) sum of k games each with exactly one head, where for $j = 1, \ldots, k$ the head in game j is at x_j. For example, the game, T H H T T H, is the sum of the three games, T H, T T H, and T T T T T H. This implies that to find the Sprague-Grundy value of a position, we only need to know the Sprague-Grundy values of positions with exactly one head. In the example, $g(\text{THHTTH}) = g(\text{TH}) \oplus g(\text{TTH}) \oplus g(\text{TTTTTH})$. We begin with some examples.

5.1 Examples

1. Subtraction Games. Many impartial combinatorial games can be put into the form of coin turning games, though the description may become rather complex. On the other hand, some games with simple descriptions as coin turning games have complex descriptions as games played with piles of counters. Subtraction Games have simple descriptions under both forms.

Take for example the simple take-away game of Sec. 1.1, where we may remove from 1 to 3 counters from a pile. As a coin-turning game, we may take the rules to be as follows. Label the positions of the coins starting from 1 for the leftmost coin, to n for the rightmost coin, where n is the number of coins. Some coin, say in position x, must be turned over from heads to tails, and a second coin in one of the positions $x-1$, $x-2$, or $x-3$, must be turned over, except when $x \leq 3$, in which case a second coin need not be turned over.

We may compute the Sprague-Grundy function of this game as we did in Sec. 3.3. A heads in position 1 can only be turned over, ending the game; so it has SG-value 1. A single heads in the second position can be moved to a single heads in the first position or to all tails; so it has SG-value 2. Continuing this analysis, we find

Position x	1	2	3	4	5	6	7	8	9	10	11	12	13	14	...
$g(x)$	1	2	3	0	1	2	3	0	1	2	3	0	1	2	...

(1)

exactly as before. A row of coins with several heads corresponds to several piles of counters.

Other subtraction games may be treated similarly. If S is the subtraction set, then as a coin-turning game it may be described as: Some x goes from heads to tails, and also some $y = x - z$ is turned over for some z in S, unless $x - z = 0$ for some z in S in which case no second coin need be reversed.

There is another way of describing subtraction games as coin-turning games. We require *exactly* two coins to be turned over at all times, the distance between the coins being y for some $y \in S$, with the rightmost coin going from heads to tails. In the analysis for this case, it is more convenient to start labelling the positions of the coins from 0 instead of 1. The first coin on the left is in position 0, the second coin from the left is in position 1 and so on.

In the analysis of the simple case where $S = \{1, 2, 3\}$, a singleton heads in position 0 is a terminal position and so has SG-value 0. A singleton heads in position 1 can only be moved to a singleton heads in position 0 and so has SG-value 1. Continuing in the manner gives the same Sprague-Grundy function as in (1) but starting at position $x = 0$ with $g(0) = 0$.

2. Twins. This method applied to the game Turning Turtles gives us an equivalent game called Twins, which is also equivalent to nim. In Turning Turtles, we may turn over one or two coins, with the rightmost coin going from heads to tails. In the game, Twins, the rules are: *Turn over exactly two coins, with the rightmost coin going from heads to tails.* If we label the positions of the coins from the left starting with 0, then the Sprague-Grundy function will be $g(x) = x$, the same as nim.

3. Mock Turtles. Here is a game that is better suited for description as a coin-turning game than as a game involving piles of counters. It is like Turning Turtles but we are allowed to turn over up to three coins. The rules are: *Turn over one, two, or three coins with the rightmost going from heads to tails.* In computing the Sprague-Grundy function, it is again useful to start labelling the positions from 0, but for a different reason.

A heads in position 0 can only move to the terminal position and so has SG-value 1. A heads in position 1 can move to heads in position 0 or to the terminal position and so has SG-value 2. A heads in position 2 can move to heads in position 1 (SG-value 2), or to heads in position 0 (SG-value 1), or to heads in positions 0 and 1 (SG-value 3) or to the terminal position, and so has SG-value 4. Continuing further we find

Position x	0	1	2	3	4	5	6	7	8	9	10	11	12	13	14	...
$g(x)$	1	2	4	7	8	11	13	14	16	19	21	22	25	26	28	...

(2)

What are these numbers? It seems as if $g(x)$ is equal to either $2x$ or $2x+1$; but which one? The answer depends on the number of 1's appearing in the binary expansion of $2x$. An nonnegative integer is called *odious* if the number of 1's in its binary expansion is odd, and it is called *evil* otherwise. Thus the numbers 1, 2, 4, and 7 are odious since the binary expansions are 1, 10, 100, and 111. The numbers 0, 3, 5, and 6 are evil (expansions 0, 11, 101, and 110). All the numbers appearing as $g(x)$ in (2) are odious. Moreover, the sequence (2) seems to contain all the odious numbers in order. Thus it seems as if $g(x)$ is $2x$ if $2x$ is odious and $2x+1$ if $2x$ is evil. How can we check this?

The key observation is to note that the nim-sum of two evil numbers or two odious numbers is evil, and that the nim-sum of an evil number and an odious number is odious. Symbolically,

$$\text{evil} \oplus \text{evil} = \text{odious} \oplus \text{odious} = \text{evil}$$
$$\text{evil} \oplus \text{odious} = \text{odious} \oplus \text{evil} = \text{odious}$$

To show that the SG-values in (2) are just the odious numbers in order, we proceed by induction. Suppose this hypothesis is true for every position to the left of x, and we want to show that $g(x)$ is the next odious number. Note that we can always move to an SG-value 0 by turning over just the coin at x. By turning over the coin at x and one other coin, we can move to all previous odious numbers. By turning over the coin at x and two other coins, we can move to all smaller evil numbers, since every non-zero evil number is the nim-sum of two smaller odious numbers; but we can never move to the next odious number because the sum of two odious numbers is evil. This completes the induction.

Suppose now that a Mock Turtles position with heads at positions $x_1, x_2, \ldots, x_n$ is a P-position, so that $g(x_1)\oplus\cdots\oplus g(x_n) = 0$. Then certainly n must be even since the sum of an odd number of odious numbers is odious and cannot be zero. Moreover, since $g(x)$ is just $2x$ with a 1 added if necessary to make $g(x)$ odious, we must have $x_1 \oplus \cdots \oplus x_n = 0$. Conversely, if $x_1 \oplus \cdots \oplus x_n = 0$ and n is even, then $g(x_1) \oplus \cdots \oplus g(x_n) = 0$. This gives us a very simple description of the P-positions in Mock Turtles.

The P-positions in Mock Turtles are exactly those P-positions in nim that have an even number of piles.

Note that $\{1, 2, 3\}$ is a P-position in nim, but not in Mock Turtles. However, the position $\{0, 1, 2, 3\}$ is a P-position in Mock Turtles. The coin at position 0 is known as the Mock Turtle. Even though the Mock Turtle has Sprague-Grundy value 0 in this interpretation of the P-positions, it still plays a role in the solution. Try out this game at http://www.tomsferguson.com/Games/mock.html.

4. Ruler. Here is another game with a simple description as a coin-turning game. *Any number of coins may be turned over but they must be consecutive, and the rightmost coin must be turned from heads to tails.* If we number the positions of the coins starting with 1 on the left, the Sprague-Grundy function for this game satisfies

$$g(n) = \text{mex}\{0, g(n-1), g(n-1) \oplus g(n-2), \ldots, g(n-1) \oplus \ldots \oplus g(1)\}.$$

From this, it is easy to show that the function satisfies

Position x	1	2	3	4	5	6	7	8	9	10	11	12	13	14	15	16	...
$g(x)$	1	2	1	4	1	2	1	8	1	2	1	4	1	2	1	16	...

(3)

Thus, $g(x)$ is the largest power of 2 dividing x. This game is called Ruler because of the way the function $g(x)$ imitates the marks on a ruler, (similar to the Sprague-Grundy function of Dim$^+$). You may play the game at http://www.tomsferguson.com/Games/ruler.html.

5. Grunt. *Turn over four symmetric coins, one of which is the leftmost coin, (e.g.* $0, x, n-x, n$ *for some* $1 \leq x < n/2$*) with the rightmost coin going from heads to tails.* This is Grundy's game in disguise. Number the positions of the coins from 0. A singleton heads in positions 0, 1, or 2, is a terminal position and so has SG-value 0. A singleton heads in position $n \geq 3$ gets replaced by heads in positions 0, x and $n-x$ for some $x < n/2$. The heads in position 0 has SG-value 0, so this is like taking a pile of n counters and splitting it into two non-empty piles of different sizes — exactly Grundy's Game.

5.2 Two-dimensional Coin-Turning Games

In the generalization of coin-turning games to two dimensions, the coins to be turned are in a rectangular array. We number the coordinates of the array starting at $(0, 0)$, with coins at coordinates (x, y) with $x \geq 0$ and $y \geq 0$.

The condition that the rightmost coin turned must be from heads to tails is replaced by the condition that one of the coins, say at (x, y) and referred to as the southeast coin, goes from heads to tails and that any other coins that are turned over must be in the rectangle $\{(a, b) : 0 \leq a \leq x, 0 \leq b \leq y\}$.

1. Acrostic Twins. A move is to turn over two coins, either in the same row or the same column, with the southeast coin going from heads to tails. The Sprague-Grundy function satisfies

$$g(x, y) = \text{mex}\{g(x, b), g(a, y) : 0 \leq b < y, 0 \leq a < x\}, \qquad (4)$$

Table 5.1. For acrostic twins, $g(x, y) = x \oplus y$.

	0	1	2	3	4	5	6	7	8	9
0	0	1	2	3	4	5	6	7	8	9
1	1	0	3	2	5	4	7	6	9	8
2	2	3	0	1	6	7	4	5	10	11
3	3	2	1	0	7	6	5	4	11	10
4	4	5	6	7	0	1	2	3	12	13
5	5	4	7	6	1	0	3	2	13	12
6	6	7	4	5	2	3	0	1	14	15
7	7	6	5	4	3	2	1	0	15	14
8	8	9	10	11	12	13	14	15	0	1
9	9	8	11	10	13	12	15	14	1	0

this being the least number not appearing earlier in the same row or column as (x, y). If $y = 0$, this game is just nim, so $g(x, 0) = x$; similarly $g(0, y) = y$, the function being symmetric. It is easy to construct the following table of the values of $g(x, y)$; each entry is just the mex of the numbers above it and to the left. When we do we find a pleasant surprise — that $g(x, y)$ is just the nim-sum, $x \oplus y$.

To fix the ideas, consider the game with layout

$$\begin{array}{ccccc} T & T & T & T & T \\ T & T & T & H & T \\ T & T & T & T & T \\ T & T & H & T & T \\ T & T & T & T & H \end{array} \qquad (5)$$

We use the convention when referring to a position at (x, y) in such a layout, that x represents row and y represents column, where rows and columns are numbered starting at 0. Thus, the three heads are at positions (1, 3), (3, 2) and (4, 4). The heads are at positions with SG-values, 2, 1 and 0. Since $2 \oplus 1 \oplus 0 = 3$, this is an N-position. It can be moved to a P-position by moving the SG-value 2 to SG-value 1. This is done by turning over the coins at positions (1, 3) and (1, 0). (There are other winning moves. Can you find them?)

2. Turning Corners. Let us examine a more typical game.

A move consists of turning over four distinct coins at the corners of a rectangle, i.e. (a, b), (a, y), (x, b) and (x, y), where $0 \le a < x$ and $0 \le b < y$, the coin at (x, y) going from heads to tails.

Table 5.2. For turning corners, $g(x, y) = x \otimes y$.

	0	1	2	3	4	5	6	7	8	9	10	11	12	13	14	15
0	0	0	0	0	0	0	0	0	0	0	0	0	0	0	0	0
1	0	1	2	3	4	5	6	7	8	9	10	11	12	13	14	15
2	0	2	3	1	8	10	11	9	12	14	15	13	4	6	7	5
3	0	3	1	2	12	15	13	14	4	7	5	6	8	11	9	10
4	0	4	8	12	6	2	14	10	11	15	3	7	13	9	5	1
5	0	5	10	15	2	7	8	13	3	6	9	12	1	4	11	14
6	0	6	11	13	14	8	5	3	7	1	12	10	9	15	2	4
7	0	7	9	14	10	13	3	4	15	8	6	1	5	2	12	11
8	0	8	12	4	11	3	7	15	13	5	1	9	6	14	10	2
9	0	9	14	7	15	6	1	8	5	12	11	2	10	3	4	13
10	0	10	15	5	3	9	12	6	1	11	14	4	2	8	13	7
11	0	11	13	6	7	12	10	1	9	2	4	15	14	5	3	8
12	0	12	4	8	13	1	9	5	6	10	2	14	11	7	15	3
13	0	13	6	11	9	4	15	2	14	3	8	5	7	10	1	12
14	0	14	7	9	5	11	2	12	10	4	13	3	15	1	8	6
15	0	15	5	10	1	14	4	11	2	13	7	8	3	12	6	9

The Sprague-Grundy function of this game satisfies the condition

$$g(x, y) = \text{mex}\{g(x, b) \oplus g(a, y) \oplus g(a, b) : 0 \leq a < x, 0 \leq b < y\}. \quad (6)$$

Since we require that the four corners of the selected rectangle be distinct, no coin along the edge of the board may be used as the southeast coin. Therefore, $g(x, 0) = g(0, y) = 0$ for all x and y. Moreover, $g(1, 1) = 1$ and $g(x, 1) = \text{mex}\{g(a, 1) : 0 \leq a < x\} = x$ by induction (or simply by noticing that the game starting at $(x, 1)$ is just nim). The rest of Table 5.2 may be constructed using Eq. (6).

As an illustration, let us compute $g(4, 4)$ using Eq. (6). There are 16 moves from position (4, 4), but using symmetry we can reduce the number to 10, say (x, y) for $0 \leq x \leq y \leq 3$. Computing the SG-values of each of these moves separately, we find $g(4, 4) = \text{mex}\{0, 4, 8, 12, 1, 14, 11, 3, 5, 2\} = 6$. With the aid of Table 5.2, you may play this game at http://www.tomsferguson.com/Games/corners.html.

If (5) represents a position in Turning Corners, we can find the SG-value easily as $3 \oplus 1 \oplus 6 = 4$. This is therefore an N-position, that can be moved to a P-position by moving the SG-value 6 to an SG-value 2. There is a unique way that this may be done, namely by turning over the four coins at (3, 3), (3, 4), (4, 3) and (4, 4).

$$\begin{array}{ccccc} T & T & T & T & T \\ T & T & T & H & T \\ T & T & T & T & T \\ T & T & H & T & T \\ T & T & T & T & H \end{array} \quad \longrightarrow \quad \begin{array}{ccccc} T & T & T & T & T \\ T & T & T & H & T \\ T & T & T & T & T \\ T & T & H & H & H \\ T & T & T & H & T \end{array}$$

5.3 Nim Multiplication

The entries in Table 5.2 may seem rather haphazard, but such an appearance is deceptive. In fact, the function g defined by (6) may be considered as a form of nim multiplication. Using the notation $x \otimes y$ to represent $g(x, y)$, we can easily see that 0 acts like a zero for multiplication since $x \otimes 0 = 0 \otimes x = 0$ for all x; and 1 acts like a unit for multiplication since $x \otimes 1 = 1 \otimes x = x$ for all x. Moreover, the commutative law obviously holds: $x \otimes y = y \otimes x$ for all x and y.

Not at all obvious is the fact that the associative law holds:

$$x \otimes (y \otimes z) = (x \otimes y) \otimes z, \quad \text{for all } x, y \text{ and } z. \tag{7}$$

Thus $3 \otimes (5 \otimes 6) = 3 \otimes 8 = 4$, and $(3 \otimes 5) \otimes 6 = 15 \otimes 6 = 4$.

Even more remarkably, when combined with nim addition, the distributive law holds:

$$x \otimes (y \oplus z) = (x \otimes y) \oplus (x \otimes z) \quad \text{for all } x, y \text{ and } z. \tag{8}$$

For example, $3 \otimes (5 \oplus 6) = 3 \otimes 3 = 2$ and $(3 \otimes 5) \oplus (3 \otimes 6) = 15 \oplus 13 = 2$.

Another significant fact is that every number other than zero has a multiplicative inverse. One sees this by noting that there is a 1 in every row except the top row of the multiplication table. Thus the nim reciprocal of 2 is 3 since $2 \otimes 3 = 1$. Using $\oslash$ to indicate nim division, this is written as $1 \oslash 2 = 3$. Check that we also have $1 \oslash 7 = 11$ and $6 \oslash 5 = 9$.

In mathematical terms, the nonnegative integers form a field under the operations of nim addition and nim multiplication. See Lenstra (1977–78). In the book of Conway (1976), it is shown that this is the "simplest" field, in some sense, that one can make out of the integers. We assume in what follows that laws (7) and (8) hold.

Aren't you glad you don't have to memorize this multiplication table? Still, it would be nice if one could find the product, say $24 \otimes 17$, without first finding all products of smaller numbers. Here is a method of accomplishing that feat. Let us call a number of the form 2^{2^n} for $n = 0, 1, 2, \ldots$, a *Fermat*

2-power. These are the numbers 2, 4, 16, 256, 65536, ... We can find the nim-product of two numbers using the following two rules, along with the associative and distributive laws, (7) and (8).

(i) The nim-product of a Fermat 2-power and any smaller number is their ordinary product.
(ii) The nim-product of a Fermat 2-power with itself is the Fermat 2-power times 3/2 in the ordinary sense.

Thus, $2 \otimes 16 = 32$, but $16 \otimes 16 = (3/2)16 = 24$. Let us find $24 \otimes 17$.

$$\begin{aligned} 24 \otimes 17 &= (16 \oplus 8) \otimes (16 \oplus 1) \\ &= (16 \otimes 16) \oplus (16 \otimes 1) \oplus (8 \otimes 16) \oplus (8 \otimes 1) \\ &= 24 \oplus 16 \oplus 128 \oplus 8 = 128. \end{aligned}$$

As another example, let us find $8 \otimes 8$. We cannot write 8 as the nim-sum of two simpler numbers, but we can write it as the nim-product of two simpler numbers, namely, $8 = 2 \otimes 4$. So,

$$\begin{aligned} 8 \otimes 8 &= (2 \otimes 4) \otimes (2 \otimes 4) = 2 \otimes 2 \otimes 4 \otimes 4 = 3 \otimes 6 = (2 \oplus 1) \otimes (4 \oplus 2) \\ &= (2 \otimes 4) \oplus (2 \otimes 2) \oplus (1 \otimes 4) \oplus (1 \otimes 2) = 8 \oplus 3 \oplus 4 \oplus 2 = 13, \end{aligned}$$

which agrees with the value given in Table 5.2.

5.4 Tartan Games

The game Turning Corners is an example of a class of games with solutions that can be found with the aid of nim multiplication. These are the Tartan Games. Given two one-dimensional coin-turning games, G_1 and G_2, one can define the tartan game $G_1 \times G_2$ to be the two-dimensional coin-turning game whose moves can be described as follows. If turning coins at positions $x_1, x_2, \ldots, x_m$ is a legal move in G_1 and turning coins at positions $y_1, y_2, \ldots, y_n$ is a legal move in G_2, then turning coins at positions (x_i, y_j) for all $1 \le i \le m$ and $1 \le j \le n$ is a legal move in $G_1 \times G_2$ (provided, of course, that the southeast coin goes from heads to tails).

With this definition, the game Turning Corners is just the game Twins $\times$ Twins. In Twins, we turn over two coins, so in Turning Corners, we turn over four coins at the corners of a rectangle. Tartan Games may be analyzed by means of the following theorem in which G_1 and G_2 represent two 1-dimensional coin turning games.

The Tartan Theorem. *If $g_1(x)$ is the Sprague-Grundy function of G_1 and $g_2(y)$ is the Sprague-Grundy function of G_2, then the Sprague-Grundy function, $g(x, y)$, of $G_1 \times G_2$ is the nim product,*

$$g(x, y) = g_1(x) \otimes g_2(y). \tag{9}$$

For the game of Twins, the Sprague-Grundy function is simply $g(x) = x$. Since Turning Corners = Twins × Twins, the Tartan Theorem says that the Sprague-Grundy function for Turning Corners is just $g(x, y) = x \otimes y$.

Turning Turtles Squared. In the game (Turning Turtles) × (Turning Turtles), we may turn over any four coins at the corners of a rectangle as in Turning Corners, but we may also turn over any two coins in the same row, or any two coins in the same column, or we may turn over any single coin. Thus it is like Turning Corners, but the rectangle may be degenerate. Although the Sprague-Grundy function of Turning Turtles is $g(x) = x$ for $x \geq 1$, remember that we started labelling the positions from 1 rather than 0. Equivalently, we may cross out all the zeros from Table 5.2 and use that as the Sprague-Grundy function, but we must remember that we are starting the labelling at (1, 1).

For example, if display (5) is a position in (Turning Turtles)^2, the SG-value is $8 \oplus 12 \oplus 7 = 3$. A winning move is to change the SG-value 7 to SG-value 4. This may be done by turning over the coin at (5, 5) (remember we begin the labelling at (1, 1)) and also the three coins at (2, 3), (2, 5) and (5, 3), whose combined SG-value is $1 \oplus 10 \oplus 15 = 4$.

```
T T T T T          T T T T T
T T T H T          T T H H H
T T T T T    →     T T T T T
T T H T T          T T H T T
T T T T H          T T H T T
```

Rugs. A more interesting example is given by the game Rugs = Ruler × Ruler. Since a move in Ruler is to turn over all coins in any set of consecutive coins, *a move in Rugs is to turn over all coins in any rectangular block of coins.* Let us construct a table of the SG-values of the game Rugs. Recall that the Sprague-Grundy sequence for Ruler is 1, 2, 1, 4, 1, 2, 1, 8, ..., starting the labelling at 1. The SG-values for Rugs are therefore just

Table 5.3. The Sprague-Grundy values for Rugs.

	1	2	3	4	5	6	7	8
1	1	2	1	4	1	2	1	8
2	2	3	2	8	2	3	2	12
3	1	2	1	4	1	2	1	8
4	4	8	4	6	4	8	4	11
5	1	2	1	4	1	2	1	8
6	2	3	2	8	2	3	2	12
7	1	2	1	4	1	2	1	8
8	8	12	8	11	8	12	8	13

the nim products of these numbers. In Table 5.3, we put the SG-values for Ruler along the edges, and the nim products in the body of the table.

As an example, consider the position of display (5) as a position in Rugs, The SG-values of the heads are 8, 4, and 1 with nim sum 13. There is a winning move that takes the SG-value 8 and replaces it with an SG-value 5. The move that achieves this is the move that turns over all the coins in the 2×4 rectangle from (1, 1) to (2, 4).

```
T T T T T          H H H H T
T T T H T          H H H T T
T T T T T    ⟶    T T T T T
T T H T T          T T H T T
T T T T H          T T T T H
```

This game is also called *Turnablock.* You may play it on the web at http:www.tomsferguson.com/Games/rugs.html.

Solving Tartan Games. One of the difficulties of dealing with tartan games is that it is not always easy to find winning moves due to the bewildering number and complexity of moves available. Consider for example the following position in a game of (Mock Turtles)$\times$(Mock Turtles) with just two heads.

```
T H T T T T
T T T T T T
T T T T T T
T T T T T T
T T T T T H
```

It is easy to find whether or not we can win if we are to move in this position. Recalling that the SG-sequence for Mock Turtles is 1, 2, 4, 7, 8, 11, we can

Table 5.4. The Sprague-Grundy values for (Mock Turtles)^2.

	0	1	2	3	4	5
0	1	2	4	7	8	11
1	2	3	8	9	12	13
2	4	8	6	10	11	7
3	7	9	10	4	15	1
4	8	12	11	15	13	9

find the corresponding SG-values for this problem using the multiplication table.

We see that the two heads in our position occur at SG-values 2 and 9, with a nim-sum of 11. Therefore we can win if it is our turn to move. Moreover, to win we may replace the heads with SG-value 9 with some move of SG-value 2, but how do we find such a move? There are 176 different moves from this position using the heads with SG-value 9 as the southeast corner, and many of them are rather complex. Here is a method that will make the search for a winning move easier. It has the advantage that if you understand why the method works, you will be able to see why the Tartan Theorem holds. (See Exercise 7.)

The method is as follows. In playing $G_1 \times G_2$, suppose you are at position (x, y) with SG-value $g_1(x) \otimes g_2(y) = v$ and suppose you desire to replace the SG-value v by an SG-value $u < v$. (In the above example, $(x, y) = (4, 5)$, $g_1(x) = 8$, $g_2(y) = 11$, $v = 9$, and $u = 2$.) The method is in three parts.

(1) Let $v_1 = g_1(x)$ and $v_2 = g_2(y)$. First, find a move in *Turning Corners* that takes (v_1, v_2) into an SG-value u. Denote the northwest corner of the move by (u_1, u_2), so that $(u_1 \otimes u_2) \oplus (u_1 \otimes v_2) \oplus (v_1 \otimes u_2) = u$.
(2) Find a move M_1 in the 1-dimensional game G_1 that moves the SG-value $g_1(x)$ at x into an SG-value u_1. (Note that $u_1 < g_1(x)$.)
(3) Find a move M_2 in the 1-dimensional game G_2 that moves the SG-value $g_2(y)$ at y into an SG-value u_2. (Note that $u_2 < g_2(y)$.)

Then the move $M_1 \times M_2$ in $G_1 \times G_2$ moves to an SG-value u as desired.

Let us use this method to find a winning move in our position in (Mock Turtles)^2. We have $(v_1, v_2) = (8, 11)$ and we want to find (u_1, u_2) with $u_1 < v_1$ and $u_2 < v_2$ such that $(u_1 \otimes u_2) \oplus (u_1 \otimes 11) \oplus (8 \otimes u_2) = 2$. One possibility is $(u_1, u_2) = (3, 10)$, since $(3 \otimes 10) \oplus (3 \otimes 11) \oplus (8 \otimes 10) = 5 \oplus 6 \oplus 1 = 2$. (A simpler possibility is $(u_1, u_2) = (1, 1)$, but it would not illustrate the method as well.) Now we want to find a move in Mock Turtles to change the SG-value 8 into an SG-value 3. This is accomplished by turning over the coins at positions 0, 1 and 4, since $g_1(0) \oplus g_1(1) = 1 \oplus 2 = 3$. We also want to find a move in Mock Turtles to change the SG-value 11 into an SG-value 10. This is accomplished by turning over coins at 1, 4 and 5, since $g_2(1) \oplus g_2(4) = 2 \oplus 8 = 10$. Therefore, the move of turning over the 9 coins at $\{0, 1, 4\} \times \{1, 4, 5\}$, resulting in position

```
T T T T H H
T H T T H H
T T T T T T
T T T T T T
T H T T H T
```

is a winning move. Check that it has SG-value 0.

There are many more aspects to this beautiful topic and many more examples that may be found in Chap. 14 of *Winning Ways.*

5.5 Exercises

1. Consider the following position in a coin turning game.

   T T H T H H T T H

   Find a winning move, if any, if the game being played is

   (a) Turning Turtles.
   (b) Twins.
   (c) The subtraction game with subtraction set $S = \{1, 3, 4\}$ when exactly two coins must be turned over.
   (d) Mock Turtles.

2. (a) How many moves can a game of Turning Turtles last, if the initial position has a just one heads as the rightmost coin of a line of n coins?
   (b) Same question but for the game of Mock Turtles.
   (c) Same question but for the game of Ruler.

3. **Triplets.** Suppose in Mock Turtles, we are not allowed to turn over just one or two coins, but are required to turn over three coins. This game is called Triplets. The rules are: *Turn over exactly three coins, with the rightmost coin going from heads to tails.* Find the Sprague-Grundy

function for Triplets. Relate it to the Sprague-Grundy function for Mock Turtles.

4. **Rulerette.** Suppose in the game Ruler, we are not allowed to turn over just one coin. The rules are: *Turn over any consecutive set of coins with at least two coins being turned over, and the rightmost coin going from heads to tails.* Find the Sprague-Grundy function for this game and relate it to the Sprague-Grundy function for Ruler.
5. Consider a coin-turning game of the subtraction game variety, but allow a single coin to be turned over also. The rules are: *Turn over one or two coins, with the rightmost coin going from heads to tails; if two coins are turned over the distance between them must be in the subtraction set S.* Relate the Sprague-Grundy function of this game to the Sprague-Grundy function of the subtraction game with subtraction set S.
6. (a) Find $6 \otimes 21$.
   (b) Find $25 \otimes 40$.
   (c) Find $15 \oslash 14$.
   (d) Find the nim square root of 8. (Note that every integer has an integer nim square root.)
   (e) Solve $x^2 \oplus x \oplus 6 = 0$. There are two roots. (Here x^2 represents $x \otimes x$.)
7. Use the notation in the statement of the Tartan Theorem.
   (a) Consider a position in Turning Corners with a single heads at (v_1, v_2). It has SG-value $v_1 \otimes v_2$. Consider also a position in $G_1 \times G_2$ with a single heads at (x, y), where $g_1(x) = v_1$ and $g_2(y) = v_2$. Assume that the Tartan Theorem holds for all positions $(x', y') \neq (x, y)$ with $x' \leq x$, $y' \leq y$. Show that there exists a move from the position in Turning Corners to a position of SG-value u if, and only if, there exists a move from the position in $G_1 \times G_2$ to a position of SG-value u.
   (b) Conclude that the position in $G_1 \times G_2$ also has SG-value $v_1 \otimes v_2$.
8. Consider the following position in the tartan game of (Mock Turtles) $\times$ Ruler.

	1	2	3	4	5	6	7	8
0	T	H	T	T	T	T	T	T
1	T	T	T	T	T	T	T	T
2	T	T	T	T	T	T	T	T
3	T	T	T	T	T	T	T	T
4	T	T	T	T	T	T	T	H

(a) Construct the table of Sprague-Grundy values for this position.
(b) Note that the SG-value of this position is 15. Find a winning move for the first player.

9. (a) Who wins the $n \times n$ game of Rugs if the initial placement of the coins is such that heads occurs at (i, j) if and only if $i + j$ is odd? Can you give a simple winning strategy?
   (b) Who wins it if heads occurs at (i, j) if and only if $i + j$ is even? Can you give a simple winning strategy?
10. Consider the tartan game, $G_1 \times G_2$, where G_1 is the subtraction game with subtraction set $S = \{1, 2, 3, 4\}$ in which exactly two coins must be turned over, and G_2 is the game Ruler. The initial possition consists of a rectangular block of 100×100 coins. (Each coin is a penny, but the winner gets all of them, worth 100 dollars.) Initially, all coins are tails, except for heads at positions (100, 100) and (4, 1). You get to move first. What move do you make to win?

6 Green Hackenbush

The game of Hackenbush is played by hacking away edges from a rooted graph and removing those pieces of the graph that are no longer connected to the ground. A rooted graph is an undirected graph with every edge attached by some path to a special vertex called **the root** or **the ground**. The ground is denoted in the figures that follow by a dotted line.

In this section, we discuss the impartial version of this game in which either player at his turn may chop any edge. This version is called Green Hackenbush where each edge is considered to be colored green. There is also the partizan version of the game, called Blue-Red Hackenbush, in which some edges are colored blue and some are colored red. Player I may only chop the blue edges and Player II may only chop the red edges so the game is no longer impartial. Blue-Red Hackenbush is the first game treated in *Winning Ways*. In the general game of Hackenbush, there may be some blue edges available only to Player I, some red edges available only to Player II, and some green edges that either player may chop.

6.1 Bamboo Stalks

As an introduction to the game of Green Hackenbush, we investigate the case where the graph consists of a number of bamboo stalks as in the left side of Fig. 6.1. A bamboo stalk with n segments is a linear graph of n edges with the bottom of the n edges rooted to the ground. A move consists of hacking away one of the segments, and removing that segment and all segments above it no longer connected to the ground. Players alternate moves and the last player to move wins. A single bamboo stalk of n segments can be moved into a bamboo stalk of any smaller number of segments from $n - 1$ to 0. So a single bamboo stalk of n segments is equivalent to a nim pile

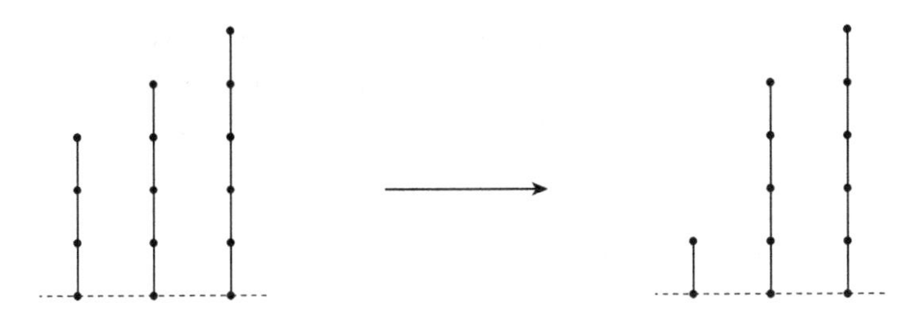

Fig. 6.1

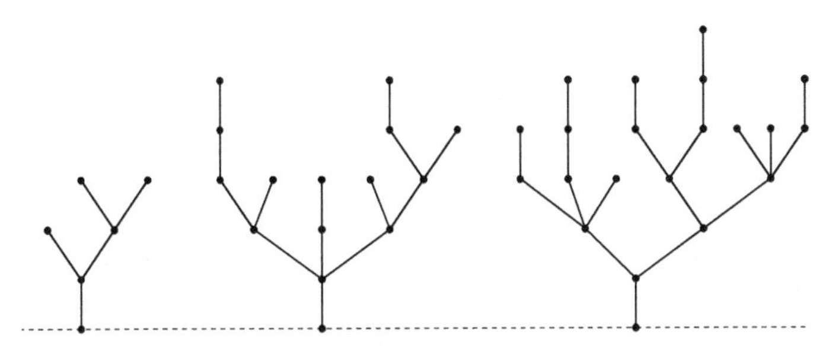

Fig. 6.2

of n chips. Playing a sum of games of bamboo stalks is thus equivalent to playing nim.

For example, the "forest" of three stalks on the left, is equivalent to the game of nim with three piles of 3, 4 and 5 chips. Since $3 \oplus 4 \oplus 5 = 2$, this is an N-position which can be moved to a P-position by hacking the second segment of the stalk with three segments, leaving a stalk of one segment. The resulting position on the right has Sprague-Grundy value 0, and so is a P-position.

6.2 Green Hackenbush on Trees

Played with bamboo stalks, Green Hackenbush is just nim in a rather transparent disguise. But what happens if we allow more general structures than these simple bamboo stalks? Suppose we have the "forest" of three rooted trees found in Fig. 6.2. A "rooted tree" is a graph with a distinguished vertex called the root, with the property that from every vertex there is a unique path (that doesn't repeat edges) to the root. Essentially this means there are no cycles.

Again a move consists of hacking away any segment and removing that segment and anything not connected to the ground. Since the game is impartial, the general theory of Chap. 4 tells us that each such tree is equivalent to some nim pile, or if you will, to some bamboo stalk. The problem is to find the Sprague-Grundy values of each of the trees.

This may be done using the following principle, known in its more general form as the **Colon Principle:** *When branches come together at a vertex, one may replace the branches by a non-branching stalk of length equal to their nim sum.*

Let us see how this principle works to find the bamboo stalk equivalent to the left tree of Fig. 6.2. There are two vertices with two branches. The higher of these vertices has two branches each with one edge. The nim sum of one and one is zero; so the two branches may be replaced by a branch with zero edges. That is to say, the two branches may be removed. This leaves a Y-shaped tree and the same reasoning may be used to show that the two branches at the Y may be removed. Thus the tree on the left of Fig. 6.2 is equivalent to a nim pile of size one.

This may have been a little too simple for illustrative purposes, so let us consider the second tree of Fig. 6.2. The leftmost branching vertex has two branches of lengths three and one. The nim sum of three and one is two, so the two branches may be replaced by a single branch of length two. The rightmost branching vertex has two branches of lengths one and two whose nim sum is three, so the two branches may be replaced by a single branch of length three. See the reduction in Fig. 6.3. Continuing in like manner we arrive at the conclusion that the second tree of Fig. 6.2 is equivalent to a nim pile of 8 chips.

Now try your luck with the third tree of Fig. 6.2. See if you can show that it is equivalent to a nim pile of size 4.

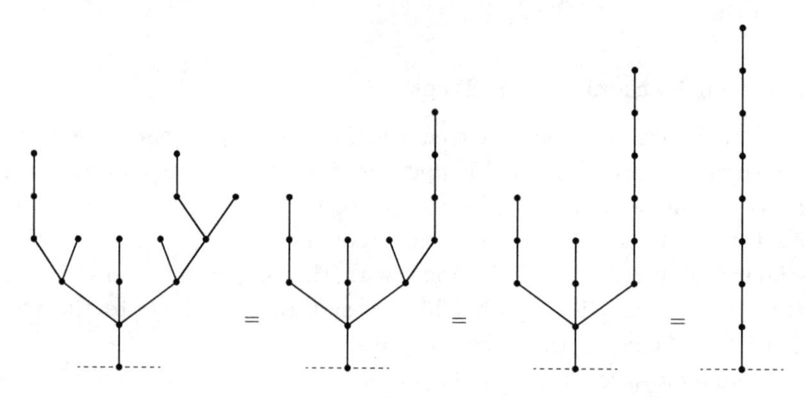

Fig. 6.3

Now we can compute the Sprague-Grundy value of the sum of the three trees of Fig. 6.2. It is $1 \oplus 8 \oplus 4 = 13$. Since this is not zero, it is a win for the next person to play. The next problem is to find a winning move. It is clear that there is a winning move using the second tree that may be obtained by chopping some edge to arrive at a tree of Sprague-Grundy value 5. But which edge must be chopped to achieve this?

The last version of the tree in Fig. 6.3 has length 8 because the three branches of the previous tree were 3, 2 and 6, whose nim-sum is $3 \oplus 2 \oplus 6 = 7$. To achieve length 5 in the last tree, we must change one of the three branches to achieve nim-sum 4. This may be done most easily by chopping the leftmost branch entirely, since $2 \oplus 6 = 4$. Alternatively, we may hack away the top edge of the middle branch leaving one edge, because $3 \oplus 1 \oplus 6 = 4$.

Each of these moves is easily translated into the corresponding chop on the tree on the left of Fig. 6.3. However, there is another way to reduce this tree to Sprague-Grundy value 5, that uses the right branch of the tree. See if you can find it.

The method of pruning trees given by the colon principle works to reduce all trees to a single bamboo stalk. One starts at the highest branches first, and then applies the principle inductively down to the root. We now show the validity of this principle for rooted graphs that may have circuits and several edges attached to the ground.

Proof of the Colon Principle. Consider a fixed but arbitrary graph, G, and select an arbitrary vertex, x, in G. Let H_1 and H_2 be arbitrary trees (or graphs) that have the same Sprague-Grundy value. Consider the two graphs $G_1 = G_x : H_1$ and $G_2 = G_x : H_2$, where $G_x : H_i$ represents the graph constructed by attaching the tree H_i to the vertex x of the graph G. The colon principle states that the two graphs G_1 and G_2 have the same Sprague-Grundy value. Consider the sum of the two games as in Fig. 6.4.

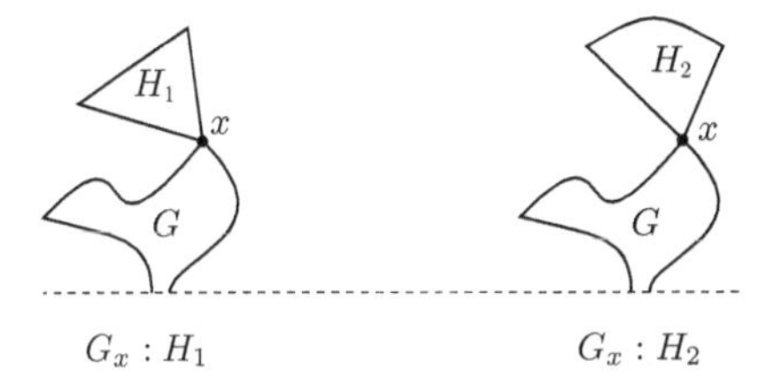

Fig. 6.4

The claim that G_1 and G_2 have the same Sprague-Grundy value is equivalent to the claim that the sum of the two games has Sprague-Grundy value 0. In other words, we are to show that the sum G_1+G_2 is a P-position.

Here is a strategy that guarantees you a win if you are the second player to move in $G_1 + G_2$. If the first player moves by chopping one of the edges in G in one of the games, then you chop the same edge in G in the other game. (Such a pair of moves may delete H_1 and H_2 from the games, but otherwise H_1 and H_2 are not disturbed.) If the first player moves by chopping an edge in H_1 or H_2, then the Sprague-Grundy values of H_1 and H_2 are no longer equal, so that there exists a move in H_1 or H_2 that keeps the Sprague-Grundy values the same. In this way you will always have a reply to every move the opponent may make. This means you will make the last move and so win. □

6.3 Green Hackenbush on General Rooted Graphs

We now consider arbitrary graphs. These graphs may have circuits and loops and several segments may be attached to the ground. Consider Fig. 6.5 as an example.

From the general theory of Chap. 4, each separate graph is equivalent to some nim pile. To find the equivalent nim pile, we look for an equivalent tree, from which the equivalent nim pile may be found. This is done using the **fusion principle**. We **fuse** two neighboring vertices by bringing them together into a single vertex and bending the edge joining them into a loop. A loop is an edge joining a vertex to itself, as for example the head of the juggler on the right of Fig. 6.5. As far as Green Hackenbush is concerned, a loop may be replaced by a leaf (an edge with one end unattached).

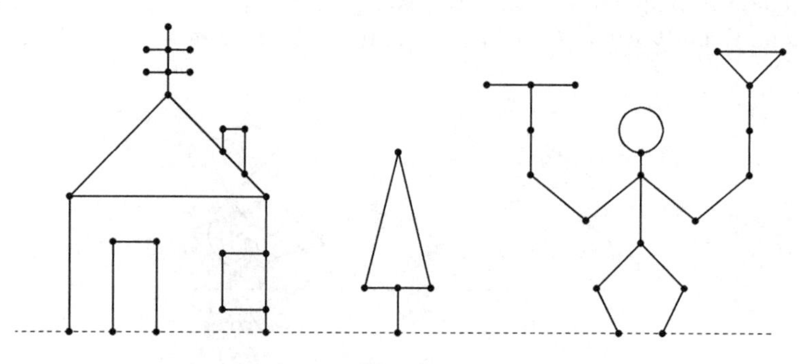

Fig. 6.5

Fig. 6.6

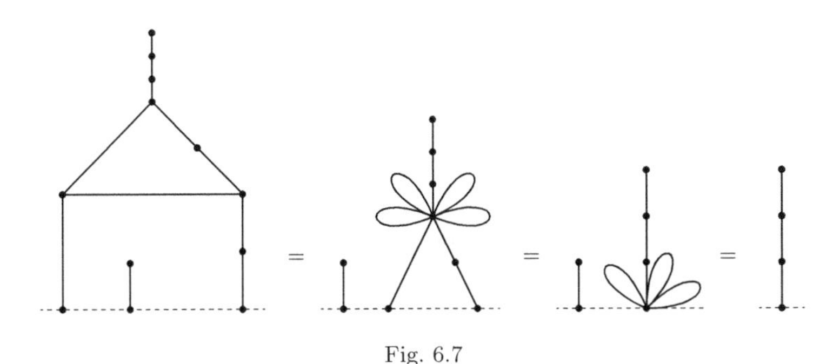

Fig. 6.7

The Fusion Principle: *The vertices on any circuit may be fused without changing the Sprague-Grundy value of the graph.*

The fusion principle allows us to reduce an arbitrary rooted graph into an equivalent tree which can be further reduced to a nim pile by the colon principle. Let us see how this works on the examples of Fig. 6.5.

Consider the door in the house on the left. The two vertices on the ground are the same vertex (remember the ground is really a single vertex) so the door is really the same as a triangle with one vertex attached to the ground. The fusion principle says that this is equivalent to a single vertex with three loops attached. Each loop is equivalent to a nim pile of size 1, and the nim sum of these is also a nim pile of size 1.

We see more generally that a circuit with an odd number of edges reduces to a single edge, and a circuit with an even number of edges reduces to a single vertex. For example, the circuit of four edges in the Christmas tree in the center of Fig. 6.5 reduces to a single vertex with four loops, which reduces to a single vertex. So the Christmas tree is equivalent to a nim pile of size 1. Similarly, the chimney on the house becomes a single vertex, and the window on the right of the house is also a single vertex. Continuing further, we find that the house (door included) has SG-value 3.

Now see if you can show that the juggler on the right of Fig. 6.5 has SG-value 4. And then see if you can find a winning move in the Hackenbush position given by Fig. 6.5.

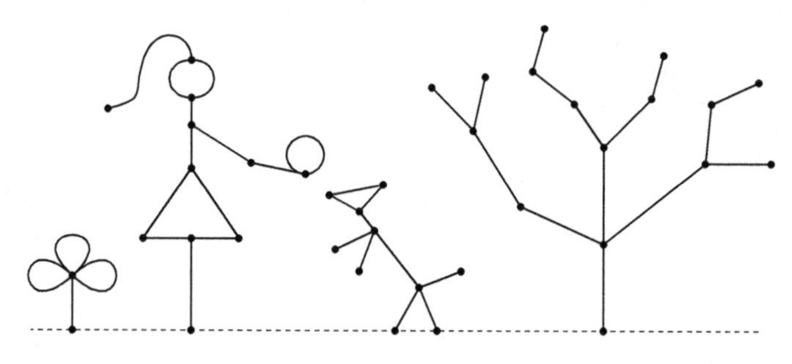

Fig. 6.8

The proof of the fusion principle is somewhat longer than the proof of the colon principle, and so is omitted. For a proof, see Chap. 7 of *Winning Ways*.

6.4 Exercise

1. (From *Fair Game* by Richard Guy.) Find the SG-values of the graphs in Fig. 6.8, and find a winning move, if any.

Part II

Two-Person Zero-Sum Games

7 The Strategic Form of a Game

The individual most closely associated with the creation of the theory of games is John von Neumann, one of the greatest mathematicians of the 20th century. Although others preceded him in formulating a theory of games — notably Émile Borel — it was von Neumann who published in 1928 the paper that laid the foundation for the theory of two-person zero-sum games. Von Neumann's work culminated in a fundamental book on game theory written in collaboration with Oskar Morgenstern entitled *Theory of Games and Economic Behavior* (1944). Other discussions of the theory of games relevant for our present purposes may be found in the textbook, *Game Theory* by Guillermo Owen, 2nd edition (1982), and the expository book, *Game Theory and Strategy* by Philip D. Straffin (1993).

The theory of von Neumann and Morgenstern is most complete for the class of games called two-person zero-sum games, i.e. games with only two players in which one player wins what the other player loses. In Part II, we restrict attention to such games. We will refer to the players as Player I and Player II.

7.1 Strategic Form

The simplest mathematical description of a game is the strategic form, mentioned in the introduction. For a two-person zero-sum game, the payoff function of Player II is the negative of the payoff of Player I, so we may restrict attention to the single payoff function of Player I, which we call here A.

Definition 1. The *strategic form*, or *normal form*, of a two-person zero-sum game is given by a triplet (X, Y, A), where

(1) X is a nonempty set, the set of strategies of Player I
(2) Y is a nonempty set, the set of strategies of Player II
(3) A is a real-valued function defined on $X \times Y$. (Thus, $A(x, y)$ is a real number for every $x \in X$ and every $y \in Y$.)

The interpretation is as follows. Simultaneously, Player I chooses $x \in X$ and Player II chooses $y \in Y$, each unaware of the choice of the other. Then their choices are made known and I wins the amount $A(x, y)$ from II. Depending on the monetary unit involved, $A(x, y)$ will be cents, dollars, pesos, beads, etc. If A is negative, I pays the absolute value of this amount to II. Thus, $A(x, y)$ represents the winnings of I and the losses of II.

This is a very simple definition of a game; yet it is broad enough to encompass the finite combinatorial games and games such as tic-tac-toe and chess. This is done by being sufficiently broadminded about the definition of a strategy. A strategy for a game of chess, for example, is a complete description of how to play the game, of what move to make in every possible situation that could occur. It is rather time-consuming to write down even one strategy, good or bad, for the game of chess. However, several different programs for instructing a machine to play chess well have been written. Each program constitutes one strategy. The program Deep Blue, that beat then world chess champion Gary Kasparov in a match in 1997, represents one strategy. The set of all such strategies for Player I is denoted by X. Naturally, in the game of chess it is physically impossible to describe all possible strategies since there are too many; in fact, there are more strategies than there are atoms in the known universe. On the other hand, the number of games of tic-tac-toe is rather small, so that it is possible to study all strategies and find an optimal strategy for each player. Later, when we study the extensive form of a game, we will see that many other types of games may be modeled and described in strategic form.

To illustrate the notions involved in games, let us consider the simplest non-trivial case when both X and Y consist of two elements. As an example, take the game called Odd-or-Even.

7.2 Example: Odd or Even

Players I and II simultaneously call out one of the numbers one or two. Player I's name is Odd; he wins if the sum of the numbers is odd. Player II's name is Even; she wins if the sum of the numbers is even. The amount paid to the winner by the loser is always the sum of the numbers in dollars. To put this game in strategic form we must specify X, Y and A. Here we may choose $X = \{1,2\}$, $Y = \{1,2\}$, and A as given in the following table.

$$\begin{array}{ccc} & \text{II (even)} & y \\ & & \begin{array}{cc} 1 & 2 \end{array} \\ \text{I (odd)} & x \begin{array}{c} 1 \\ 2 \end{array} & \begin{pmatrix} -2 & +3 \\ +3 & -4 \end{pmatrix} \end{array}$$

$$A(x,y) = \text{I's winnings} = \text{II's losses.}$$

It turns out that one of the players has a distinct advantage in this game. Can you tell which one it is?

Let us analyze this game from Player I's point of view. Suppose he calls "one" 3/5ths of the time and "two" 2/5ths of the time at random. In this case,

1. If II calls "one", I loses 2 dollars 3/5ths of the time and wins 3 dollars 2/5ths of the time; on the average, he wins $-2(3/5) + 3(2/5) = 0$ (he breaks even in the long run).
2. If II call "two", I wins 3 dollars 3/5ths of the time and loses 4 dollars 2/5ths of the time; on the average he wins $3(3/5) - 4(2/5) = 1/5$.

That is, if I mixes his choices in the given way, the game is even every time II calls "one", but I wins 20¢ on the average every time II calls "two". By employing this simple strategy, I is assured of at least breaking even on the average no matter what II does. Can Player I fix it so that he wins a positive amount no matter what II calls?

Let p denote the proportion of times that Player I calls "one". Let us try to choose p so that Player I wins the same amount on the average whether II calls "one" or "two". Then since I's average winnings when II calls "one" is $-2p+3(1-p)$, and his average winnings when II calls "two" is $3p-4(1-p)$ Player I should choose p so that

$$\begin{aligned} -2p + 3(1-p) &= 3p - 4(1-p) \\ 3 - 5p &= 7p - 4 \\ 12p &= 7 \\ p &= 7/12. \end{aligned}$$

Hence, I should call "one" with probability $7/12$, and "two" with probability $5/12$. On the average, I wins $-2(7/12)+3(5/12) = 1/12$, or $8\frac{1}{3}$ cents every time he plays the game, no matter what II does. Such a strategy that produces the same average winnings no matter what the opponent does is called an **equalizing strategy**.

Therefore, the game is clearly in I's favor. Can he do better than $8\frac{1}{3}$ cents per game on the average? The answer is: Not if II plays properly. In fact, II could use the same procedure:

call "one" with probability $7/12$

call "two" with probability $5/12$.

If I calls "one", II's average loss is $-2(7/12) + 3(5/12) = 1/12$. If I calls "two", II's average loss is $3(7/12) - 4(5/12) = 1/12$.

Hence, I has a procedure that guarantees him at least $1/12$ on the average, and II has a procedure that keeps her average loss to at most $1/12$. $1/12$ is called the **value** of the game, and the procedure each uses to insure this return is called an **optimal strategy** or a **minimax strategy**.

If instead of playing the game, the players agree to call in an arbitrator to settle this conflict, it seems reasonable that the arbitrator should require II to pay $8\frac{1}{3}$ cents to I. For I could argue that he should receive at least $8\frac{1}{3}$ cents since his optimal strategy guarantees him that much on the average no matter what II does. On the other hand, II could argue that she should not have to pay more than $8\frac{1}{3}$ cents since she has a strategy that keeps her average loss to at most that amount no matter what I does.

7.3 Pure Strategies and Mixed Strategies

It is useful to make a distinction between a pure strategy and a mixed strategy. We refer to elements of X or Y as pure strategies. The more complex entity that chooses among the pure strategies at random in various proportions is called a mixed strategy. Thus, I's optimal strategy in the game of Odd-or-Even is a mixed strategy; it mixes the pure strategies one and two with probabilities $7/12$ and $5/12$ respectively. Of course every pure strategy, $x \in X$, can be considered as the mixed strategy that chooses the pure strategy x with probability 1.

In our analysis, we made a rather subtle assumption. We assumed that when a player uses a mixed strategy, he is only interested in his average return. He does not care about his maximum possible winnings or losses — only the average. This is actually a rather drastic assumption. We are evidently assuming that a player is indifferent between receiving

5 million dollars outright, and receiving 10 million dollars with probability 1/2 and nothing with probability 1/2. I think nearly everyone would prefer the $5,000,000 outright. This is because the utility of having 10 megabucks is not twice the utility of having 5 megabucks.

The main justification for this assumption comes from **utility theory** and is treated in Appendix 1. The basic premise of utility theory is that one should evaluate a payoff by its utility to the player rather than on its numerical monetary value. Generally a player's utility of money will not be linear in the amount. The main theorem of utility theory states that under certain reasonable assumptions, a player's preferences among outcomes are consistent with the existence of a utility function and the player judges an outcome only on the basis of the *average* utility of the outcome.

However, utilizing utility theory to justify the above assumption raises a new difficulty. Namely, the two players may have different utility functions. The same outcome may be perceived in quite different ways. This means that the game is no longer zero-sum. We need an assumption that says the utility functions of two players are the same (up to change of location and scale). This is a rather strong assumption, but for moderate to small monetary amounts, we believe it is a reasonable one.

A mixed strategy may be implemented with the aid of a suitable outside random mechanism, such as tossing a coin, rolling dice, drawing a number out of a hat and so on. The seconds indicator of a watch provides a simple personal method of randomization provided it is not used too frequently. For example, Player I of Odd-or-Even wants an outside random event with probability $7/12$ to implement his optimal strategy. Since $7/12 = 35/60$, he could take a quick glance at his watch; if the seconds indicator showed a number between 0 and 35, he would call "one", while if it were between 35 and 60, he would call "two".

7.4 The Minimax Theorem

A two-person zero-sum game (X, Y, A) is said to be a **finite game** if both strategy sets X and Y are finite sets. The fundamental theorem of game theory due to von Neumann states that the situation encountered in the game of Odd-or-Even holds for all finite two-person zero-sum games. Specifically,

The Minimax Theorem. *For every finite two-person zero-sum game,*

(1) *there is a number V, called the value of the game,*

(2) *there is a mixed strategy for Player I such that I's average gain is at least V no matter what II does, and*
(3) *there is a mixed strategy for Player II such that II's average loss is at most V no matter what I does.*

This is one form of the minimax theorem to be stated more precisely and discussed in greater depth later. If V is zero we say the game is fair. If V is positive, we say the game favors Player I, while if V is negative, we say the game favors Player II.

7.5 Exercises

1. Consider the game of Odd-or-Even with the sole change that the loser pays the winner the product, rather than the sum, of the numbers chosen (who wins still depends on the sum). Find the table for the payoff function A, and analyze the game to find the value and optimal strategies of the players. Is the game fair?
2. Player I holds a black Ace and a red 8. Player II holds a red 2 and a black 7. The players simultaneously choose a card to play. If the chosen cards are of the same color, Player I wins. Player II wins if the cards are of different colors. The amount won is a number of dollars equal to the number on the winner's card (Ace counts as 1). Set up the payoff function, find the value of the game and the optimal mixed strategies of the players.
3. Sherlock Holmes boards the train from London to Dover in an effort to reach the continent and so escape from Professor Moriarty. Moriarty can take an express train and catch Holmes at Dover. However, there is an intermediate station at Canterbury at which Holmes may detrain to avoid such a disaster. But of course, Moriarty is aware of this too and may himself stop instead at Canterbury. Von Neumann and Morgenstern (loc. cit.) estimate the value to Moriarty of these four possibilities to be given in the following matrix (in some unspecified units).

		Holmes	
		Canterbury	Dover
Moriarty	Canterbury	100	−50
	Dover	0	100

What are the optimal strategies for Holmes and Moriarty, and what is the value? (Historically, as related by Dr. Watson in "The Final Problem" in Arthur Conan Doyle's *The Memoires of Sherlock Holmes*, Holmes detrained at Canterbury and Moriarty went on to Dover.)

4. The entertaining book *The Compleat Strategyst* by John Williams contains many simple examples and informative discussion of strategic form games. Here is one of his problems.

> *"I know a good game," says Alex. "We point fingers at each other; either one finger or two fingers. If we match with one finger, you buy me one Daiquiri, If we match with two fingers, you buy me two Daiquiris. If we don't match I let you off with a payment of a dime. It'll help pass the time."*
>
> *Olaf appears quite unmoved. "That sounds like a very dull game — at least in its early stages." His eyes glaze on the ceiling for a moment and his lips flutter briefly; he returns to the conversation with: "Now if you'd care to pay me 42 cents before each game, as a partial compensation for all those 55-cent drinks I'll have to buy you, then I'd be happy to pass the time with you.*

Olaf could see that the game was inherently unfair to him so he insisted on a side payment as compensation. Does this side payment make the game fair? What are the optimal strategies and the value of the game?

8 Matrix Games — Domination

A finite two-person zero-sum game in strategic form, (X, Y, A), is sometimes called a matrix game because the payoff function A can be represented by a matrix. If $X = \{x_1, \ldots, x_m\}$ and $Y = \{y_1, \ldots, y_n\}$, then by the *game matrix* or *payoff matrix* we mean the matrix

$$\boldsymbol{A} = \begin{pmatrix} a_{11} & \cdots & a_{1n} \\ \vdots & & \vdots \\ a_{m1} & \cdots & a_{mn} \end{pmatrix} \quad \text{where } a_{ij} = A(x_i, y_j),$$

In this form, Player I chooses a row, Player II chooses a column, and II pays I the entry in the chosen row and column. Note that the entries of the matrix are the winnings of the row chooser and losses of the column chooser.

A mixed strategy for Player I may be represented by an m-tuple, $\boldsymbol{p} = (p_1, p_2, \ldots, p_m)^T$ of probabilities that add to 1. If I uses the mixed strategy $\boldsymbol{p} = (p_1, p_2, \ldots, p_m)^T$ and II chooses column j, then the (average) payoff to I is $\sum_{i=1}^m p_i a_{ij}$. Similarly, a mixed strategy for Player II is an n-tuple $\boldsymbol{q} = (q_1, q_2, \ldots, q_n)^T$. If II uses $\boldsymbol{q}$ and I uses row i the payoff to I is $\sum_{j=1}^n a_{ij} q_j$. More generally, if I uses the mixed strategy $\boldsymbol{p}$ and II uses the mixed strategy $\boldsymbol{q}$, the (average) payoff to I is $\boldsymbol{p}^T \boldsymbol{A} \boldsymbol{q} = \sum_{i=1}^m \sum_{j=1}^n p_i a_{ij} q_j$.

Note that the pure strategy for Player I of choosing row i may be represented as the mixed strategy $\boldsymbol{e}_i$, the unit vector with a 1 in the ith position and 0's elsewhere. Similarly, the pure strategy for II of choosing the jth column may be represented by $\boldsymbol{e}_j$. In the following, we shall be attempting to "solve" games. This means finding the value, and at least one optimal strategy for each player. Occasionally, we shall be interested in finding all optimal strategies for a player.

8.1 Saddle Points

Occasionally it is easy to solve the game. If some entry a_{ij} of the matrix $\boldsymbol{A}$ has the property that

(1) a_{ij} is the minimum of the ith row, and
(2) a_{ij} is the maximum of the jth column,

then we say a_{ij} is a saddle point. If a_{ij} is a saddle point, then Player I can then win at least a_{ij} by choosing row i, and Player II can keep her loss to at most a_{ij} by choosing column j. Hence a_{ij} is the value of the game.

Example 1.

$$\boldsymbol{A} = \begin{pmatrix} 4 & 1 & -3 \\ 3 & 2 & 5 \\ 0 & 1 & 6 \end{pmatrix}.$$

The central entry, 2, is a saddle point, since it is a minimum of its row and maximum of its column. Thus it is optimal for I to choose the second row, and for II to choose the second column. The value of the game is 2, and $(0, 1, 0)$ is an optimal mixed strategy for both players.

For large $m \times n$ matrices it is tedious to check each entry of the matrix to see if it has the saddle point property. It is easier to compute the minimum of each row and the maximum of each column to see if there is a match. Here is an example of the method.

$$\begin{array}{cc} & \text{row min} \\ \boldsymbol{A} = \begin{pmatrix} 3 & 2 & 1 & 0 \\ 0 & 1 & 2 & 0 \\ 1 & 0 & 2 & 1 \\ 3 & 1 & 2 & 2 \end{pmatrix} & \begin{array}{c} 0 \\ 0 \\ 0 \\ 1 \end{array} \\ \text{col max} \quad 3 \quad 2 \quad 2 \quad 2 & \end{array} \qquad \begin{array}{cc} & \text{row min} \\ \boldsymbol{B} = \begin{pmatrix} 3 & 1 & 1 & 0 \\ 0 & 1 & 2 & 0 \\ 1 & 0 & 2 & 1 \\ 3 & 1 & 2 & 2 \end{pmatrix} & \begin{array}{c} 0 \\ 0 \\ 0 \\ 1 \end{array} \\ \text{col max} \quad 3 \quad 1 \quad 2 \quad 2 & \end{array}$$

In matrix $\boldsymbol{A}$, no row minimum is equal to any column maximum, so there is no saddle point. However, if the 2 in position a_{12} were changed to a 1, then we have matrix $\boldsymbol{B}$. Here, the minimum of the fourth row is equal to the maximum of the second column; so b_{42} is a saddle point.

8.2 Solution of All 2 × 2 Matrix Games

Consider the general 2×2 game matrix

$$\boldsymbol{A} = \begin{pmatrix} a & b \\ d & c \end{pmatrix}.$$

To solve this game (i.e. to find the value and at least one optimal strategy for each player) we proceed as follows.

1. Test for a saddle point.

2. If there is no saddle point, solve by finding equalizing strategies.

We now prove the method of finding equalizing strategies of Sec. 7.2 works whenever there is no saddle point by deriving the value and the optimal strategies.

Assume there is no saddle point. If $a \geq b$, then $b < c$, as otherwise b is a saddle point. Since $b < c$, we must have $c > d$, as otherwise c is a saddle point. Continuing thus, we see that $d < a$ and $a > b$. In other words, if $a \geq b$, then $a > b < c > d < a$. By symmetry, if $a \leq b$, then $a < b > c < d > a$. This shows that *if there is no saddle point, then either* $a > b$, $b < c$, $c > d$ *and* $d < a$, *or* $a < b$, $b > c$, $c < d$ *and* $d > a$.

In Eqs. (1), (2) and (3) below, we develop formulas for the optimal strategies and value of the general 2×2 game. If I chooses the first row with probability p (i.e. uses the mixed strategy $(p, 1-p)$), we equate his average return when II uses columns 1 and 2.

$$ap + d(1-p) = bp + c(1-p).$$

Solving for p, we find

$$p = \frac{c-d}{(a-b)+(c-d)}. \tag{1}$$

Since there is no saddle point, $(a-b)$ and $(c-d)$ are either both positive or both negative; hence, $0 < p < 1$. Player I's average return using this strategy is

$$v = ap + d(1-p) = \frac{ac-bd}{a-b+c-d}.$$

If II chooses the first column with probability q (i.e. uses the strategy $(q, 1-q)$), we equate his average losses when I uses rows 1 and 2.

$$aq + b(1-q) = dq + c(1-q).$$

Hence,

$$q = \frac{c-b}{a-b+c-d}. \tag{2}$$

Again, since there is no saddle point, $0 < q < 1$. Player II's average loss using this strategy is

$$aq + b(1-q) = \frac{ac - bd}{a - b + c - d} = v, \tag{3}$$

the same value achievable by I. This shows that the game has a value, and that the players have optimal strategies (something the minimax theorem says holds for all finite games).

Example 2.

$$\boldsymbol{A} = \begin{pmatrix} -2 & 3 \\ 3 & -4 \end{pmatrix} \qquad \begin{aligned} p &= \frac{-4-3}{-2-3-4-3} = 7/12 \\ q &= \text{same} \\ v &= \frac{8-9}{-2-3-4-3} = 1/12 \end{aligned}$$

Example 3.

$$\boldsymbol{A} = \begin{pmatrix} 0 & -10 \\ 1 & 2 \end{pmatrix} \qquad \begin{aligned} p &= \frac{2-1}{0+10+2-1} = 1/11 \\ q &= \frac{2+10}{0+10+2-1} = 12/11. \end{aligned}$$

But q must be between zero and one. What happened? The trouble is we "forgot to test this matrix for a saddle point, so of course it has one" [J. D. Williams (1966) p. 56]. The lower left corner is a saddle point. So $p = 0$ and $q = 1$ are optimal strategies, and the value is $v = 1$.

8.3 Removing Dominated Strategies

Sometimes, large matrix games may be reduced in size (hopefully to the 2×2 case) by deleting rows and columns that are obviously bad for the player who uses them.

Definition. We say the ith row of a matrix $\boldsymbol{A} = (a_{ij})$ *dominates* the kth row if $a_{ij} \geq a_{kj}$ for all j. We say the ith row of $\boldsymbol{A}$ *strictly dominates* the kth row if $a_{ij} > a_{kj}$ for all j. Similarly, the jth column of $\boldsymbol{A}$ dominates (strictly dominates) the kth column if $a_{ij} \leq a_{ik}$ (resp. $a_{ij} < a_{ik}$) for all i.

Anything Player I can achieve using a dominated row can be achieved at least as well using the row that dominates it. Hence dominated rows may be deleted from the matrix. A similar argument shows that dominated columns may be removed. To be more precise, *removal of a dominated row or column does not change the value of a game.* However, there may exist an optimal strategy that uses a dominated row or column (see Exercise 9). If so, removal of that row or column will also remove the use of that optimal strategy (although there will still be at least one optimal strategy left). However, in the case of removal of a *strictly dominated* row or column, the set of optimal strategies does not change.

We may iterate this procedure and successively remove several rows and columns. As an example, consider the matrix, $\boldsymbol{A}$.

The last column is dominated by the middle column. Deleting the last column we obtain:

$$\boldsymbol{A} = \begin{pmatrix} 2 & 0 & 4 \\ 1 & 2 & 3 \\ 4 & 1 & 2 \end{pmatrix}$$

Now the top row is dominated by the bottom row. (Note that this is not the case in the original matrix.) Deleting the top row we obtain:

$$\begin{pmatrix} 2 & 0 \\ 1 & 2 \\ 4 & 1 \end{pmatrix}$$

This 2×2 matrix does not have a saddle point, so $p = 3/4$, $q = 1/4$ and $v = 7/4$. I's optimal strategy in the original game is $(0, 3/4, 1/4)$; II's is $(1/4, 3/4, 0)$.

$$\begin{pmatrix} 1 & 2 \\ 4 & 1 \end{pmatrix}$$

A row (column) may also be removed if it is dominated by a probability combination of other rows (columns).

If for some $0 < p < 1$, $pa_{i_1 j} + (1-p)a_{i_2 j} \geq a_{kj}$ for all j, then the kth row is dominated by the mixed strategy that chooses row i_1 with probability p and row i_2 with probability $1-p$. Player I can do at least as well using this mixed strategy instead of choosing row k. (In addition, any mixed strategy choosing row k with probability p_k may be replaced by the one in which k's probability is split between i_1 and i_2. That is, i_1's probability is increased by pp_k and i_2's probability is increased by $(1-p)p_k$.) A similar argument may be used for columns.

Consider the matrix $\boldsymbol{A} = \begin{pmatrix} 0 & 4 & 6 \\ 5 & 7 & 4 \\ 9 & 6 & 3 \end{pmatrix}$.

The middle column is dominated by the outside columns taken with probability 1/2 each. With the central column deleted, the middle row is dominated by the combination of the top row with probability $1/3$ and the bottom row with probability $2/3$. The reduced matrix, $\begin{pmatrix} 0 & 6 \\ 9 & 3 \end{pmatrix}$, is easily solved. The value is $V = 54/12 = 9/2$.

Of course, mixtures of more than two rows (columns) may be used to dominate and remove other rows (columns). For example, the mixture of columns 1, 2 and 3 with probabilities $1/3$ each in matrix $\boldsymbol{B} = \begin{pmatrix} 1 & 3 & 5 & 3 \\ 4 & 0 & 2 & 2 \\ 3 & 7 & 3 & 5 \end{pmatrix}$ dominates the last column, and so the last column may be removed.

Not all games may be reduced by dominance. In fact, even if the matrix has a saddle point, there may not be any dominated rows or columns. The 3×3 game with a saddle point found in Example 1 demonstrates this.

8.4 Solving $2 \times n$ and $m \times 2$ Games

Games with matrices of size $2 \times n$ or $m \times 2$ may be solved with the aid of a graphical interpretation. Take the following example.

$$\begin{matrix} p \\ 1-p \end{matrix} \quad \begin{pmatrix} 2 & 3 & 1 & 5 \\ 4 & 1 & 6 & 0 \end{pmatrix}$$

Suppose Player I chooses the first row with probability p and the second row with probability $1-p$. If II chooses Column 1, I's average payoff is $2p+4(1-p)$. Similarly, choices of Columns 2, 3 and 4 result in average payoffs of $3p+(1-p)$, $p+6(1-p)$, and $5p$ respectively. We graph these four linear functions of p for $0 \le p \le 1$. For a fixed value of p, Player I can be sure that his average winnings is at least the minimum of these four functions evaluated at p. This is known as the lower envelope of these functions. Since I wants to maximize his guaranteed average winnings, he wants to find p that achieves the maximum of this lower envelope. According to the drawing, this should occur at the intersection of the lines for Columns 2 and 3. This essentially, involves solving the game in which II is restricted to Columns 2 and 3. The value of the game $\begin{pmatrix} 3 & 1 \\ 1 & 6 \end{pmatrix}$ is $v = 17/7$, I's optimal strategy is $(5/7, 2/7)$, and II's optimal strategy is $(5/7, 2/7)$. Subject to the accuracy of the drawing, we conclude therefore that in the original game I's optimal strategy is $(5/7, 2/7)$, II's is $(0, 5/7, 2/7, 0)$ and the value is $17/7$.

The accuracy of the drawing may be checked: *Given any guess at a solution to a game, there is a sure-fire test to see if the guess is correct,* as follows. If I uses the strategy $(5/7, 2/7)$, his average payoff if II uses Columns 1, 2, 3 and 4, is $18/7$, $17/7$, $17/7$, and $25/7$ respectively. Thus his average payoff is at least $17/7$ no matter what II does. Similarly, if II

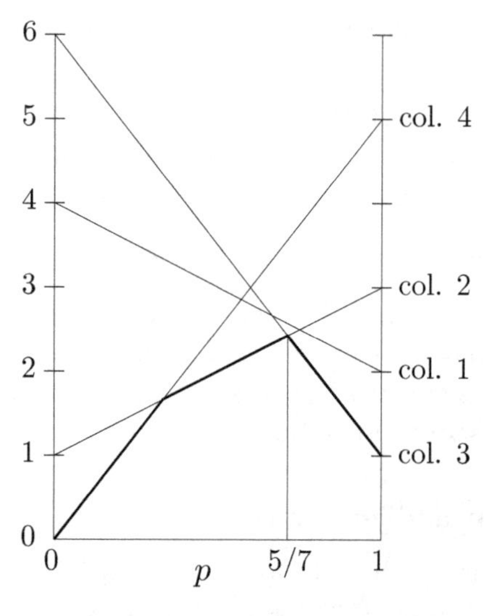

Fig. 8.1.

uses $(0, 5/7, 2/7, 0)$, her average loss is (at most) $17/7$. Thus, $17/7$ is the value, and these strategies are optimal.

We note that the line for Column 1 plays no role in the lower envelope (that is, the lower envelope would be unchanged if the line for Column 1 were removed from the graph). *This is a test for domination.* Column 1 is, in fact, dominated by Columns 2 and 3 taken with probability 1/2 each. The line for Column 4 does appear in the lower envelope, and hence Column 4 cannot be dominated.

As an example of an $m \times 2$ game, consider the matrix associated with Fig. 8.2. If q is the probability that II chooses Column 1, then II's average loss for I's three possible choices of rows is given in the accompanying graph. Here, Player II looks at the largest of her average losses for a given q. This is the upper envelope of the function. II wants to find q that minimizes this upper envelope. From the graph, we see that any value of q between $1/4$ and $1/3$ inclusive achieves this minimum. The value of the game is 4, and I has an optimal pure strategy: row 2.

These techniques work just as well for $2 \times \infty$ and $\infty \times 2$ games.

8.5 Latin Square Games

A Latin square is an $n \times n$ array of n different letters such that each letter occurs once and only once in each row and each column. The 5×5

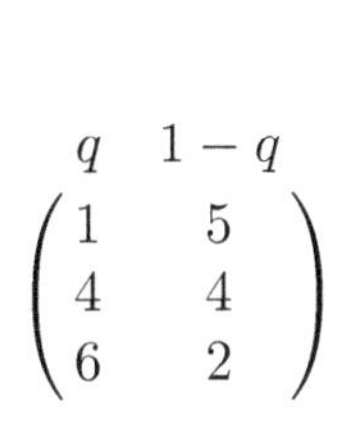

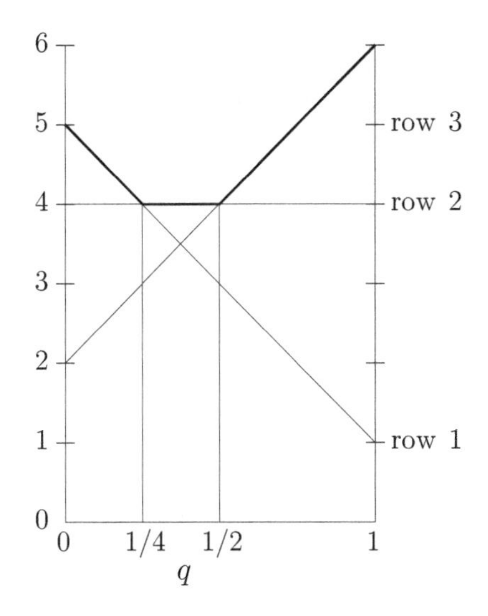

Fig. 8.2.

array at the right is an example. If in a Latin square each letter is assigned a numerical value, the resulting matrix is the matrix of a Latin square game. Such games have simple solutions. The value is the average of the numbers in a row, and the strategy that chooses each pure strategy with equal probability $1/n$ is optimal for both players. The reason is not very deep. The conditions for optimality are satisfied.

$$\begin{pmatrix} a & b & c & d & e \\ b & e & a & c & d \\ c & a & d & e & b \\ d & c & e & b & a \\ e & d & b & a & c \end{pmatrix} \qquad \begin{matrix} a=1,\ b=2,\ c=d=3,\ e=6 \\ \begin{pmatrix} 1 & 2 & 3 & 3 & 6 \\ 2 & 6 & 1 & 3 & 3 \\ 3 & 1 & 3 & 6 & 2 \\ 3 & 3 & 6 & 2 & 1 \\ 6 & 3 & 2 & 1 & 3 \end{pmatrix} \end{matrix}$$

In the example above, the value is $V = (1+2+3+3+6)/5 = 3$, and the mixed strategy $\boldsymbol{p} = \boldsymbol{q} = (1/5, 1/5, 1/5, 1/5, 1/5)$ is optimal for both players. The game of matching pennies is a Latin square game. Its value is zero and $(1/2, 1/2)$ is optimal for both players.

8.6 Exercises

1. Solve the game with matrix $\begin{pmatrix} -1 & -3 \\ -2 & 2 \end{pmatrix}$, that is find the value and an optimal (mixed) strategy for both players.

2. Solve the game with matrix $\begin{pmatrix} 0 & 2 \\ t & 1 \end{pmatrix}$ for an arbitrary real number t. (Don't forget to check for a saddle point!) Draw the graph of $v(t)$, the value of the game, as a function of t, for $-\infty < t < \infty$.
3. Show that if a game with $m \times n$ matrix has two saddle points, then they have equal values.
4. Reduce by dominance to 2×2 games and solve.

$$\text{(a)} \quad \begin{pmatrix} 5 & 4 & 1 & 0 \\ 4 & 3 & 2 & -1 \\ 0 & -1 & 4 & 3 \\ 1 & -2 & 1 & 2 \end{pmatrix} \qquad \text{(b)} \quad \begin{pmatrix} 10 & 0 & 7 & 1 \\ 2 & 6 & 4 & 7 \\ 6 & 3 & 3 & 5 \end{pmatrix}.$$

5. (a) Solve the game with matrix $\begin{pmatrix} 3 & 2 & 4 & 0 \\ -2 & 1 & -4 & 5 \end{pmatrix}$.

   (b) Reduce by dominance to a 3×2 matrix game and solve: $\begin{pmatrix} 0 & 8 & 5 \\ 8 & 4 & 6 \\ 12 & -4 & 3 \end{pmatrix}$.

6. Players I and II choose integers i and j respectively from the set $\{1, 2, \ldots, n\}$ for some $n \geq 2$. Player I wins 1 if $|i - j| = 1$. Otherwise there is no payoff. If $n = 7$, for example, the game matrix is

$$\begin{pmatrix} 0 & 1 & 0 & 0 & 0 & 0 & 0 \\ 1 & 0 & 1 & 0 & 0 & 0 & 0 \\ 0 & 1 & 0 & 1 & 0 & 0 & 0 \\ 0 & 0 & 1 & 0 & 1 & 0 & 0 \\ 0 & 0 & 0 & 1 & 0 & 1 & 0 \\ 0 & 0 & 0 & 0 & 1 & 0 & 1 \\ 0 & 0 & 0 & 0 & 0 & 1 & 0 \end{pmatrix}$$

   (a) Using dominance to reduce the size of the matrix, solve the game for $n = 7$ (i.e. find the value and one optimal strategy for each player).

   (b) See if you can solve the game for arbitrary n.

7. In general, the sure-fire test may be stated thus: For a given game, conjectured optimal strategies $(p_1, \ldots, p_m)$ and $(q_1, \ldots, q_n)$ are indeed optimal if the minimum of I's average payoffs using $(p_1, \ldots, p_m)$ is equal to the maximum of II's average payoffs using $(q_1, \ldots, q_n)$. Show that for the game with the following matrix the mixed strategies

$p = (6/37, 20/37, 0, 11/37)$ and $q = (14/37, 4/37, 0, 19/37, 0)$ are optimal for I and II respectively. What is the value?

$$\begin{pmatrix} 5 & 8 & 3 & 1 & 6 \\ 4 & 2 & 6 & 3 & 5 \\ 2 & 4 & 6 & 4 & 1 \\ 1 & 3 & 2 & 5 & 3 \end{pmatrix}.$$

8. Given that $p = (52/143, 50/143, 41/143)$ is optimal for I in the game with the following matrix, what is the value?

$$\begin{pmatrix} 0 & 5 & -2 \\ -3 & 0 & 4 \\ 6 & -4 & 0 \end{pmatrix}$$

9. Player I secretly chooses one of the numbers, 1, 2 and 3, and Player II tries to guess which. If II guesses correctly, she loses nothing; otherwise, she loses the absolute value of the difference of I's choice and her guess. Set up the matrix and reduce it by dominance to a 2×2 game and solve. Note that II has an optimal pure strategy that was eliminated by dominance. Moreover, this strategy dominates the optimal mixed strategy in the 2×2 game.
10. **Magic Square Games.** A magic square is an $n \times n$ array of the first n integers with the property that all row and column sums are equal. Show how to solve all games with magic square game matrices. Solve the example,

$$\begin{pmatrix} 16 & 3 & 2 & 13 \\ 5 & 10 & 11 & 8 \\ 9 & 6 & 7 & 12 \\ 4 & 15 & 14 & 1 \end{pmatrix}.$$

(This is the magic square that appears in Albrecht Dürer's engraving, *Melencolia.* See http://freemasonry.bcy.ca/art/melencolia.html)
11. In an article, "Normandy: Game and Reality" by W. Drakert in *Moves*, No. 6 (1972), an analysis is given of the invasion of Europe at Normandy in World War II. Six possible attacking configurations (1 to 6) by the Allies and six possible defensive strategies (A to F) by the Germans were simulated and evaluated, 36 simulations in all. The following table

gives the estimated value to the Allies of each hypothetical battle in some numerical units.

$$\begin{array}{c} \\ 1 \\ 2 \\ 3 \\ 4 \\ 5 \\ 6 \end{array} \begin{array}{c} \begin{array}{cccccc} A & B & C & D & E & F \end{array} \\ \begin{pmatrix} 13 & 29 & 8 & 12 & 16 & 23 \\ 18 & 22 & 21 & 22 & 29 & 31 \\ 18 & 22 & 31 & 31 & 27 & 37 \\ 11 & 22 & 12 & 21 & 21 & 26 \\ 18 & 16 & 19 & 14 & 19 & 28 \\ 23 & 22 & 19 & 23 & 30 & 34 \end{pmatrix} \end{array}$$

(a) Assuming this is a matrix of a 6×6 game, reduce by dominance and solve.

(b) The historical defense by the Germans was B, and the historical attack by the Allies was 1. Criticize these choices.

9 The Principle of Indifference

For a matrix game with $m \times n$ matrix $\boldsymbol{A}$, if Player I uses the mixed strategy $\boldsymbol{p} = (p_1, \ldots, p_m)^T$ and Player II uses column j, Player I's average payoff is $\sum_{i=1}^{m} p_i a_{ij}$. If V is the value of the game, an optimal strategy, $\boldsymbol{p}$, for I is characterized by the property that Player I's average payoff is at least V no matter what column j Player II uses, i.e.

$$\sum_{i=1}^{m} p_i a_{ij} \geq V \quad \text{for all } j = 1, \ldots, n. \tag{1}$$

Similarly, a strategy $\boldsymbol{q} = (q_1, \ldots, q_n)^T$ is optimal for II if and only if

$$\sum_{j=1}^{n} a_{ij} q_j \leq V \quad \text{for all } i = 1, \ldots, m. \tag{2}$$

When both players use their optimal strategies the average payoff, $\sum_i \sum_j p_i a_{ij} q_j$, is exactly V. This may be seen from the inequalities

$$\begin{aligned} V = \sum_{j=1}^{n} V q_j &\leq \sum_{j=1}^{n} \left(\sum_{i=1}^{m} p_i a_{ij} \right) q_j = \sum_{i=1}^{m} \sum_{j=1}^{n} p_i a_{ij} q_j \\ &= \sum_{i=1}^{m} p_i \left(\sum_{j=1}^{n} a_{ij} q_j \right) \leq \sum_{i=1}^{m} p_i V = V. \end{aligned} \tag{3}$$

Since this begins and ends with V we must have equality throughout.

9.1 The Equilibrium Theorem

The following simple theorem — the Equilibrium Theorem — gives conditions for equality to be achieved in (1) for certain values of j, and in (2) for certain values of i.

Theorem 9.1. *Consider a game with* $m \times n$ *matrix* $\boldsymbol{A}$ *and value* V. *Let* $\boldsymbol{p} = (p_1, \ldots, p_m)^T$ *be any optimal strategy for I and* $\boldsymbol{q} = (q_1, \ldots, q_n)^T$ *be any optimal strategy for II. Then*

$$\sum_{j=1}^{n} a_{ij} q_j = V \quad \text{for all } i \text{ for which } p_i > 0 \tag{4}$$

and

$$\sum_{i=1}^{m} p_i a_{ij} = V \quad \text{for all } j \text{ for which } q_j > 0. \tag{5}$$

Proof. Suppose there is a k such that $p_k > 0$ and $\sum_{j=1}^{n} a_{kj} q_j \neq V$. Then from (2), $\sum_{j=1}^{n} a_{kj} q_j < V$. But then from (3) with equality throughout

$$V = \sum_{i=1}^{m} p_i \left(\sum_{j=1}^{n} a_{ij} q_j \right) < \sum_{i=1}^{m} p_i V = V.$$

The inequality is strict since it is strict for the kth term of the sum. This contradiction proves the first conclusion. The second conclusion follows analogously. □

Another way of stating the first conclusion of this theorem is: If there exists an optimal strategy for I giving positive probability to row i, then every optimal strategy of II gives I the value of the game if he uses row i.

This theorem is useful in certain classes of games for helping to direct us toward the solution. The procedure this theorem suggests for Player 1 is to try to find a solution to the set of equations (5) formed by those j for which you think it likely that $q_j > 0$. One way of saying this is that Player 1 searches for a strategy that makes Player 2 *indifferent* as to which of the (good) pure strategies to use. Similarly, Player 2 should play in such a way to make Player 1 indifferent among his (good) strategies. This is called the **Principle of Indifference**.

Example. As an example of this consider the game of Odd-or-Even in which both players simultaneously call out one of the numbers zero, one, or two. The matrix is

$$\begin{array}{cc} & \text{Even} \\ \text{Odd} & \begin{pmatrix} 0 & 1 & -2 \\ 1 & -2 & 3 \\ -2 & 3 & -4 \end{pmatrix}. \end{array}$$

Again it is difficult to guess who has the advantage. If we play the game a few times we might become convinced that Even's optimal strategy gives positive weight (probability) to each of the columns. If this assumption is true, Odd should play to make Player 2 indifferent; that is, Odd's optimal strategy $\boldsymbol{p}$ must satisfy

$$\begin{aligned} p_2 - 2p_3 &= V, \\ p_1 - 2p_2 + 3p_3 &= V, \\ -2p_1 + 3p_2 - 4p_3 &= V, \end{aligned} \tag{6}$$

for some number, V — three equations in four unknowns. A fourth equation that must be satisfied is

$$p_1 + p_2 + p_3 = 1. \tag{7}$$

This gives four equations in four unknowns. This system of equations is solved as follows. First we work with (6); add the first equation to the second.

$$p_1 - p_2 + p_3 = 2V. \tag{8}$$

Then add the second equation to the third.

$$-p_1 + p_2 - p_3 = 2V. \tag{9}$$

Taken together (8) and (9) imply that $V = 0$. Adding (7) to (9), we find $2p_2 = 1$, so that $p_2 = 1/2$. The first equation of (6) implies $p_3 = 1/4$ and (7) implies $p_1 = 1/4$. Therefore

$$\boldsymbol{p} = (1/4, 1/2, 1/4)^T \tag{10}$$

is a strategy for I that keeps his average gain to zero no matter what II does. Hence the value of the game is at least zero, and $V = 0$ if our assumption that II's optimal strategy gives positive weight to all columns is correct.

To complete the solution, we note that if the optimal $\boldsymbol{p}$ for I gives positive weight to all rows, then II's optimal strategy $\boldsymbol{q}$ must satisfy the same set of equations (6) and (7) with $\boldsymbol{p}$ replaced by $\boldsymbol{q}$ (because the game matrix here is symmetric). Therefore,

$$\boldsymbol{q} = (1/4, 1/2, 1/4)^{\mathsf{T}} \tag{11}$$

is a strategy for II that keeps his average loss to zero no matter what I does. Thus the value of the game is zero and (10) and (11) are optimal for I and II respectively. The game is fair.

9.2 Nonsingular Game Matrices

Let us extend the method used to solve this example to arbitrary nonsingular square matrices. Let the game matrix $\boldsymbol{A}$ be $m \times m$, and suppose that $\boldsymbol{A}$ is nonsingular. *Assume that I has an optimal strategy giving positive weight to each of the rows.* (This is called the *all-strategies-active* case.) Then by the principle of indifference, every optimal strategy $\boldsymbol{q}$ for II satisfies (4), or

$$\sum_{j=1}^{m} a_{ij} q_j = V \quad \text{for } i = 1, \ldots, m. \tag{12}$$

This is a set of m equations in m unknowns, and since $\boldsymbol{A}$ is nonsingular, we may solve for q_i. Let us write this set of equations in vector notation using $\boldsymbol{q}$ to represent the column vector of II's strategy, and $\mathbf{1} = (1, 1, \ldots, 1)^{\mathsf{T}}$ to represent the column vector of all 1's:

$$\boldsymbol{A}\boldsymbol{q} = V\mathbf{1}. \tag{13}$$

We note that V cannot be zero since (13) would imply that $\boldsymbol{A}$ was singular. Since $\boldsymbol{A}$ is non-singular, $\boldsymbol{A}^{-1}$ exists. Multiplying both sides of (13) on the left by $\boldsymbol{A}^{-1}$ yields

$$\boldsymbol{q} = V\boldsymbol{A}^{-1}\mathbf{1}. \tag{14}$$

If the value of V were known, this would give the unique optimal strategy for II. To find V, we may use the equation $\sum_{j=1}^{m} q_j = 1$, or in vector notation $\mathbf{1}^{\mathsf{T}}\boldsymbol{q} = 1$. Multiplying both sides of (14) on the left by $\mathbf{1}^{\mathsf{T}}$ yields $1 = \mathbf{1}^{\mathsf{T}}\boldsymbol{q} = V\mathbf{1}^{\mathsf{T}}\boldsymbol{A}^{-1}\mathbf{1}$. This shows that $\mathbf{1}^{\mathsf{T}}\boldsymbol{A}^{-1}\mathbf{1}$ cannot be zero so we can solve for V:

$$V = 1/\mathbf{1}^T \boldsymbol{A}^{-1}\mathbf{1}. \tag{15}$$

The unique optimal strategy for II is therefore

$$\boldsymbol{q} = \boldsymbol{A}^{-1}\mathbf{1}/\mathbf{1}^T \boldsymbol{A}^{-1}\mathbf{1}. \tag{16}$$

However, if some component, q_j, turns out to be negative, then our assumption that I has an optimal strategy giving positive weight to each row is false.

However, if $q_j \geq 0$ for all j, we may seek an optimal strategy for I by the same method. The result would be

$$\boldsymbol{p}^T = \mathbf{1}^T \boldsymbol{A}^{-1}/\mathbf{1}^T \boldsymbol{A}^{-1}\mathbf{1}. \tag{17}$$

Now, if in addition $p_i \geq 0$ for all i, then both $\boldsymbol{p}$ and $\boldsymbol{q}$ are optimal since both guarantee an average payoff of V no matter what the other player does. Note that we do not require the p_i to be strictly positive as was required by our original "all-strategies-active" assumption.

We summarize this discussion as a theorem.

Theorem 9.2. *Assume the square matrix $\boldsymbol{A}$ is nonsingular and $\mathbf{1}^T \boldsymbol{A}^{-1}\mathbf{1} \neq 0$. Then the game with matrix $\boldsymbol{A}$ has value $V = 1/\mathbf{1}^T \boldsymbol{A}^{-1}\mathbf{1}$ and optimal strategies $\boldsymbol{p}^T = V\mathbf{1}^T \boldsymbol{A}^{-1}$ and $\boldsymbol{q} = V\boldsymbol{A}^{-1}\mathbf{1}$, provided both $\boldsymbol{p} \geq \mathbf{0}$ and $\boldsymbol{q} \geq \mathbf{0}$.*

If the value of a game is zero, this method cannot work directly since (13) implies that $\boldsymbol{A}$ is singular. However, the addition of a positive constant to all entries of the matrix to make the value positive, *may* change the game matrix into being nonsingular. The previous example of Odd-or-Even is a case in point. The matrix is singular so it would seem that the above method would not work. Yet if 1, say, were added to each entry of the matrix to obtain the matrix $\boldsymbol{A}$ below, then $\boldsymbol{A}$ is nonsingular and we may apply the above method. Let us carry through the computations. By some method or another $\boldsymbol{A}^{-1}$ is obtained.

$$\boldsymbol{A} = \begin{pmatrix} 1 & 2 & -1 \\ 2 & -1 & 4 \\ -1 & 4 & -3 \end{pmatrix}, \quad \boldsymbol{A}^{-1} = \frac{1}{16}\begin{pmatrix} 13 & -2 & -7 \\ -2 & 4 & 6 \\ -7 & 6 & 5 \end{pmatrix}.$$

Then $\mathbf{1}^T \boldsymbol{A}^{-1}\mathbf{1}$, the sum of the elements of $\boldsymbol{A}^{-1}$, is found to be 1, so from (15), $V = 1$. Therefore, we compute $\boldsymbol{p}^T = \mathbf{1}^T \boldsymbol{A}^{-1} = (1/4, 1/2, 1/4)^T$ and

$\boldsymbol{q} = \boldsymbol{A}^{-1}\boldsymbol{1} = (1/4, 1/2, 1/4)^T$. Since both are nonnegative, both are optimal and 1 is the value of the game with matrix $\boldsymbol{A}$.

What do we do if either $\boldsymbol{p}$ or $\boldsymbol{q}$ has negative components? A complete answer to questions of this sort is given in the comprehensive theorem of Shapley and Snow (1950). This theorem shows that an arbitrary $m \times n$ matrix game whose value is not zero may be solved by choosing some suitable square submatrix $\boldsymbol{A}$, and applying the above methods and checking that the resulting optimal strategies are optimal for the whole matrix, $\boldsymbol{A}$. Optimal strategies obtained in this way are called **basic**, and it is noted that every optimal strategy is a probability mixture of basic optimal strategies. Such a submatrix, $\boldsymbol{A}$, is called an *active submatrix* of the game. See Karlin (1959, Vol. I, Sec. 2.4) for a discussion and proof. The problem is to determine which square submatrix to use. The simplex method of linear programming is simply an efficient method not only for solving equations of the form (13), but also for finding which square submatrix to use. This is described in Sec. 10.4.

9.3 Diagonal Games

We apply these ideas to the class of diagonal games — games whose game matrix $\boldsymbol{A}$ is square and diagonal,

$$\boldsymbol{A} = \begin{pmatrix} d_1 & 0 & \dots & 0 \\ 0 & d_2 & \dots & 0 \\ \vdots & \vdots & \ddots & \vdots \\ 0 & 0 & \dots & d_m \end{pmatrix}. \tag{18}$$

Suppose all diagonal terms are positive, $d_i > 0$ for all i. (The other cases are treated in Exercise 2.) One may apply Theorem 9.2 to find the solution, but it is as easy to proceed directly. The set of equations (12) becomes

$$p_i d_i = V \quad \text{for } i = 1, \dots, m \tag{19}$$

whose solution is simply

$$p_i = V/d_i \quad \text{for } i = 1, \dots, m. \tag{20}$$

To find V, we sum both sides over i to find

$$1 = V \sum_{i=1}^{m} 1/d_i \quad \text{or} \quad V = \left(\sum_{i=1}^{m} 1/d_i \right)^{-1}. \tag{21}$$

Similarly, the equations for Player II yield

$$q_i = V/d_i \quad \text{for } i = 1, \ldots, m. \tag{22}$$

Since V is positive from (21), we have $p_i > 0$ and $q_i > 0$ for all i, so that (20) and (22) give optimal strategies for I and II respectively, and (21) gives the value of the game.

As an example, consider the game with matrix $\boldsymbol{C}$.

$$\boldsymbol{C} = \begin{pmatrix} 1 & 0 & 0 & 0 \\ 0 & 2 & 0 & 0 \\ 0 & 0 & 3 & 0 \\ 0 & 0 & 0 & 4 \end{pmatrix}.$$

From (20) and (22) the optimal strategy is proportional to the reciprocals of the diagonal elements. The sum of these reciprocals is $1+1/2+1/3+1/4 = 25/12$. Therefore, the value is $V = 12/25$, and the optimal strategies are $\boldsymbol{p} = \boldsymbol{q} = (12/25, 6/25, 4/25, 3/25)^T$.

9.4 Triangular Games

Another class of games for which Eq. (12) are easy to solve are the games with triangular matrices — matrices with zeros above or below the main diagonal. Unlike for diagonal games, the method does not always work to solve triangular games because the resulting $\boldsymbol{p}$ or $\boldsymbol{q}$ may have negative components. Nevertheless, it works often enough to merit special mention. Consider the game with triangular matrix $\boldsymbol{T}$.

$$\boldsymbol{T} = \begin{pmatrix} 1 & -2 & 3 & -4 \\ 0 & 1 & -2 & 3 \\ 0 & 0 & 1 & -2 \\ 0 & 0 & 0 & 1 \end{pmatrix}.$$

Equations (12) become

$$\begin{aligned} p_1 \qquad\qquad\qquad\qquad\qquad &= V, \\ -2p_1 + p_2 \qquad\qquad\qquad &= V, \\ 3p_1 - 2p_2 + p_3 \qquad\quad &= V, \\ -4p_1 + 3p_2 - 2p_3 + p_4 &= V. \end{aligned}$$

These equations may be solved one at a time from the top down to give

$$p_1 = V, \quad p_2 = 3V, \quad p_3 = 4V, \quad p_4 = 4V.$$

Since $\sum p_i = 1$, we find $V = 1/12$ and $\boldsymbol{p} = (1/12, 1/4, 1/3, 1/3)$. The equations for the q's are

$$\begin{aligned} q_1 - 2q_2 + 3q_3 - 4q_4 &= V, \\ q_2 - 2q_3 + 3q_4 &= V, \\ q_3 - 2q_4 &= V, \\ q_4 &= V. \end{aligned}$$

The solution is

$$q_1 = 4V, \quad q_2 = 4V, \quad q_3 = 3V, \quad q_4 = V.$$

Since the p's and q's are nonnegative, $V = 1/12$ is the value, $\boldsymbol{p} = (1/12, 1/4, 1/3, 1/3)$ is optimal for I, and $\boldsymbol{q} = (1/3, 1/3, 1/4, 1/12)$ is optimal for II.

9.5 Symmetric Games

A game is symmetric if the rules do not distinguish between the players. For symmetric games, both players have the same options (the game matrix is square), and the payoff if I uses i and II uses j is the negative of the payoff if I uses j and II uses i. This means that the game matrix should be **skew-symmetric**: $\boldsymbol{A} = -\boldsymbol{A}^T$, or $a_{ij} = -a_{ji}$ for all i and j.

Definition 9.1. A finite game is said to be *symmetric* if its game matrix is square and skew-symmetric.

Speaking more generally, we may say that a game is symmetric if after some rearrangement of the rows or columns the game matrix is skew-symmetric.

The game of paper-scissors-rock is an example. In this game, Players I and II simultaneously display one of the three objects: paper, scissors, or rock. If they both choose the same object to display, there is no payoff. If they choose different objects, then scissors win over paper (scissors cut paper), rock wins over scissors (rock breaks scissors), and paper wins over rock (paper covers rock). If the payoff upon winning or losing is one unit, then the matrix of the game is as follows.

		II		
		paper	scissors	rock
I	paper	0	−1	1
	scissors	1	0	−1
	rock	−1	1	0

This matrix is skew-symmetric so the game is symmetric. The diagonal elements of the matrix are zero. This is true of any skew-symmetric matrix, since $a_{ii} = -a_{ii}$ implies $a_{ii} = 0$ for all i.

A contrasting example is the game of matching pennies. The two players simultaneously choose to show a penny with either the heads or the tails side facing up. One of the players, say Player I, wins if the choices match. The other player, Player II, wins if the choices differ. Although there is a great deal of symmetry in this game, we do not call it a symmetric game. Its matrix is

$$\begin{array}{ccc} & & \text{II} \\ & & \begin{array}{cc}\text{heads} & \text{tails}\end{array} \\ \text{I} & \begin{array}{c}\text{heads}\\ \text{tails}\end{array} & \begin{pmatrix} 1 & -1 \\ -1 & 1 \end{pmatrix} \end{array}$$

This matrix is not skew-symmetric.

We expect a symmetric game to be fair, that is to have value zero, $V = 0$. This is indeed the case.

Theorem 9.3. *A finite symmetric game has value zero. Any strategy optimal for one player is also optimal for the other.*

Proof. Let $\boldsymbol{p}$ be an optimal strategy for I. If II uses the same strategy the average payoff is zero, because

$$\begin{aligned} \boldsymbol{p}^T\boldsymbol{A}\boldsymbol{p} &= \sum\sum p_i a_{ij} p_j = \sum\sum p_i(-a_{ji})p_j \\ &= -\sum\sum p_j a_{ji} p_i = -\boldsymbol{p}^T\boldsymbol{A}\boldsymbol{p} \end{aligned} \tag{23}$$

implies that $\boldsymbol{p}^T\boldsymbol{A}\boldsymbol{p} = 0$. This shows that the value $V \leq 0$. A symmetric argument shows that $V \geq 0$. Hence $V = 0$. Now suppose $\boldsymbol{p}$ is optimal for I. Then $\sum_{i=1}^m p_i a_{ij} \geq 0$ for all j. Hence $\sum_{j=1}^m a_{ij} p_j = -\sum_{j=1}^m p_j a_{ji} \leq 0$ for all i, so that $\boldsymbol{p}$ is also optimal for II. By symmetry, if $\boldsymbol{q}$ is optimal for II, it is optimal for I also. □

Mendelsohn Games. (N. S. Mendelsohn (1946)) In Mendelsohn games, two players simultaneously choose a positive integer. Both players want to choose an integer larger but not too much larger than the opponent. Here is a simple example. The players choose an integer between 1 and 100. If the numbers are equal, there is no payoff. The player that chooses a number one larger than that chosen by his opponent wins 1. The player that chooses a number two or more larger than his opponent loses 2. Find the game matrix and solve the game.

Solution. The payoff matrix is

$$
\begin{array}{c} \\ 1 \\ 2 \\ 3 \\ 4 \\ 5 \\ \vdots \end{array}
\begin{array}{c}
\begin{array}{cccccc} 1 & 2 & 3 & 4 & 5 & \cdots \end{array} \\
\begin{pmatrix}
0 & -1 & 2 & 2 & 2 & \cdots \\
1 & 0 & -1 & 2 & 2 & \cdots \\
-2 & 1 & 0 & -1 & 2 & \cdots \\
-2 & -2 & 1 & 0 & -1 & \cdots \\
-2 & -2 & -2 & 1 & 0 & \cdots \\
\vdots & & & & & \ddots
\end{pmatrix}
\end{array}
\qquad (24)
$$

The game is symmetric so the value is zero and the players have identical optimal strategies. We see that row 1 dominates rows $4, 5, 6, \ldots$ so we may restrict attention to the upper left 3×3 submatrix. We suspect that there is an optimal strategy for I with $p_1 > 0$, $p_2 > 0$ and $p_3 > 0$. If so, it would follow from the principle of indifference (since $q_1 = p_1 > 0$, $q_2 = p_2 > 0$ $q_3 = p_3 > 0$ is optimal for II) that

$$
\begin{aligned}
p_2 - 2p_3 &= 0, \\
-p_1 \quad + p_3 &= 0, \\
2p_1 - p_2 \quad &= 0.
\end{aligned}
\qquad (25)
$$

We find $p_2 = 2p_3$ and $p_1 = p_3$ from the first two equations, and the third equation is redundant. Since $p_1 + p_2 + p_3 = 1$, we have $4p_3 = 1$; so $p_1 = 1/4$, $p_2 = 1/2$, and $p_3 = 1/4$. Since p_1, p_2 and p_3 are positive, this gives the solution: $\boldsymbol{p} = \boldsymbol{q} = (1/4, 1/2, 1/4, 0, 0, \ldots)^T$ is optimal for both players.

9.6 Invariance

Consider the game of matching pennies: Two players simultaneously choose heads or tails. Player I wins if the choices match and Player II wins otherwise.

There doesn't seem to be much of a reason for either player to choose heads instead of tails. In fact, the problem is the same if the names of heads and tails are interchanged. In other words, the problem is *invariant* under interchanging the names of the pure strategies. In this section, we make the notion of invariance precise. We then define the notion of an invariant strategy and show that in the search for a minimax strategy, a player may restrict attention to invariant strategies. Use of this result greatly simplifies the search for minimax strategies in many games. In the game of matching pennies for example, there is only one invariant strategy for either player, namely, choose heads or tails with probability 1/2 each. Therefore this strategy is minimax without any further computation.

We look at the problem from Player II's viewpoint. Let Y denote the pure strategy space of Player II. A transformation, g of Y into Y is said to be *onto* Y if the range of g is the whole of Y, that is, if for every $y_1 \in Y$ there is $y_2 \in Y$ such that $g(y_2) = y_1$. A transformation, g, of Y into itself is said to be *one-to-one* if $g(y_1) = g(y_2)$ implies $y_1 = y_2$. For finite Y, g is one-to-one if and only if g is onto.

Definition 9.2. Let $G = (X, Y, A)$ be a finite game, and let g be a one-to-one transformation of Y onto itself. The game G is said to be *invariant* under g if for every $x \in X$ there is a unique $x' \in X$ such that

$$A(x, y) = A(x', g(y)) \quad \text{for all } y \in Y. \tag{26}$$

The requirement that x' be unique is not restrictive, for if there were another point $x'' \in X$ such that

$$A(x, y) = A(x'', g(y)) \quad \text{for all } y \in Y, \tag{27}$$

then, we would have $A(x', g(y)) = A(x'', g(y))$ for all $y \in Y$, and since g is onto,

$$A(x', y) = A(x'', y) \quad \text{for all } y \in Y. \tag{28}$$

Thus the strategies x' and x'' have identical payoffs and we could remove one of them from X without changing the problem at all.

To keep things simple, we assume without loss of generality that all duplicate pure strategies have been eliminated. That is, we assume

$$\begin{aligned} A(x', y) = A(x'', y) &\quad \text{for all } y \in Y \text{ implies that } x' = x'', \quad \text{and} \\ A(x, y') = A(x, y'') &\quad \text{for all } x \in X \text{ implies that } y' = y''. \end{aligned} \tag{29}$$

Unicity of x' in Definition 9.2 follows from this assumption.

The given x' in Definition 9.2 depends on g and x only. We denote it by $x' = \overline{g}(x)$. We may write Eq. (26) defining invariance as

$$A(x, y) = A(\overline{g}(x), g(y)) \quad \text{for all } x \in X \text{ and } y \in Y. \tag{26$'$}$$

The mapping $\overline{g}$ is a one-to-one transformation of X since if $\overline{g}(x_1) = \overline{g}(x_2)$, then

$$A(x_1, y) = A(\overline{g}(x_1), g(y)) = A(\overline{g}(x_2), g(y)) = A(x_2, y) \tag{30}$$

for all $y \in Y$, which implies $x_1 = x_2$ from assumption (29). Therefore the inverse, g^{-1}, of g, defined by $g^{-1}(g(x)) = g(g^{-1}(x)) = x$, exists. Moreover, $\overline{g}$ is one-to-one and onto.

Lemma 9.1. *If a finite game, $G = (X, Y, A)$, is invariant under a one-to-one transformation, g, then G is also invariant under g^{-1}.*

Proof. We are given $A(x, y) = A(\overline{g}(x), g(y))$ for all $x \in X$ and all $y \in Y$. Since true for all x and y, it is true if y is replaced by $g^{-1}(y)$ and x is replaced by $\overline{g}^{-1}(x)$. This gives $A(\overline{g}^{-1}(x), g^{-1}(y)) = A(x, y)$ for all $x \in X$ and all $y \in Y$. This shows that G is invariant under g^{-1}. □

Lemma 9.2. *If a finite game, $G = (X, Y, A)$, is invariant under two one-to-one transformations, g_1 and g_2, then G is also invariant under the composition transformation, $g_2 g_1$, defined by $g_2 g_1(y) = g_2(g_1(y))$.*

Proof. We are given $A(x, y) = A(\overline{g}_1(x), g_1(y))$ for all $x \in X$ and all $y \in Y$, and $A(x, y) = A(\overline{g}_2(x), g_2(y))$ for all $x \in X$ and all $y \in Y$. Therefore,

$$\begin{aligned} A(x, y) &= A(\overline{g}_2(\overline{g}_1(x)), g_2(g_1(y))) \\ &= A(\overline{g}_2(\overline{g}_1(x)), g_2 g_1(y)) \quad \text{for all } y \in Y \text{ and } x \in X, \end{aligned} \tag{31}$$

which shows that G is invarant under $g_2 g_1$. □

Furthermore, these proofs show that

$$\overline{g_2 g_1} = \overline{g}_2\, \overline{g}_1 \quad \text{and} \quad \overline{g^{-1}} = \overline{g}^{-1}. \tag{32}$$

Thus the class of transformations, g on Y, under which the problem is invariant forms a group, $\mathcal{G}$, with composition as the multiplication operator. The identity element, e of the group is the identity transformation, $e(y) = y$ for all $y \in Y$. The set, $\overline{\mathcal{G}}$, of corresponding transformations $\overline{g}$ on X is also a group, with identity $\overline{e}(x) = x$ for all $x \in X$. Equations (32) say that $\overline{\mathcal{G}}$ is isomorphic to $\mathcal{G}$; as groups, they are indistinguishable.

This shows that we could have analyzed the problem from Player I's viewpoint and arrived at the same groups $\overline{\mathcal{G}}$ and $\mathcal{G}$.

Definition 9.3 A finite game $G = (X, Y, A)$ is said to be invariant under a group, $\mathcal{G}$ of transformations, if $(26')$ holds for all $g \in \mathcal{G}$.

We now define what it means for a mixed strategy, $\boldsymbol{q}$, for Player II to be invariant under a group $\mathcal{G}$. Let m denote the number of elements in X and n denote the number of elements in Y.

Definition 9.4. Given that a finite game $G = (X, Y, A)$ is invariant under a group, $\mathcal{G}$, of one-to-one transformations of Y, a mixed strategy, $\boldsymbol{q} = (q(1), \ldots, q(n))$, for Player II is said to be invariant under $\mathcal{G}$ if

$$q(g(y)) = q(y) \quad \text{for all } y \in Y \text{ and all } g \in \mathcal{G}. \tag{33}$$

Similarly a mixed strategy $\boldsymbol{p} = (p(1), \ldots, p(m))$, for Player I is said to be invariant under $\mathcal{G}$ (or $\overline{\mathcal{G}}$) if

$$p(\overline{g}(x)) = p(x) \quad \text{for all } x \in X \text{ and all } \overline{g} \in \overline{\mathcal{G}}. \tag{34}$$

Two points y_1 and y_2 in Y are said to be equivalent if there exists a g in $\mathcal{G}$ such that $g(y_2) = y_1$. It is an easy exercise to show that this is an equivalence relation. The set of points, $E_y = \{y' : g(y') = y \text{ for some } g \in \mathcal{G}\}$, is called an equivalence class, or an orbit. Thus, y_1 and y_2 are equivalent if they lie in the same orbit. Definition 9.4 says that a mixed strategy $\boldsymbol{q}$ for Player II is invariant if it is constant on orbits, that is, if it assigns the same probability to all pure strategies in the orbit. The power of this notion is contained in the following theorem.

Theorem 9.4. *If a finite game $G = (X, Y, A)$ is invariant under a group $\mathcal{G}$, then there exist invariant optimal strategies for the players.*

Proof. It is sufficient to show that Player II has an invariant optimal strategy. Since the game is finite, there exist a value, V, and an optimal mixed strategy for player II, $\boldsymbol{q}^*$. This is to say that

$$\sum_{y \in Y} A(x, y) q^*(y) \leq V \quad \text{for all } x \in X. \tag{35}$$

We must show that there is an invariant strategy $\tilde{\boldsymbol{q}}$ that satisfies this same condition. Let $N = |\mathcal{G}|$ be the number of elements in the group $\mathcal{G}$. Define

$$\tilde{q}(y) = \frac{1}{N} \sum_{g \in \mathcal{G}} q^*(g(y)). \tag{36}$$

(This takes each orbit and replaces each probability by the average of the probabilities in the orbit.) Then $\tilde{\boldsymbol{q}}$ is invariant since for any $g' \in \mathcal{G}$,

$$\begin{aligned} \tilde{q}(g'(y)) &= \frac{1}{N}\sum_{g\in\mathcal{G}} q^*(g(g'(y))) \\ &= \frac{1}{N}\sum_{g\in\mathcal{G}} q^*(g(y)) = \tilde{q}(y) \end{aligned} \tag{37}$$

since applying g' to $Y = \{1, 2, \ldots, n\}$ is just a reordering of the points of Y. Moreover, $\tilde{\boldsymbol{q}}$ satisfies (35) since

$$\begin{aligned} \sum_{y\in Y} A(x,y)\tilde{q}(y) &= \sum_{y\in Y} A(x,y)\frac{1}{N}\sum_{g\in\mathcal{G}} q^*(g(y)) \\ &= \frac{1}{N}\sum_{g\in\mathcal{G}}\sum_{y\in Y} A(x,y)q^*(g(y)) \\ &= \frac{1}{N}\sum_{g\in\mathcal{G}}\sum_{y\in Y} A(\overline{g}(x),g(y))q^*(g(y)) \\ &= \frac{1}{N}\sum_{g\in\mathcal{G}}\sum_{y\in Y} A(\overline{g}(x),y)q^*(y) \\ &\le \frac{1}{N}\sum_{g\in\mathcal{G}} V = V. \end{aligned} \tag{38}$$

□

In matching pennies, $X = Y = \{1, 2\}$, and $A(1,1) = A(2,2) = 1$ and $A(1,2) = A(2,1) = -1$. The Game $G = (X, Y, A)$ is invariant under the group $\mathcal{G} = \{e, g\}$, where e is the identity transformation, and g is the transformation, $g(1) = 2$, $g(2) = 1$. The (mixed) strategy $(q(1), q(2))$ is invariant under $\mathcal{G}$ if $q(1) = q(2)$. Since $q(1) + q(2) = 1$, this implies that $q(1) = q(2) = 1/2$ is the only invariant strategy for Player II. It is therefore minimax. Similarly, $p(1) = p(2) = 1/2$ is the only invariant, and hence minimax, strategy for Player I.

Similarly, the game of paper-scissors-rock is invariant under the group $\mathcal{G} = \{e, g, g^2\}$, where $g(\text{paper}) = \text{scissors}$, $g(\text{scissors}) = \text{rock}$ and $g(\text{rock}) = \text{paper}$. The unique invariant, and hence minimax, strategy gives probability $1/3$ to each of paper, scissors and rock.

Colonel Blotto Games. For more interesting games reduced by invariance, we consider a class of tactical military games called Blotto

Games, introduced by Tukey (1949). There are many variations of these games; just google "Colonel Blotto Games" to get a sampling. Here, we describe the discrete version treated in Williams (1966), Karlin (1959) and Dresher (1961).

Colonel Blotto has four regiments with which to occupy two posts. The famous Lieutenant Kije has three regiments with which to occupy the same posts. The payoff is defined as follows. The army sending the most units to either post captures it and all the regiments sent by the other side, scoring one point for the captured post and one for each captured regiment. If the players send the same number of regiments to a post, both forces withdraw and there is no payoff.

Colonel Blotto must decide how to split his forces between the two posts. There are five pure strategies he may employ, namely, $X = \{(4,0), (3,1), (2,2), (1,3), (0,4)\}$, where (n_1, n_2) represents the strategy of sending n_1 units to post number 1, and n_2 units to post number two. Lieutenant Kije has four pure strategies, $Y = \{(3,0), (2,1), (1,2), (0,3)\}$. The payoff matrix is

$$\begin{array}{c} \\ (4,0) \\ (3,1) \\ (2,2) \\ (1,3) \\ (0,4) \end{array} \begin{array}{c} \begin{array}{cccc} (3,0) & (2,1) & (1,2) & (0,3) \end{array} \\ \begin{pmatrix} 4 & 2 & 1 & 0 \\ 1 & 3 & 0 & -1 \\ -2 & 2 & 2 & -2 \\ -1 & 0 & 3 & 1 \\ 0 & 1 & 2 & 4 \end{pmatrix} \end{array}. \tag{39}$$

Unfortunately, the 5×4 matrix game cannot be reduced by removing dominated strategies. So it seems that to solve it, we must use the simplex method. However, there is an invariance in this problem that simplifies it considerably. This involves the symmetry between the posts. This leads to the group, $\mathcal{G} = \{e, g\}$, where

$$g((3,0)) = (0,3), \quad g((0,3)) = (3,0), \quad g((2,1)) = (1,2), \quad g((1,2)) = (2,1)$$

and the corresponding group, $\overline{\mathcal{G}} = \{\overline{e}, \overline{g}\}$, where

$$\overline{g}((4,0)) = (0,4), \quad \overline{g}((0,4)) = (4,0), \quad \overline{g}((3,1)) = (1,3),$$
$$\overline{g}((1,3)) = (3,1) \quad \text{and} \quad \overline{g}((2,2)) = (2,2).$$

The orbits for Kije are $\{(3,0), (0,3)\}$ and $\{(2,1), (1,2)\}$. Therefore a strategy, $\boldsymbol{q}$, is invariant if $q((3,0)) = q((0,3))$ and $q((2,1)) = q((1,2))$.

Similarly, the orbits for Blotto are $\{(4,0),(0,4)\}$, $\{(3,1),(1,3)\}$ and $\{(2,2)\}$. So a strategy, $\boldsymbol{p}$, for Blotto is invariant if $p((4,0)) = p((0,4))$ and $p((3,1)) = p((1,3))$.

We may reduce Kije's strategy space to two elements, defined as follows:

$(3,0)^*$: use $(3,0)$ and $(0,3)$ with probability $1/2$ each.
$(2,1)^*$: use $(2,1)$ and $(1,2)$ with probability $1/2$ each.

Similarly, Blotto's strategy space reduces to three elements:

$(4,0)^*$: use $(4,0)$ and $(0,4)$ with probability $1/2$ each.
$(3,1)^*$: use $(3,1)$ and $(1,3)$ with probability $1/2$ each.
$(2,2)$: use $(2,2)$.

With these strategy spaces, the payoff matrix becomes

$$\begin{array}{c} \\ (4,0)^* \\ (3,1)^* \\ (2,2) \end{array} \begin{array}{c} \begin{array}{cc} (3,0)^* & (2,1)^* \end{array} \\ \begin{pmatrix} 2 & 1.5 \\ 0 & 1.5 \\ -2 & 2 \end{pmatrix} \end{array}. \tag{40}$$

As an example of the computations used to arrive at these payoffs, consider the upper left entry. If Blotto uses (4,0) and (0,4) with probability 1/2 each, and if Kije uses (3,0) and (0,3) with probability 1/2 each, then the four corners of the matrix (39) occur with probability 1/4 each, so the expected payoff is the average of the four numbers, 4, 0, 0, 4, namely 2.

To complete the analysis, we solve the game with matrix (40). We first note that the middle row is dominated by the top row (even though there was no domination in the original matrix). Removal of the middle row reduces the game to a 2×2 matrix game whose solution is easily found. The mixed strategy (8/9,0,1/9) is optimal for Blotto, the mixed strategy (1/9,8/9) is optimal for Kije, and the value is $V = 14/9$.

Returning now to the original matrix (39), we find that (4/9,0,1/9,0,4/9) is optimal for Blotto, (1/18,4/9,4/9,1/18) is optimal for Kije, and $V = 14/9$ is the value.

9.7 Exercises

1. Consider the game with matrix $\begin{pmatrix} -2 & 2 & -1 \\ 1 & 1 & 1 \\ 3 & 0 & 1 \end{pmatrix}$.

   (a) Note that this game has a saddle point.

(b) Show that the inverse of the matrix exists.
(c) Show that II has an optimal strategy giving positive weight to each of his columns.
(d) Why then, don't Eqs. (16) give an optimal strategy for II?

2. Consider the diagonal matrix game with matrix (18).

(a) Suppose one of the diagonal terms is zero. What is the value of the game?
(b) Suppose one of the diagonal terms is positive and another is negative. What is the value of the game?
(c) Suppose all diagonal terms are negative. What is the value of the game?

3. Player II chooses a number $j \in \{1,2,3,4\}$, and Player I tries to guess what number II has chosen. If he guesses correctly and the number was j, he wins 2^j dollars from II. Otherwise there is no payoff. Set up the matrix of this game and solve.
4. Player II chooses a number $j \in \{1,2,3,4\}$ and I tries to guess what it is. If he guesses correctly, he wins 1 from II. If he overestimates he wins 1/2 from II. If he underestimates, there is no payoff. Set up the matrix of this game and solve.
5. Player II chooses a number $j \in \{1,2,\ldots,n\}$ and I tries to guess what it is. If he guesses correctly, he wins 1. If he guesses too high, he loses 1. If he guesses too low, there is no payoff. Set up the matrix and solve.
6. Player II chooses a number $j \in \{1,2,\ldots,n\}$, $n \geq 2$, and Player I tries to guess what it is by guessing some $i \in \{1,2,\ldots,n\}$. If he guesses correctly, i.e. $i=j$, he wins 1. If $i>j$, he wins b^{i-j} for some number $b<1$. Otherwise, if $i<j$, he wins nothing. Set up the matrix and solve. Hint: If $\boldsymbol{A}_n = (a_{ij})$ denotes the game matrix, then show the inverse matrix is $\boldsymbol{A}_n^{-1} = (a^{ij})$, where $a^{ij} = \begin{cases} 1 & \text{if } i=j \\ -b & \text{if } i=j+1 \\ 0 & \text{otherwise} \end{cases}$, and use Theorem 9.2.
7. **The Pascal Matrix Game.** The Pascal matrix of order n is the $n \times n$ matrix $\boldsymbol{B}_n$ of elements b_{ij}, where

$$b_{ij} = \binom{i-1}{j-1} \quad \text{if } i \geq j, \quad \text{and} \quad b_{ij} = 0 \quad \text{if } i<j.$$

The ith row of $\boldsymbol{B}_n$ consists of the binomial coefficients in the expansion of $(x+y)^i$. Call and Velleman (1993) show that the inverse of $\boldsymbol{B}_n$ is the matrix $\boldsymbol{A}_n$ with entries a_{ij}, where $a_{ij} = (-1)^{i+j}b_{ij}$. Using this, find the value and optimal strategies for the matrix game with matrix $\boldsymbol{A}_n$.

8. Solve the games with the following matrices.

$$\text{(a)}\quad \begin{pmatrix} 1 & -1 & -1 \\ 0 & 2 & 1 \\ 0 & 0 & 3 \end{pmatrix}, \quad \text{(b)}\quad \begin{pmatrix} 2 & 1 & 1 & 1 \\ 1 & 3/2 & 1 & 1 \\ 1 & 1 & 4/3 & 1 \\ 1 & 1 & 1 & 5/4 \end{pmatrix},$$

$$\text{(c)}\quad \begin{pmatrix} 2 & 0 & 0 & 2 \\ 0 & 3 & 0 & 0 \\ 0 & 0 & 4 & 3 \\ 1 & 1 & 0 & 1 \end{pmatrix}.$$

9. **Another Mendelsohn game.** Two players simultaneously choose an integer between 1 and n inclusive, where $n \geq 5$. If the numbers are equal there is no payoff. The player that chooses a number one larger than that chosen by his opponent wins 2. The player that chooses a number two or more larger than that chosen by his opponent loses 1.

   (a) Set up the game matrix.
   (b) It turns out that the optimal strategy satisfies $p_i > 0$ for $i = 1, \ldots, 5$, and $p_i = 0$ for all other i. Solve for the optimal $\boldsymbol{p}$. (It is not too difficult since you can argue that $p_1 = p_5$ and $p_2 = p_4$ by symmetry of the equations.) Check that in fact the strategy you find is optimal.

10. **Silverman Games.** (See R. T. Evans (1979) and Heuer and Leopold-Wildburger (1991).) Two players simultaneously choose positive integers. As in Mendelsohn games, a player wants to choose an integer larger but not too much larger than the opponent, but in Silverman games "too much larger" is determined multiplicatively rather than additively. Solve the following example: The player whose number is larger but less than three times as large as the opponent's wins 1. But the player whose number is three times as large or larger loses 2. If the numbers are the same, there is no payoff.

(a) Note that this is a symmetric game, and show that dominance reduces the game to a 3×3 matrix.
(b) Solve.

11. Solve the following games.

$$\text{(a)} \quad \begin{pmatrix} 0 & 1 & -2 \\ -1 & 0 & 3 \\ 2 & -3 & 0 \end{pmatrix}, \quad \text{(b)} \quad \begin{pmatrix} 0 & 1 & -2 \\ -2 & 0 & 1 \\ 1 & -2 & 0 \end{pmatrix},$$

$$\text{(c)} \quad \begin{pmatrix} 1 & 4 & -1 & 5 \\ 4 & -1 & 5 & 1 \\ -1 & 5 & 1 & 4 \\ 5 & 1 & 4 & -1 \\ 2 & 2 & 2 & 2 \end{pmatrix}.$$

12. Run the original Blotto matrix (39) through the Matrix Game Solver, on the web at: http://www.tomsferguson.com/gamesolve.html, and note that it gives different optimal strategies than those found in the text. What does this mean? Show that $(3,1)^*$ is strictly dominated in (40). This means that no optimal strategy can give weight to $(3,1)^*$. Is this true for the solution found?
13. (a) Suppose Blotto has 2 units and Kije just 1 unit, with 2 posts to capture. Solve.
    (b) Suppose Blotto has 3 units and Kije 2 units, with 2 posts to capture. Solve.
14. (a) Suppose there are 3 posts to capture. Blotto has 4 units and Kije has 3. Solve. (Reduction by invariance leads to a 4×3 matrix, reducible further by domination to 2×2.)
    (b) Suppose there are 4 posts to capture. Blotto has 4 units and Kije has 3. Solve. (A 5×3 reduced matrix, reducible by domination to 4×3. But you may as well use the Matrix Game Solver to solve it.)
15. **Battleship.** The game of Battleship, sometimes called Salvo, is played on two square boards, usually 10×10. Each player hides a fleet of ships on his own board and tries to sink the opponent's ships before the opponent sinks his. (For one set of rules, see http://www.kielack.de/games/destroya.htm, and while you are there, have a game.)

    For simplicity, consider a 3×3 board and suppose that Player I hides a destroyer (length 2 squares) horizontally or vertically on this board. Then Player II shoots by calling out squares of the board, one at a time. After each shot, Player I says whether the shot was a hit or a miss. Player II continues until both squares of the destroyer have been hit. The payoff to Player I is the number of shots that Player II has made. Let us label the squares from 1 to 9 as follows.

1	2	3
4	5	6
7	8	9

The problem is invariant under rotations and reflections of the board. In fact, of the 12 possible positions for the destroyer, there are only two distinct invariant choices available to Player I: the strategy, $[1,2]^*$, that chooses one of [1,2], [2,3], [3,6], [6,9], [8,9], [7,8], [4,7], and [1,4], at random with probability 1/8 each, and the strategy, $[2,5]^*$, that chooses one of [2,5], [5,6], [5,8], and [4,5], at random with probability 1/4 each. This means that invariance reduces the game to a $2 \times n$ game where n is the number of invariant strategies of Player II. Domination may reduce it somewhat further. Solve the game.

16. **Dresher's Guessing Game.** Player I secretly writes down one of the numbers $1, 2, \ldots, n$. Player II must repeatedly guess what I's number is until she guesses correctly, losing 1 for each guess. After each guess, Player I must say whether the guess is correct, too high, or too low. Solve this game for $n = 3$. (This game was investigated by Dresher (1961) and solved for $n \le 11$ by Johnson (1964). A related problem is treated in Gal (1974).)
17. **Thievery.** Player I wants to steal one of $m \ge 2$ items from Player II. Player II can only guard one item at a time. Item i is worth $u_i > 0$, for $i = 1, \ldots, m$. This leads to a matrix game with $m \times m$ matrix,

$$\boldsymbol{A} = \begin{pmatrix} 0 & u_1 & u_1 & \ldots & u_1 \\ u_2 & 0 & u_2 & \ldots & u_2 \\ u_3 & u_3 & 0 & \ldots & u_3 \\ \vdots & \vdots & \vdots & \ddots & \vdots \\ u_m & u_m & u_m & \ldots & 0 \end{pmatrix}$$

Solve!

Hint: It might be expected that for some $k \le m$ Player I will give all his probability to stealing one of the k most expensive items. Order the items from most expensive to least expensive, $u_1 \ge u_2 \ge \cdots \ge u_m > 0$, and use the principle of indifference on the upper left $k \times k$ submatrix of $\boldsymbol{A}$ for some k.

18. Player II chooses a number $j \in \{1, 2, \ldots, n\}$, $n \geq 2$, and Player I tries to guess what it is by guessing some $i \in \{1, 2, \ldots, n\}$. If he guesses correctly, i.e. $i = j$, he wins 2. If he misses by exactly 1, i.e. $|i - j| = 1$, then he loses 1. Otherwise there is no payoff. Solve. Hint: Let $\boldsymbol{A}_n$ denote the $n \times n$ payoff matrix, and show that $\boldsymbol{A}_n^{-1} = \boldsymbol{B}_n = (b_{ij})$, where $b_{ij} = i(n+1-j)/(n+1)$ for $i \leq j$, and $b_{ij} = b_{ji}$ for $i > j$.
19. **The Number Hides Game.** The Number Hides Game, introduced by Ruckle (1983) and solved by Baston, Bostock and Ferguson (1989), may be described as follows. From the set $S = \{1, 2, \ldots, k\}$, Player I chooses an interval of m_1 consecutive integers and Player II chooses an interval of m_2 consecutive integers. The payoff to Player I is the number of integers in the intersection of the two intervals. When $k = n+1$ and $m_1 = m_2 = 2$, this game is equivalent to the game with $n \times n$ matrix $\boldsymbol{A}_n = (a_{ij})$, where

$$a_{ij} = \begin{cases} 2 & \text{if } i = j, \\ 1 & \text{if } |i - j| = 1, \\ 0 & \text{otherwise.} \end{cases}$$

[In this form, the game is also a special case of the **Helicopter versus Submarine Game**, solved in the book of Garnaev (2000), in which the payoff for $|i - j| = 1$ is allowed to be an arbitrary number a, $0 \leq a \leq 1$.] Since $\boldsymbol{A}_n^{-1}$ is just $\boldsymbol{B}_n$ of the previous exercise with b_{ij} replaced by $(-1)^{i+j} b_{ij}$, the solution can be derived as in that exercise. Instead, just show the following.

(a) For n odd, the value is $V_n = 4/(n+1)$. There is an optimal equalizing strategy (the same for both players) that is proportional to $(1, 0, 1, 0, \ldots, 0, 1)$.

(b) For n even, the value is $4(n+1)/(n(n+2))$. There is an optimal equalizing strategy (the same for both players) that is proportional to $(k, 1, k-1, 2, k-2, 3, \ldots, 2, k-1, 1, k)$, where $k = n/2$.

10 Solving Finite Games

Consider an arbitrary finite two-person zero-sum game, (X, Y, A), with $m \times n$ matrix, $\boldsymbol{A}$. Let us take the strategy space X to be the first m integers, $X = \{1, 2, \ldots, m\}$, and similarly, $Y = \{1, 2, \ldots, n\}$. A mixed strategy for Player I may be represented by a column vector, $(p_1, p_2, \ldots, p_m)^{\mathsf{T}}$ of probabilities that add to 1. Similarly, a mixed strategy for Player II is an n-tuple $\boldsymbol{q} = (q_1, q_2, \ldots, q_n)^{\mathsf{T}}$. The sets of mixed strategies of players I and II will be denoted respectively by X^* and Y^*,

$$X^* = \left\{ \boldsymbol{p} = (p_1, \ldots, p_m)^{\mathsf{T}} : \; p_i \geq 0, \quad \text{for } i = 1, \ldots, m \quad \text{and} \quad \sum_1^m p_i = 1 \right\},$$

$$Y^* = \left\{ \boldsymbol{q} = (q_1, \ldots, q_n)^{\mathsf{T}} : \; q_j \geq 0, \quad \text{for } j = 1, \ldots, n \quad \text{and} \quad \sum_1^n q_j = 1 \right\}.$$

The m-dimensional unit vector $\mathbf{e}_k \in X^*$ with a one for the kth component and zeros elsewhere may be identified with the pure strategy of choosing row k. Thus, we may consider the set of Player I's pure strategies, X, to be a subset of X^*. Similarly, Y may be considered to be a subset of Y^*. We could if we like consider the game (X, Y, A) in which the players are allowed to use mixed strategies as a new game (X^*, Y^*, A), where $A(\boldsymbol{p}, \boldsymbol{q}) = \boldsymbol{p}^{\mathsf{T}} \boldsymbol{A} \boldsymbol{q}$, though we would no longer call this game a finite game.

In this section, we give an algorithm for solving finite games; that is, we show how to find the value and at least one optimal strategy for each player. Occasionally, we shall be interested in finding all optimal strategies for a player.

10.1 Best Responses

Suppose that Player II chooses a column at random using $\boldsymbol{q} \in Y^*$. If Player I chooses row i, the average payoff to I is

$$\sum_{j=1}^{n} a_{ij} q_j = (\boldsymbol{A}\boldsymbol{q})_i, \tag{1}$$

the ith component of the vector $\boldsymbol{A}\boldsymbol{q}$. Similarly, if Player I uses $\boldsymbol{p} \in X^*$ and Player II chooses column j, Then I's average payoff is

$$\sum_{i=1}^{n} p_i a_{ij} = (\boldsymbol{p}^T \boldsymbol{A})_j, \tag{2}$$

the jth component of the vector $\boldsymbol{p}^T\boldsymbol{A}$. More generally, if Player I uses $\boldsymbol{p} \in X^*$ and Player II uses $\boldsymbol{q} \in Y^*$, the average payoff to I becomes

$$\sum_{i=1}^{m} \left(\sum_{j=1}^{n} a_{ij} q_j \right) p_i = \sum_{i=1}^{m} \sum_{j=1}^{n} p_i a_{ij} q_j = \boldsymbol{p}^T \boldsymbol{A}\boldsymbol{q}. \tag{3}$$

Suppose it is known that Player II is going to use a particular strategy $\boldsymbol{q} \in Y^*$. Then Player I would choose that row i that maximizes (1); or, equivalently, he would choose that $\boldsymbol{p} \in X^*$ that maximizes (3). His average payoff would be

$$\max_{1 \le i \le m} \sum_{j=1}^{n} a_{ij} q_j = \max_{p \in X^*} \boldsymbol{p}^T \boldsymbol{A}\boldsymbol{q}. \tag{4}$$

To see that these quantities are equal, note that the left side is the maximum of $\boldsymbol{p}^T\boldsymbol{A}\boldsymbol{q}$ over $\boldsymbol{p} \in X^*$, and so, since $X \subset X^*$, must be less than or equal to the right side. The reverse inequality follows since (3) is an average of the quantities in (1) and so must be less than or equal to the largest of the values in (1).

Any $\boldsymbol{p} \in X^*$ that achieves the maximum of (3) is called a *best response* or a *Bayes strategy* against $\boldsymbol{q}$. In particular, any row i that achieves the maximum of (1) is a (pure) Bayes strategy against $\boldsymbol{q}$. There always exist pure Bayes strategies against $\boldsymbol{q}$ for every $\boldsymbol{q} \in Y^*$ in finite games.

Similarly, if it is known that Player I is going to use a particular strategy $\boldsymbol{p} \in X^*$, then Player II would choose that column j that minimizes (2), or, equivalently, that $\boldsymbol{q} \in Y^*$ that minimizes (3). Her average payoff would be

$$\min_{1 \le j \le n} \sum_{i=1}^{m} p_i a_{ij} = \min_{\boldsymbol{q} \in Y^*} \boldsymbol{p}^T \boldsymbol{A} \boldsymbol{q}. \tag{5}$$

Any $\boldsymbol{q} \in Y^*$ that achieves the minimum in (5) is called a *best response* or a *Bayes strategy* for Player II against $\boldsymbol{p}$.

The notion of a best response presents a practical way of playing a game: Make a guess at the probabilities that you think your opponent will play his/her various pure strategies, and choose a best response against this. This method is available in quite complex situations. In addition, it allows a player to take advantage of an opponent's perceived weaknesses. Of course this may be a dangerous procedure. Your opponent may be better at this type of guessing than you. (See Exercise 1.)

10.2 Upper and Lower Values of a Game

Suppose now that II is required to announce her choice of a mixed strategy $\boldsymbol{q} \in Y^*$ before I makes his choice. This changes the game to make it apparently more favorable to I. If II announces $\boldsymbol{q}$, then certainly I would use a Bayes strategy against $\boldsymbol{q}$ and II would lose the quantity (4) on the average. Therefore, II would choose to announce that $\boldsymbol{q}$ that minimizes (4). The minimum of (4) over all $\boldsymbol{q} \in Y^*$ is denoted by $\overline{V}$ and called the *upper value of the game* (X, Y, A).

$$\overline{V} = \min_{\boldsymbol{q} \in Y^*} \max_{1 \le i \le m} \sum_{j=1}^{n} a_{ij} q_j = \min_{\boldsymbol{q} \in Y^*} \max_{\boldsymbol{p} \in X^*} \boldsymbol{p}^T \boldsymbol{A} \boldsymbol{q}. \tag{6}$$

Any $\boldsymbol{q} \in Y^*$ that achieves the minimum in (6) is called a *minimax strategy for II.* It minimizes her maximum loss. There always exists a minimax strategy in finite games: the quantity (4), being the maximum of m linear functions of $\boldsymbol{q}$, is a continuous function of $\boldsymbol{q}$ and since Y^* is a closed bounded set, this function assumes its minimum over Y^* at some point of Y^*.

In words, $\overline{V}$ as the smallest average loss that Player II can assure for herself no matter what I does.

A similar analysis may be carried out assuming that I must announce his choice of a mixed strategy $\boldsymbol{p} \in X^*$ before II makes her choice. If I announces $\boldsymbol{p}$, then II would choose that column with the smallest average payoff, or equivalently that $\boldsymbol{q} \in Y^*$ that minimizes the average payoff (5).

Given that (5) is the average payoff to I if he announces $\boldsymbol{p}$, he would therefore choose $\boldsymbol{p}$ to maximize (5) and obtain on the average

$$\underline{V} = \max_{p \in X^*} \min_{1 \le j \le n} \sum_{i=1}^{m} p_i a_{ij} = \max_{p \in X^*} \min_{q \in Y^*} \boldsymbol{p}^T \boldsymbol{A} \boldsymbol{q}. \tag{7}$$

The quantity $\underline{V}$ is called the *lower value of the game*. It is the maximum amount that I can guarantee himself no matter what II does. Any $\boldsymbol{p} \in X^*$ that achieves the maximum in (7) is called a *minimax strategy for I*. Perhaps *maximin strategy* would be more appropriate terminology in view of (7), but from symmetry (either player may consider himself Player II for purposes of analysis) the same word to describe the same idea may be preferable and it is certainly the customary terminology. As in the analysis for Player II, we see that Player I always has a minimax strategy. The existence of minimax strategies in matrix games is worth stating as a lemma.

Lemma 10.1. *In a finite game, both players have minimax strategies.*

It is easy to argue that the lower value is less than or equal to the upper value. For if $\overline{V} < \underline{V}$ and if I can assure himself of winning at least $\underline{V}$, Player II cannot assure herself of not losing more than $\overline{V}$, an obvious contradiction. It is worth stating this fact as a lemma too.

Lemma 10.2. *The lower value is less than or equal to the upper value,*

$$\underline{V} \le \overline{V}.$$

This lemma also follows from the general mathematical principle that for *any* real-valued function, $f(x, y)$, and *any* sets, X^* and Y^*,

$$\max_{x \in X^*} \min_{y \in Y^*} f(x, y) \le \min_{y \in Y^*} \max_{x \in X^*} f(x, y).$$

To see this general principle, note that $\min_{y'} f(x, y') \le f(x, y) \le \max_{x'} f(x', y)$ for every fixed x and y. Then, taking $\max_x$ on the left does not change the inequality, nor does taking $\min_y$ on the right, which gives the result.

If $\underline{V} < \overline{V}$, the average payoff should fall between $\underline{V}$ and $\overline{V}$. Player II can keep it from getting larger than $\overline{V}$ and Player I can keep it from getting smaller than $\underline{V}$. When $\underline{V} = \overline{V}$, a very nice stable situation exists.

Definition. If $\underline{V} = \overline{V}$, we say the value of the game exists and is equal to the common value of $\underline{V}$ and $\overline{V}$, denoted simply by V. If the value of the game exists, we refer to minimax strategies as optimal strategies.

The Minimax Theorem, stated in Chap. 7, may be expressed simply by saying that for finite games, $\underline{V} = \overline{V}$.

The Minimax Theorem. *Every finite game has a value, and both players have minimax strategies.*

There are several nice proofs of the Minimax Theorem. There is one completely elementary proof due to Owen (1967). This is presented in Appendix 2. Another proof of the Minimax Theorem is presented in Sec. 10.4 based on the Duality Theorem of linear programming. In Sec. 10.5, it is shown how linear programming may be used to calculate the value and optimal strategies of a matrix game.

We note one remarkable corollary of this theorem. If the rules of the game are changed so that Player II is required to announce her choice of a mixed strategy before Player I makes his choice, then the apparent advantage given to Player I by this is illusory. Player II can simply announce her minimax strategy.

10.3 Invariance under Change of Location and Scale

Another simple observation is useful in this regard. This concerns the *invariance* of the minimax strategies under the operations of adding a constant to each entry of the game matrix, and of multiplying each entry of the game matrix by a positive constant. The game having matrix $\boldsymbol{A} = (a_{ij})$ and the game having matrix $\boldsymbol{A}' = (a'_{ij})$ with $a'_{ij} = a_{ij} + b$, where b is an arbitrary real number, are very closely related. In fact, the game with matrix $\boldsymbol{A}'$ is equivalent to the game in which II pays I the amount b, and then I and II play the game with matrix $\boldsymbol{A}$. Clearly any strategies used in the game with matrix $\boldsymbol{A}'$ give Player I b plus the payoff using the same strategies in the game with matrix $\boldsymbol{A}$. Thus, any minimax strategy for either player in one game is also minimax in the other, and the upper (lower) value of the game with matrix $\boldsymbol{A}'$ is b plus the upper (lower) value of the game with matrix $\boldsymbol{A}$.

Similarly, the game having matrix $\boldsymbol{A}'' = (a''_{ij})$ with $a''_{ij} = ca_{ij}$, where c is a positive constant, may be considered as the game with matrix $\boldsymbol{A}$ with a change of scale (a change of monetary unit if you prefer). Again, minimax strategies do not change, and the upper (lower) value of $\boldsymbol{A}''$ is c times the upper (lower) value of $\boldsymbol{A}$. We combine these observations as follows. (See Exercise 2.)

Lemma 10.3. *If $\boldsymbol{A} = (a_{ij})$ and $\boldsymbol{A}' = (a'_{ij})$ are matrices with $a'_{ij} = ca_{ij}+b$, where $c > 0$, then the game with matrix $\boldsymbol{A}$ has the same minimax strategies for I and II as the game with matrix $\boldsymbol{A}'$. Also, if V denotes the value of the game with matrix $\boldsymbol{A}$, then the value V' of the game with matrix $\boldsymbol{A}'$ satisfies $V' = cV + b$.*

10.4 Reduction to a Linear Programming Problem

Here we introduce the area of linear programming. This has the advantage of leading to a simple algorithm for solving finite games. For a background in linear programming, the book by Chvátal (1983) can be recommended. A short course on Linear Programming more in tune with the material as it is presented here may be found on the web at http://www.tomsferguson.com/LP.pdf.

A Linear Program is defined as the problem of choosing real variables to maximize or minimize a linear function of the variables, called the objective function, subject to linear constraints on the variables. The constraints may be equalities or inequalities. A standard form of this problem is to choose $y_1, \ldots, y_n$, to

$$\text{maximize} \quad b_1 y_1 + \cdots + b_n y_n, \tag{8}$$

subject to the constraints

$$\begin{aligned} a_{11}y_1 + \cdots + a_{1n}y_n &\leq c_1 \\ &\vdots \\ a_{m1}y_1 + \cdots + a_{mn}y_n &\leq c_m \end{aligned} \tag{9}$$

and

$$y_j \geq 0 \quad \text{for } j = 1, \ldots, n.$$

Let us consider the game problem from Player I's point of view. He wants to choose $p_1, \ldots, p_m$ to maximize (5) subject to the constraint $\boldsymbol{p} \in X^*$. This becomes the mathematical program: choose $p_1, \ldots, p_m$ to

$$\text{maximize} \quad \min_{1 \leq j \leq n} \sum_{i=1}^{m} p_i a_{ij} \tag{10}$$

subject to the constraints

$$p_1 + \cdots + p_m = 1 \tag{11}$$

and

$$p_i \geq 0 \quad \text{for } i = 1, \ldots, m.$$

Although the constraints are linear, the objective function is not a linear function of the p's because of the min operator, so this is not a linear program. However, it can be changed into a linear program through a trick. Add one new variable v to Player I's list of variables, restrict it to be less than the objective function, $v \leq \min_{1\leq j \leq n} \sum_{i=1}^{m} p_i a_{ij}$, and try to make v as large as possible subject to this new constraint. The problem becomes:

Choose v and $p_1, \ldots, p_m$ to

$$\text{maximize} \quad v \tag{12}$$

subject to the constraints

$$\begin{aligned} v &\leq \sum_{i=1}^{m} p_i a_{i1} \\ &\vdots \\ v &\leq \sum_{i=1}^{m} p_i a_{in} \\ p_1 + \cdots &+ p_m = 1 \end{aligned} \tag{13}$$

and

$$p_i \geq 0 \quad \text{for } i = 1, \ldots, m.$$

This is indeed a linear program. For solving such problems, there exists a simple algorithm known as the simplex method.

In a similar way, one may view the problem from Player II's point of view and arrive at a similar linear program. II's problem is: choose w and $q_1, \ldots, q_n$ to

$$\text{minimize} \quad w \tag{14}$$

subject to the constraints

$$
\begin{aligned}
w &\geq \sum_{j=1}^{n} a_{1j} q_j \\
&\vdots \\
w &\geq \sum_{j=1}^{n} a_{mj} q_j \\
q_1 + \cdots + q_n &= 1
\end{aligned}
\tag{15}
$$

and

$$q_j \geq 0 \quad \text{for } j = 1, \ldots, n.$$

In Linear Programming, there is a theory of duality that says these two programs, (12)–(13), and (14)–(15), are dual programs. And there is a remarkable theorem, called the Duality Theorem, that says dual programs have the same value. The maximum Player I can achieve in (14) is equal to the minimum that Player II can achieve in (12). But this is exactly the claim of the Minimax Theorem. In other words, the Duality Theorem implies the Minimax Theorem.

There is another way to transform the linear program, (12)–(13), into a linear program that is somewhat simpler for computations when it is known that the value of the game is positive. So suppose $v > 0$ and let $x_i = p_i/v$. Then the constraint $p_1 + \cdots + p_m = 1$ becomes $x_1 + \cdots + x_m = 1/v$, which looks nonlinear. But maximizing v is equivalent to minimizing $1/v$, so we can remove v from the problem by minimizing $x_1 + \cdots + x_m$ instead. The problem, (12)–(13), becomes: choose $x_1, \ldots, x_m$ to

$$\text{minimize} \quad x_1 + \cdots + x_m \tag{16}$$

subject to the constraints

$$
\begin{aligned}
1 &\leq \sum_{i=1}^{m} x_i a_{i1} \\
&\vdots \\
1 &\leq \sum_{i=1}^{m} x_i a_{in}
\end{aligned}
\tag{17}
$$

and

$$x_i \geq 0 \quad \text{for } i = 1, \ldots, m.$$

When we have solved this problem, the solution of the original game may be easily found. The value will be $v = 1/(x_1 + \cdots + x_m)$ and the optimal strategy for Player I will be $p_i = vx_i$ for $i = 1, \ldots, m$.

10.5 Description of the Pivot Method for Solving Games

The following algorithm for solving finite games is essentially the simplex method for solving (16)–(17) as described in Williams (1966).

Step 1. Add a constant to all elements of the game matrix if necessary to ensure that the value is positive. (If you do, you must remember at the end to subtract this constant from the value of the new matrix game to get the value of the original matrix game.)
Step 2. Create a *tableau* by augmenting the game matrix with a border of -1's along the lower edge, $+1$'s along the right edge, and zero in the lower right corner. Label I's strategies on the left from x_1 to x_m and II's strategies on the top from y_1 to y_n.

	y_1	y_2	$\cdots$	y_n	
x_1	a_{11}	a_{12}	$\cdots$	a_{1n}	1
x_2	a_{21}	a_{22}	$\cdots$	a_{2n}	1
$\vdots$	$\vdots$	$\vdots$		$\vdots$	$\vdots$
x_m	a_{m1}	a_{m2}	$\cdots$	a_{mn}	1
	-1	-1	$\cdots$	-1	0

Step 3. Select any entry in the interior of the tableau to be the *pivot*, say row p column q, subject to the properties:

a. The border number in the pivot column, $a(m+1, q)$, must be negative.
b. The pivot, $a(p, q)$, itself must be positive.
c. The pivot row, p, must be chosen to give the smallest of the ratios the border number in the pivot row to the pivot, $a(p, n+1)/a(p, q)$, among all positive pivots for that column.

Step 4. Pivot as follows:

a. Replace each entry, $a(i, j)$, not in the row or column of the pivot by $a(i, j) - a(p, j) \cdot a(i, q)/a(p, q)$.

b. Replace each entry in the pivot row, except for the pivot, by its value divided by the pivot value.
c. Replace each entry in the pivot column, except for the pivot, by the negative of its value divided by the pivot value.
d. Replace the pivot value by its reciprocal.

This may be represented symbolically by

$$\boxed{\begin{matrix} \textcircled{p} & r \\ c & q \end{matrix}} \longrightarrow \boxed{\begin{matrix} 1/p & r/p \\ -c/p & q-(rc/p) \end{matrix}}$$

where p stands for the pivot, r represents any number in the same row as the pivot, c represents any number in the same column as the pivot, and q is an arbitrary entry not in the same row or column as the pivot.

Step 5. Exchange the label on the left of the pivot row with the label on the top of the pivot column.
Step 6. If there are any negative numbers remaining in the lower border row, go back to Step 3.
Step 7. Otherwise, a solution may now be read out:

a. The value, v, is the reciprocal of the number in the lower right corner. (If you subtracted a number from each entry of the matrix in Step 1, it must be added to v here.)
b. I's optimal strategy is constructed as follows. Those variables of Player I that end up on the left side receive probability zero. Those that end up on the top receive the value of the bottom edge in the same column divided by the lower right corner.
c. II's optimal strategy is constructed as follows. Those variables of Player II that end up on the top receive probability zero. Those that end up on the left receive the value of the right edge in the same row divided by the lower right corner.

10.6 A Numerical Example

Let us illustrate these steps using an example. Let us take a 3×3 matrix since that is the simplest example we cannot solve using previous methods. Consider the matrix game with the following matrix,

$$B = \begin{pmatrix} 2 & -1 & 6 \\ 0 & 1 & -1 \\ -2 & 2 & 1 \end{pmatrix}.$$

We might check for a saddle point (there is none) and we might check for domination (there is none). Is the value positive? We might be able to guess by staring at the matrix long enough, but why don't we simply make the first row positive by adding 2 to each entry of the matrix:

$$B' = \begin{pmatrix} 4 & 1 & 8 \\ 2 & 3 & 1 \\ 0 & 4 & 3 \end{pmatrix}.$$

The value of this game is at least one since Player I can guarantee at least 1 by using the first (or second) row. We will have to remember to subtract 2 from the value of B' to get the value of B. This completes Step 1 of the algorithm.

In Step 2, we set up the tableau for the matrix B' as follows:

	y_1	y_2	y_3	
x_1	4	1	8	1
x_2	2	3	1	1
x_3	0	4	3	1
	−1	−1	−1	0

In Step 3, we must choose the pivot. Since all three columns have a negative number in the lower edge, we may choose any of these columns as the pivot column. Suppose we choose column 1. The pivot row must have a positive number in this column, so it must be one of the top two rows. To decide which row, we compute the ratios of border numbers to pivot. For the first row it is $1/4$; for the second row it is $1/2$. The former is smaller, so the pivot is in row 1. We pivot about the 4 in the upper left corner.

Step 4 tells us how to pivot. The pivot itself gets replaced by its reciprocal, namely $1/4$. The rest of the numbers in the pivot row are simply divided by the pivot, giving $1/4$, 2, and $1/4$. Then the rest of the numbers in the pivot column are divided by the pivot and changed in sign. The remaining nine numbers are modified by subtracting $r{\cdot}c/p$ for the corresponding r and c. For example, from the 1 in second row third column we subtract $8 \times 2/4 = 4$, leaving -3. The complete pivoting operation is

	y_1	y_2	y_3	
x_1	④	1	8	1
x_2	2	3	1	1
x_3	0	4	3	1
	−1	−1	−1	0

$\longrightarrow$

	x_1	y_2	y_3	
y_1	1/4	1/4	2	1/4
x_2	−1/2	5/2	−3	1/2
x_3	0	4	3	1
	1/4	−3/4	1	1/4

In Step 5, we interchange the labels of the pivot row and column. Here we interchange x_1 and y_1. This has been done in the display.

For Step 6, we check for negative entries in the lower edge. Since there is one, we return to Step 3.

This time, we must pivot in column 2 since it has the unique negative number in the lower edge. All three numbers in this column are positive. We find the ratios of border numbers to pivot for rows 1, 2, and 3 to be 1, 1/5, and 1/4 respectively. The smallest occurs in the second row, so we pivot about the 5/2 in the second row, second column. Completing Steps 4 and 5, we obtain

	x_1	y_2	y_3	
y_1	1/4	1/4	2	1/4
x_2	−1/2	(5/2)	−3	1/2
x_3	0	4	3	1
	1/4	−3/4	1	1/4

$\longrightarrow$

	x_1	x_2	y_3	
y_1	0.3	−0.1	2.3	0.2
y_2	−0.2	0.4	−1.2	0.2
x_3	0.8	−1.6	7.8	0.2
	0.1	0.3	0.1	0.4

At Step 6 this time, all values on the lower edge are nonnegative so we pass to Step 7. We may now read the solution to the game with matrix B'.

The value is the reciprocal of 0.4, namely 5/2.

Since x_3 is on the left in the final tableau, the optimal p_3 is zero. The optimal p_1 and p_2 are the ratios, 0.1/0.4 and 0.3/0.4, namely 1/4 and 3/4. Therefore, I's optimal mixed strategy is $(p_1, p_2, p_3) = (0.25, 0.75, 0)$.

Since y_3 is on top in the final tableau, the optimal q_3 is zero. The optimal q_1 and q_2 are the ratios, 0.2/0.4 and 0.2/0.4, namely 1/2 and 1/2. Therefore, II's optimal mixed strategy is $(q_1, q_2, q_3) = (0.5, 0.5, 0)$.

The game with matrix B has the same optimal mixed strategies but the value is $5/2 - 2 = 1/2$.

Remarks.

1. The reason that the pivot row is chosen according to the rule in Step 3(c) is that the numbers in the resulting right edge of the tableau stay nonnegative. If after pivoting you find a negative number in the last column, you have made a mistake, either in the choice of the pivot row, or in your numerical calculations for pivoting.
2. There may be ties in comparing the ratios to choose the pivot row. The rule given allows you to choose among those rows with the smallest ratios. The smallest ratio may be zero.

3. The value of the number in the lower right corner never decreases. (Can you see why this is?) In fact, the lower right corner is always equal to the sum of the values in the lower edge corresponding to Player I's labels along the top. Similarly, it is also the sum of the values on the right edge corresponding to Player II's labels on the left. This gives another small check on your arithmetic.
4. One only pivots around numbers in the main body of the tableau, never in the lower or right edges.
5. This method gives one optimal strategy for each player. If other optimal strategies exist, there will be one or more zeros in the bottom edge or right edge in the final tableau. Other optimal basic strategies can be found by pivoting further, in a column with a zero in the bottom edge or a row with a zero in the right edge, in such a way that the bottom row and right edge stay nonnegative.

10.7 Approximating the Solution: Fictitious Play

As an alternative to the simplex method, the method of fictitious play may be used to approximate the value and optimal strategies of a finite game. It is a sequential procedure that approximates the value of a game as closely as desired, giving upper and lower bounds that converge to the value and strategies for the players that achieve these bounds.

The advantage of the simplex method is that it gives answers that are accurate, generally to machine accuracy, and for small size problems is extremely fast. The advantage of the method of fictitious play is its simplicity, both to program and understand, and the fact that you can stop it at any time and obtain answers whose accuracy you know. The simplex method only gives answers when it is finished. For large size problems, say a matrix 50×50 or greater, the method of fictitious play will generally give a sufficiently accurate answer in a shorter time than the simplex method. For very large problems, it may be the only way to proceed.

Let $A(i, j)$ be an $m \times n$ payoff matrix. The method starts with an arbitrary initial pure strategy $1 \leq i_1 \leq m$ for Player I. Alternatively from then on, each player chooses his next pure strategy as a best reply assuming the other player chooses among his previous choices at random equally likely. For example, if $i_1, \ldots, i_k$ have already been chosen by Player I for some $k \geq 1$, then j_k is chosen as that j that minimizes the expectation $(1/k)\sum_{\ell=1}^{k} A(i_\ell, j)$. Similarly, if $j_1, \ldots, j_k$ have already been chosen, i_{k+1}

is then chosen as that i that maximizes the expectation $(1/k)\sum_{\ell=1}^{k} A(i, j_\ell)$. To be specific, we define

$$s_k(j) = \sum_{\ell=1}^{k} A(i_\ell, j) \quad \text{and} \quad t_k(i) = \sum_{\ell=1}^{k} A(i, j_\ell) \tag{1}$$

and then define

$$j_k = \text{argmin } s_k(j) \quad \text{and} \quad i_{k+1} = \text{argmax } t_k(i). \tag{2}$$

If the maximum of $t_k(i)$ is assumed at several different values of i, then it does not matter which of these is taken as i_{k+1}. To be specific, we choose i_{k+1} as the smallest value of i that maximizes $t_k(i)$. Similarly j_k is taken as the smallest j that minimizes $s_k(j)$. In this way, the sequences i_k and j_k are defined deterministically once i_1 is given.

If we define $\overline{V}_k = (1/k)t_k(i_{k+1})$, then $\overline{V}_k$ is an upper bound to the value of the game since Player II can use the strategy that chooses j randomly and equally likely from $j_1, \ldots, j_k$ and keep Player I's expected return to be at most $\overline{V}_k$. Similarly, the quamtity $\underline{V}_k = (1/k)s_k(j_k)$ is a lower bound to the value of the game. It is rather surprising that these upper and lower bounds to the value converge to the value of the game as k tends to infinity.

Theorem 1. *If V denotes the value of the game, then $\underline{V}_k \to V$, $\overline{V}_k \to V$, and $\underline{V}_k \leq V \leq \overline{V}_k$, for all k.*

This approximation method was suggested by Brown (1951), and the proof of convergence was provided by Robinson (1951). The convergence of $\underline{V}_k$ and $\overline{V}_k$ to V is slow. It is thought to be of order at least $1/\sqrt{k}$. In addition, the convergence is not monotone. See the example below.

A modification of this method in which i_1 and j_1 are initially arbitrarily chosen, and then the selection of future i_k and j_k is made simultaneously by the players rather than sequentially, is often used, but it is not as fast.

It should be mentioned that as a practical matter, choosing at each stage a best reply to an opponent's imagined strategy of choosing among his previous choices at random is not a good idea. See Exercise 7. On the other hand, Baños (1968) and Megiddo (1980) describe sequential methods for Player I, say, to choose mixed strategies such that liminf of the average payoff is at least the value of the game no matter what

Player II does. This choice of mixed strategies is based only upon Player I's past pure strategy choices and the past observed payoffs, *but not otherwise on the payoff matrix or upon the opponent's pure strategy choices.* Indeed, Player II's strategy set may be infinite.

Example 1. Take as an example the game with matrix

$$A = \begin{pmatrix} 2 & -1 & 6 \\ 0 & 1 & -1 \\ -2 & 2 & 1 \end{pmatrix}$$

This is the game solved in Sec. 10.6. It has value 0.5, and optimal mixed strategies, (0.25, 0.75, 0) and (0.5, 0.5, 0) for Player I and Player II respectively. It is easy to set up a program to perform the calculations. In particular, the computations, (1), may be made recursively in the simpler form

$$s_k(j) = s_{k-1}(j) + A(i_k, j) \quad \text{and} \quad t_k(i) = t_{k-1}(i) + A(i, j_k). \tag{3}$$

We take the initial $i_1 = 1$, and find

k	i_k	$s_k(1)$	$s_k(2)$	$s_k(3)$	$\underline{V}_k$	j_k	$t_k(1)$	$t_k(2)$	$t_k(3)$	$\overline{V}_k$
1	1	2	**−1**	6	−1	2	−1	1	**2**	2
2	3	**0**	1	7	0	1	**1**	1	0	0.5
3	1	2	**0**	13	0	2	0	**2**	2	0.6667
4	2	2	**1**	12	0.25	2	−1	3	**4**	1
5	3	**0**	3	13	0	1	1	**3**	2	0.6
6	2	**0**	4	12	0	1	**3**	3	0	0.5
7	1	**2**	3	18	0.2857	1	**5**	3	−2	0.7143
8	1	4	**2**	24	0.25	2	**4**	4	0	0.5
9	1	6	**1**	30	0.1111	2	3	**5**	2	0.5556
10	2	6	**2**	29	0.2	2	2	**6**	4	0.6
11	2	6	**3**	28	0.2727	2	1	**7**	6	0.6364
12	2	6	**4**	27	0.3333	2	0	**8**	8	0.6667
13	2	6	**5**	26	0.3846	2	−1	9	**10**	0.7692
14	3	**4**	7	27	0.2857	1	1	**9**	8	0.6429
15	2	**4**	8	26	0.2667	1	3	**9**	6	0.6

The initial choice of $i_1 = 1$ gives $(s_1(1), s_1(2), s_1(3))$ as the first row of A, which has a minimum at $s_1(2)$. Therefore, $j_1 = 2$. The second column of A has $t_1(3)$ as the maximum, so $i_2 = 3$. Then the third row of A is added to the s_1 to produce the s_2 and so on. The minima of the s_k and the maxima of the t_k are indicated in boldface. The largest of the $\underline{V}_k$ found so far occurs at $k = 13$ and has value $s_k(j_k)/k = 5/13 = 0.3846\ldots$. This value can be guaranteed to Player I by using the mixed strategy (5/13,6/13,2/13), since in the first 13 of the i_k there are 5 1's, 6 2's and 2 3's. The smallest of the $\overline{V}_k$ occurs several times and has value 0.5. It can be achieved by Player II using the first and second columns equally likely. So far we know that $0.3846 \le V \le 0.5$, although we know from Sec. 10.6 that $V = 0.5$.

Computing further, we can find that $\underline{V}_{91} = 44/91 = 0.4835\ldots$ and is achieved by the mixed strategy $(25/91, 63/91, 3/91)$. From row 9 on, the difference between the boldface numbers in each row seems to be bounded between 4 and 6. This implies that the convergence is of order $1/k$.

10.8 Exercises

1. Consider the game with matrix $\boldsymbol{A}$. Past experience in playing the game with Player II enables Player I to arrive at a set of probabilities reflecting his belief of the column that II will choose. I thinks that with probabilities 1/5, 1/5, 1/5, and 2/5, II will choose columns 1, 2, 3, and 4 respectively.

$$\boldsymbol{A} = \begin{pmatrix} 0 & 7 & 2 & 4 \\ 1 & 4 & 8 & 2 \\ 9 & 3 & -1 & 6 \end{pmatrix}.$$

   (a) Find for I a Bayes strategy (best response) against $(1/5, 1/5, 1/5, 2/5)$.
   (b) Suppose II guesses correctly that I is going to use a Bayes strategy against $(1/5, 1/5, 1/5,\ 2/5)$. Instruct II on the strategy she should use — that is, find II's Bayes strategy against I's Bayes strategy against $(1/5, 1/5, 1/5, 2/5)$.

2. The game with matrix $\boldsymbol{A}$ has value zero, and $(6/11, 3/11, 2/11)$ is optimal for I.

$$\boldsymbol{A} = \begin{pmatrix} 0 & -1 & 1 \\ 2 & 0 & -2 \\ -3 & 3 & 0 \end{pmatrix}, \quad \boldsymbol{B} = \begin{pmatrix} 5 & 3 & 7 \\ 9 & 5 & 1 \\ -1 & 11 & 5 \end{pmatrix}.$$

   (a) Find the value of the game with matrix $\boldsymbol{B}$ and an optimal strategy for I.
   (b) Find an optimal strategy for II in both games.

3. **A game without a value.** Let $X = \{1, 2, 3, \ldots\}$, let $Y = \{1, 2, 3, \ldots\}$ and

$$A(i,j) = \begin{cases} +1 & \text{if } i > j, \\ 0 & \text{if } i = j, \\ -1 & \text{if } i < j. \end{cases}$$

This is the game "The player that chooses the larger integer wins". Here we may take for the space X^* of mixed strategies of Player I

$$X^* = \left\{ (p_1, p_2, \ldots) : p_i \geq 0 \text{ for all } i, \text{ and } \sum_{i=1}^{\infty} p_i = 1 \right\}.$$

Similarly,

$$Y^* = \left\{ (q_1, q_2, \ldots) : q_j \geq 0 \text{ for all } j, \text{ and } \sum_{j=1}^{\infty} q_j = 1 \right\}.$$

The payoff for given $\boldsymbol{p} \in X^*$ and $\boldsymbol{q} \in Y^*$ is

$$A(\boldsymbol{p}, \boldsymbol{q}) = \sum_{i=1}^{\infty} \sum_{j=1}^{\infty} p_i A(i,j) q_j.$$

(a) Show that for all $\boldsymbol{q} \in Y^*$, $\sup_{1 \leq i < \infty} \sum_{j=1}^{\infty} A(i,j) q_j = +1$.
(b) Conclude that $\overline{V} = +1$.
(c) Using symmetry, argue that $\underline{V} = -1$.
(d) What are I's minimax strategies in this game?

4. Use the method presented in Sec. 10.5 to solve the game with matrix

$$\boldsymbol{A} = \begin{pmatrix} 0 & 1 & 2 \\ 2 & -1 & -2 \\ 3 & -3 & 0 \end{pmatrix}.$$

Either argue that the value is positive, or add $+1$ to the elements of the matrix. To go easy on the homework grader, make the first pivot in the second column.

5. **An Example in Which the Lower Value is Greater than the Upper Value?** Consider the infinite game with strategy spaces

$X = Y = \{0, 1, 2, \ldots\}$, and payoff function,

$$A(i,j) = \begin{cases} 0 & \text{if } i = j, \\ 4^j & \text{if } i > j, \\ -4^i & \text{if } i < j. \end{cases}$$

Note that the game is symmetric. Let $\boldsymbol{p} = (p_0, p_1, p_2, \ldots)^T = (1/2, 1/4, 1/8, \ldots)^T$ be a mixed strategy for Player I, $p_i = 2^{-(i+1)}$.

(a) Show that if Player I uses this strategy, his average return, $\sum_{i=0}^{\infty} p_i A(i,j)$, is equal to $1/2$ for all pure strategies j for Player II.

(b) So $\boldsymbol{p}$ is an equalizer strategy that guarantees Player I at least $1/2$. So the lower value is at least $1/2$. Perhaps he can do better. In fact he can, but ... Wait a minute! The game is symmetric. Shouldn't the value be zero? Worse, suppose Player II uses the same strategy. By symmetry, she can keep Player I's winnings down to $-1/2$ no matter what pure strategy he chooses. So the upper value is at most $-1/2$. What is wrong? What if both players use the mixed strategy, $\boldsymbol{p}$? We haven't talked much about infinite games, but what restrictions would you place on infinite games to avoid such absurd examples? Should the restrictions be placed on the payoff function, A, or on the notion of a mixed strategy?

6. Carry out the fictitious play algorithm on the matrix $A = \begin{pmatrix} 1 & -1 \\ 0 & 2 \end{pmatrix}$ through step $k = 4$. Find the upper and lower bounds on the value of the game that this gives.

7. Suppose the game with matrix, $\begin{pmatrix} \sqrt{2} & 0 \\ 0 & 1 \end{pmatrix}$ is played repeatedly. On the first round the players make any choices.

   (a) Thereafter Player I makes a best response to his opponent's imagined strategy of choosing among her previous choice at random. If Player II knows this, what should she do? What are the limiting average frequencies of the choices of the players?

   (b) Suppose Player II is required to play a best response to her opponent's previous choices. What should Player I do, and what would his limiting average payoff be?

11 The Extensive Form of a Game

The strategic form of a game is a compact way of describing the mathematical aspects of a game. In addition, it allows a straightforward method of analysis, at least in principle. However, the flavor of many games is lost in such a simple model. Another mathematical model of a game, called the extensive form, is built on the basic notions of position and move, concepts not apparent in the strategic form of a game. In the extensive form, we may speak of other characteristic notions of games such as bluffing, signaling, sandbagging, and so on. Three new concepts make their appearance in the extensive form of a game: the game tree, chance moves, and information sets.

11.1 The Game Tree

The extensive form of a game is modeled using a directed graph. A *directed graph* is a pair (T, F) where T is a nonempty set of vertices and F is a function that gives for each $x \in T$ a subset, $F(x)$ of T called the followers of x. When a directed graph is used to represent a game, the vertices represent positions of the game. The followers, $F(x)$, of a position, x, are those positions that can be reached from x in one move.

A *path from a vertex* t_0 *to a vertex* t_1 is a sequence, $x_0, x_1, \ldots, x_n$, of vertices such that $x_0 = t_0$, $x_n = t_1$ and x_i is a follower of x_{i-1} for $i = 1, \ldots, n$. For the extensive form of a game, we deal with a particular type of directed graph called a tree.

Definition. A *tree* is a directed graph, (T, F) in which there is a special vertex, t_0, called the root or the initial vertex, such that for every other vertex $t \in T$, there is a unique path beginning at t_0 and ending at t.

The existence and uniqueness of the path implies that a tree is connected, has a unique initial vertex, and has no circuits or loops.

In the extensive form of a game, play starts at the initial vertex and continues along one of the paths eventually ending in one of the terminal vertices. At terminal vertices, the rules of the game specify the payoff. For n-person games, this would be an n-tuple of payoffs. Since we are dealing with two-person zero-sum games, we may take this payoff to be the amount Player I wins from Player II. For the nonterminal vertices there are three possibilities. Some nonterminal vertices are assigned to Player I who is to choose the move at that position. Others are assigned to Player II. However, some vertices may be singled out as positions from which a chance move is made.

Chance Moves. Many games involve chance moves. Examples include the rolling of dice in board games like monopoly or backgammon or gambling games such as craps, the dealing of cards as in bridge or poker, the spinning of the wheel of fortune, or the drawing of balls out of a cage in lotto. In these games, chance moves play an important role. Even in chess, there is generally a chance move to determine which player gets the white pieces (and therefore the first move which is presumed to be an advantage). It is assumed that the players are aware of the probabilities of the various outcomes resulting from a chance move.

Information. Another important aspect we must consider in studying the extensive form of games is the amount of information available to the players about past moves of the game. In poker for example, the first move is the chance move of shuffling and dealing the cards, each player is aware of certain aspects of the outcome of this move (the cards he received) but he is not informed of the complete outcome (the cards received by the other players). This leads to the possibility of "bluffing."

11.2 Basic Endgame in Poker

One of the simplest and most useful mathematical models of a situation that occurs in poker is called the "classical betting situation" by Friedman (1971) and "basic endgame" by Cutler (1976). These papers provide explicit situations in the game of stud poker and of lowball stud for which the model gives a very accurate description. This model is also found in the exercises of the book of Ferguson (1967). Since this is a model of a situation that occasionally arises in the last round of betting when there are two players left, we adopt the terminology of Cutler and call it Basic Endgame in poker. This will also emphasize what we feel is an important feature of the game of poker, that like chess, go, backgammon and other games, there is a distinctive phase of the game that occurs at the close, where special strategies and tactics that are analytically tractable become important.

Basic Endgame is played as follows. Both players put \$1, called the ante, in the center of the table. The money in the center of the table, so far \$2, is called the pot. Then Player I is dealt a card from a deck. It is a winning card with probability 1/4 and a losing card with probability 3/4. Player I sees this card but keeps it hidden from Player II. (Player II does not get a card.) Player I then checks or bets. If he checks, then his card is inspected; if he has a winning card he wins the pot and hence wins the \$1 ante from II, and otherwise he loses the \$1 ante to II. If I bets, he puts \$2 more into the pot. Then Player II — not knowing what card Player I has — must fold or call. If she folds, she loses the \$1 ante to I no matter what card I has. If II calls, she adds \$2 to the pot. Then Player I's card is exposed and I wins \$3 (the ante plus the bet) from II if he has a winning card, and loses \$3 to II otherwise.

Let us draw the tree for this game. There are at most three moves in this game: (1) the chance move that chooses a card for I, (2) I's move in which he checks or bets, and (3) II's move in which she folds or calls. To each vertex of the game tree, we attach a label indicating which player is to move from that position. Chance moves we generally refer to as moves by nature and use the label N. See Fig. 11.1.

Each edge is labelled to identify the move. (The arrows are omitted for the sake of clarity. Moves are assumed to proceed down the page.) Also,

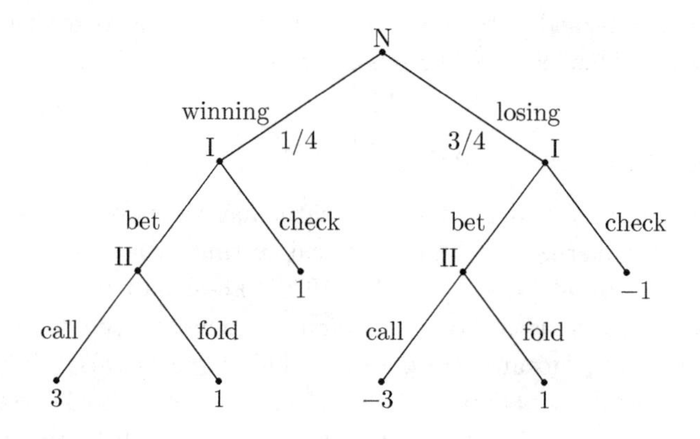

Fig. 11.1.

the moves leading from a vertex at which nature moves are labelled with the probabilities with which they occur. At each terminal vertex, we write the numerical value of I's winnings (II's losses).

There is only one feature lacking from the above figure. From the tree we should be able to reconstruct all the essential rules of the game. That is not the case with the tree of Fig. 11.1 since we have not indicated that at the time II makes her decision she does not know which card I has received. That is, when it is II's turn to move, she does not know at which of her two possible positions she is. We indicate this on the diagram by encircling the two positions in a closed curve, and we say that these two vertices constitute an information set. The two vertices at which I is to move constitute two separate information sets since he is told the outcome of the chance move. To be complete, this must also be indicated on the diagram by drawing small circles about these vertices. We may delete one of the labels indicating II's vertices since they belong to the same information set. It is really the information set that must be labelled. The completed game tree becomes

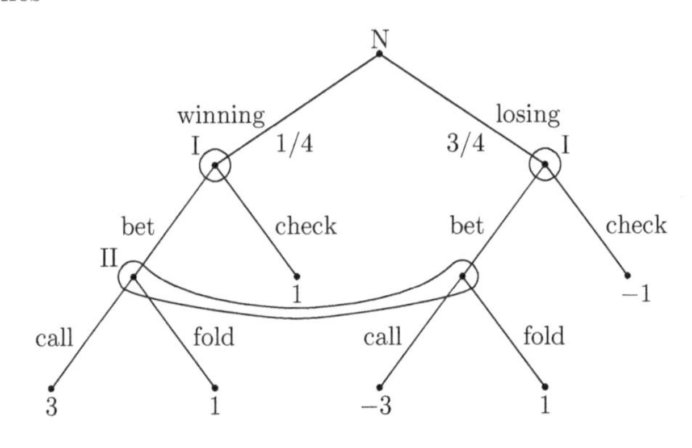

Fig. 11.2.

The diagram now contains all the essential rules of the game.

11.3 The Kuhn Tree

The game tree with all the payoffs, information sets, and labels for the edges and vertices included is known as the Kuhn Tree. We now give the formal definition of a Kuhn tree.

Not every set of vertices can form an information set. In order for a player not to be aware of which vertex of a given information set the game has come to, each vertex in that information set must have the same number of edges leaving it. Furthermore, it is important that the edges from each vertex of an information set have the same set of labels. The player moving from such an information set really chooses a label. It is presumed that a player makes just one choice from each information set.

Definition. A finite two-person zero-sum game in extensive form is given by

1) a finite tree with vertices T,
2) a payoff function that assigns a real number to each terminal vertex,
3) a set T_0 of non-terminal vertices (representing positions at which chance moves occur) and for each $t \in T_0$, a probability distribution on the edges leading from t,
4) a partition of the rest of the vertices (not terminal and not in T_0) into two groups of information sets $T_{11}, T_{12}, \ldots, T_{1k_1}$ (for Player I) and $T_{21}, T_{22}, \ldots, T_{2k_2}$ (for Player II), and
5) for each information set T_{jk} for player j, a set of labels L_{jk}, and for each $t \in T_{jk}$, a one-to-one mapping of L_{jk} onto the set of edges leading from t.

The information structure in a game in extensive form can be quite complex. It may involve lack of knowledge of the other player's moves or of some of the chance moves. It may indicate a lack of knowledge of how many moves have already been made in the game (as is the case With Player II in Fig. 11.3).

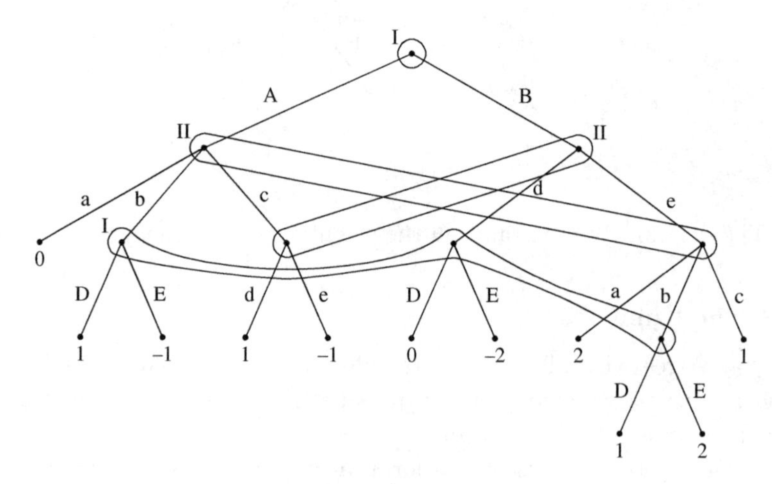

Fig. 11.3.

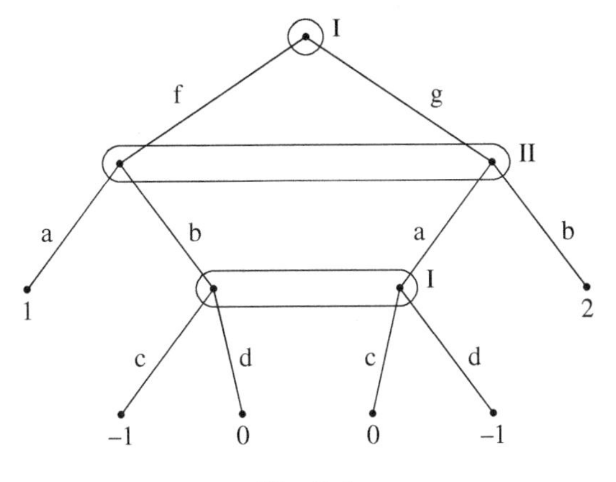

Fig. 11.4.

It may describe situations in which one player has forgotten a move he has made earlier (as is the case With Player I in Fig. 11.3 or Fig. 11.4). In fact, one way to try to model the game of bridge as a two-person zero-sum game involves the use of this idea. In bridge, there are four individuals forming two teams or partnerships of two players each. The interests of the members of a partnership are identical, so it makes sense to describe this as a two-person game. But the members of one partnership make bids alternately based on cards that one member knows and the other does not. This may be described as a single player who alternately remembers and forgets the outcomes of some of the previous random moves. Games in which players remember all past information they once knew and all past moves they made are called games of *perfect recall.*

A kind of degenerate situation exists when an information set contains two vertices which are joined by a path, as is the case with I's information set in Fig. 11.5.

We take it as a convention that a player makes one choice from each information set during a game. That choice is used no matter how many times the information set is reached. In Fig. 11.5, if I chooses option a there is no problem. If I chooses option b, then in the lower of I's two vertices the a is superfluous, and the tree is really equivalent to Fig. 11.6. Instead of using the above convention, we may if we like assume in the definition of a game in extensive form that no information set contains two vertices joined by a path.

Games in which both players know the rules of the game, that is, in which both players know the Kuhn tree, are called *games of complete*

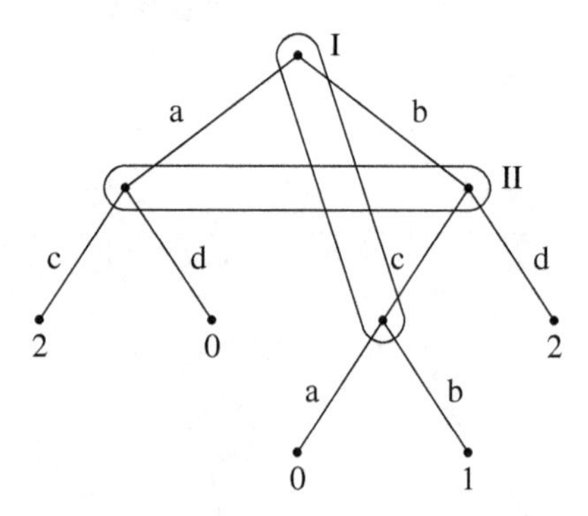

Fig. 11.5.

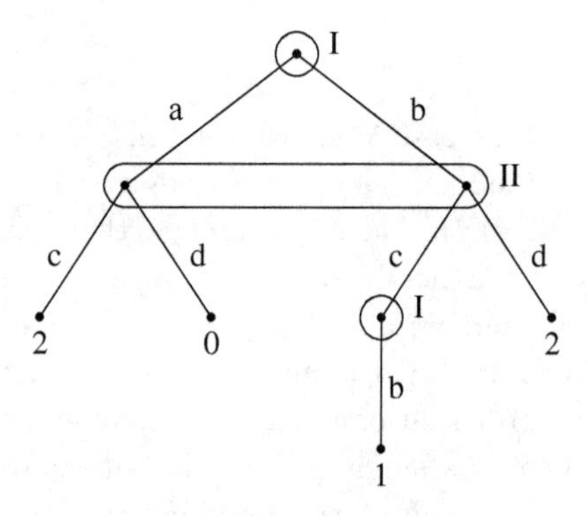

Fig. 11.6.

information. Games in which one or both of the players do not know some of the payoffs, or some of the probabilities of chance moves, or some of the information sets, or even whole branches of the tree, are called *games with incomplete information*, or *pseudogames*. We assume in the following that we are dealing with games of complete information.

11.4 The Representation of a Strategic Form Game in Extensive Form

The notion of a game in strategic form is quite simple. It is described by a triplet (X, Y, A) as in Sec. 7.1. The extensive form of a game on the other hand is quite complex. It is described by the game tree with each non-terminal vertex labelled as a chance move or as a move of one of the players, with all information sets specified, with probability distributions given for all chance moves, and with a payoff attached to each terminal vertex. It would seem that the theory of games in extensive is much more comprehensive than the theory of games in strategic form. However, by taking a game in extensive form and considering only the strategies and average payoffs, we may reduce the game to strategic form.

First, let us check that a game in strategic form can be put into extensive form. In the strategic form of a game, the players are considered to make their choices simultaneously, while in the extensive form of a game simultaneous moves are not allowed. However, simultaneous moves may be made sequentially as follows. We let one of the players, say Player I, move first, and then let player II move without knowing the outcome of I's move. This lack of knowledge may be described by the use of an appropriate information set. The example below illustrates this.

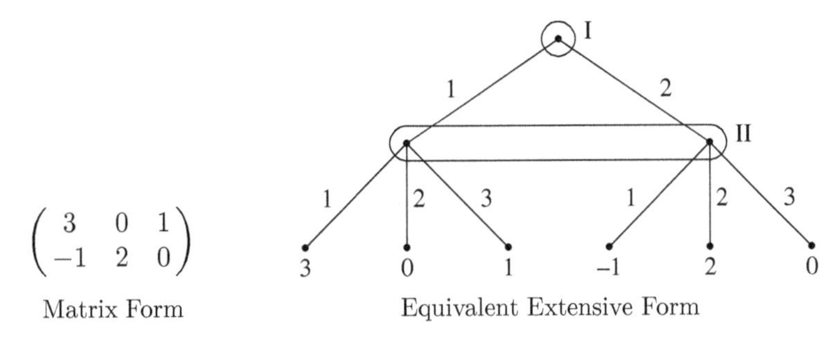

Matrix Form Equivalent Extensive Form

Player I has 2 pure strategies and Player II has 3. We pretend that Player I moves first by choosing row 1 or row 2. Then Player II moves, not knowing the choice of Player I. This is indicated by the information set for Player II. Then Player II moves by choosing column 1, 2 or 3, and the appropriate payoff is made.

11.5 Reduction of a Game in Extensive Form to Strategic Form

To go in the reverse direction, from the extensive form of a game to the strategic form, requires the consideration of pure strategies and the usual convention regarding random payoffs.

Pure strategies. Given a game in extensive form, we first find X and Y, the sets of pure strategies of the players to be used in the strategic form. A pure strategy for Player I is a rule that tells him exactly what move to make in each of his information sets. Let $T_{11}, \ldots, T_{1k_1}$ be the information sets for Player I and let $L_{11}, \ldots, L_{1k_1}$ be the corresponding sets of labels.

A pure strategy for I is a k_1-tuple $\boldsymbol{x} = (x_l, ..., x_{k_1})$ where for each i, x_i is one of the elements of L_{1i}. If there are m_i elements in L_{1i}, the number of such k_l-tuples and hence the number of Is pure strategies is the product $m_1 m_2 \cdots m_{k_1}$. The set of all such strategies is X. Similarly, if $T_{21}, \ldots, T_{2k_2}$ represent II's information sets and $L_{21}, \ldots, L_{2k_2}$ the corresponding sets of labels, a pure strategy for II is a k_2-tuple, $\boldsymbol{y} = (y_1, \ldots, y_{k_2})$ where $y_j \in L_{2j}$ for each j. Player II has $n_1 n_2 \cdots n_{k_2}$ pure strategies if there are n_j elements in L_{2j}. Y denotes the set of these strategies.

Random payoffs. A referee, given $\boldsymbol{x} \in X$ and $\boldsymbol{y} \in Y$, could play the game, playing the appropriate move from $\boldsymbol{x}$ whenever the game enters one of I's information sets, playing the appropriate move from $\boldsymbol{y}$ whenever the game enters one of II's information sets, and playing the moves at random with the indicated probabilities at each chance move. The actual outcome of the game for given $\boldsymbol{x} \in X$ and $\boldsymbol{y} \in Y$ depends on the chance moves selected, and is therefore a random quantity. Strictly speaking, random payoffs were not provided for in our definition of games in normal form. However, we are quite used to replacing random payoffs by their average values (expected values) when the randomness is due to the use of mixed strategies by the players. We adopt the same convention in dealing with random payoffs when the randomness is due to the chance moves. The justification of this comes from utility theory.

Convention. *If for fixed pure strategies of the players, $\boldsymbol{x} \in X$ and $\boldsymbol{y} \in Y$, the payoff is a random quantity, we replace the payoff by the average value, and denote this average value by $A(\boldsymbol{x}, \boldsymbol{y})$.*

For example, if for given strategies $\boldsymbol{x} \in X$ and $\boldsymbol{y} \in Y$, Player I wins 3 with probability $1/4$, wins 1 with probability $1/4$, and loses 1 with probability $1/2$, then his average payoff is $\frac{1}{4}(3) + \frac{1}{4}(1) + \frac{1}{2}(-1) = 1/2$ so we let $A(\boldsymbol{x}, \boldsymbol{y}) = 1/2$.

Therefore, given a game in extensive form, we say (X, Y, A) is the equivalent strategic form of the game if X and Y are the pure strategy spaces of players I and II respectively, and if $A(\boldsymbol{x}, \boldsymbol{y})$ is the average payoff for $\boldsymbol{x} \in X$ and $\boldsymbol{y} \in Y$.

11.6 Example

Let us find the equivalent strategic form to Basic Endgame in Poker described in Sec. 11.2, whose tree is given in Fig. 11.2. Player I has two information sets. In each set he must make a choice from among two options. He therefore has $2 \cdot 2 = 4$ pure strategies. We may denote them by

(b, b): bet with a winning card or a losing card.
(b, c): bet with a winning card, check with a losing card.
(c, b): check with a winning card, bet with a losing card.
(c, c): check with a winning card or a losing card.

Therefore, $X = \{(b, b), (b, c), (c, b), (c, c)\}$. We include in X all pure strategies whether good or bad (in particular, (c, b) seems a rather perverse sort of strategy).

Player II has only one information set. Therefore, $Y = \{c, f\}$, where
c: if I bets, call.
f: if I bets, fold.

Now we find the payoff matrix. Suppose I uses (b, b) and II uses c. Then if I gets a winning card (which happens with probability $1/4$), he bets, II calls, and I wins \$3. But if I gets a losing card (which happens with probability $3/4$), he bets, II calls, and I loses \$3. I's average or expected winnings is

$$A((b, b), c) = \frac{1}{4}(3) + \frac{3}{4}(-3) = -\frac{3}{2}.$$

This gives the upper left entry in the following matrix. The other entries may be computed similarly and are left as exercises.

$$\begin{array}{c} \\ (b,b) \\ (b,c) \\ (c,b) \\ (c,c) \end{array} \begin{array}{c} \begin{array}{cc} c & \quad f \end{array} \\ \begin{pmatrix} -3/2 & 1 \\ 0 & -1/2 \\ -2 & 1 \\ -1/2 & -1/2 \end{pmatrix} \end{array}$$

Let us solve this 4×2 game. The third row is dominated by the first row, and the fourth row is dominated by the second row. In terms of the original form of the game, this says something you may already have suspected: that if I gets a winning card, it cannot be good for him to check. By betting he will win at least as much, and maybe more. With the bottom two rows eliminated the matrix becomes $\begin{pmatrix} -3/2 & 1 \\ 0 & -1/2 \end{pmatrix}$, whose solution is easily found. The value is $V = -1/4$. I's optimal strategy is to mix (b, b) and (b, c) with probabilities $1/6$ and $5/6$ respectively, while II's optimal strategy is to mix c and f with equal probabilities $1/2$ each. The strategy (b, b) is Player I's bluffing strategy. Its use entails betting with a losing hand. The strategy (b, c) is Player I's "honest" strategy, bet with a winning hand and check with a losing hand. I's optimal strategy requires some bluffing and some honesty.

In Exercise 4, there are six information sets for I each with two choices. The number of I's pure strategies is therefore $2^6 = 64$. II has two information sets each with two choices. Therefore, II has $2^2 = 4$ pure strategies. The game matrix for the equivalent strategic form has dimension 64×4. Dominance can help reduce the dimension to a 2×3 game! (See Exercise 10(d).)

11.7 Games of Perfect Information

Now that a game in extensive form has been defined, we may make precise the notion of a game of perfect information.

Definition. A game of perfect information is a game in extensive form in which each information set of every player contains a single vertex.

In a game of perfect information, each player when called upon to make a move knows the exact position in the tree. In particular, each player knows all the past moves of the game including the chance ones. Examples include tic-tac-toe, chess, backgammon, craps, etc.

Games of perfect information have a particularly simple mathematical structure. The main result is that *every game of perfect information when reduced to strategic form has a saddle point*; both players have optimal pure strategies. Moreover, *the saddle point can be found by removing dominated rows and columns.* This has an interesting implication for the game of chess for example. Since there are no chance moves, every entry of the game matrix for chess must be either $+1$ (a win for Player I), or -1 (a win for Player II), or 0 (a draw). A saddle point must be one of these numbers. Thus, either Player I can guarantee himself a win, or Player II can guarantee herself a win, or both players can assure themselves at least a draw. From the game-theoretic viewpoint, chess is a very simple game. One needs only to write down the matrix of the game. If there is a row of all $+1$'s, Player I can win. If there is a column of all -1's, then Player II can win. Otherwise, there is a row with all $+1$'s and 0's and a column with all -1's and 0's, and so the game is drawn with best play. Of course, the real game of chess is so complicated, there is virtually no hope of ever finding an optimal strategy. In fact, it is not yet understood how humans can play the game so well.

11.8 Behavioral Strategies

For games in extensive form, it is useful to consider a different method of randomization for choosing among pure strategies. All a player really needs to do is to make one choice of an edge for each of his information sets in the game. A *behavioral strategy* is a strategy that assigns to each information set a probability distributions over the choices of that set.

For example, suppose the first move of a game is the deal of one card from a deck of 52 cards to Player I. After seeing his card, Player I either bets or passes, and then Player II takes some action. Player I has 52 information sets each with 2 choices of action, and so he has 2^{52} pure strategies. Thus, a mixed strategy for I is a vector of 2^{52} components adding to 1. On the other hand, a behavioral strategy for I simply given by the probability of betting for each card he may receive, and so is specified by only 52 numbers.

In general, the dimension of the space of behavioral strategies is much smaller than the dimension of the space of mixed strategies. The question arises – Can we do as well with behavioral strategies as we can with mixed strategies? The answer is we can if both players in the game have *perfect recall*. The basic theorem, due to Kuhn in 1953 says that in finite games with perfect recall, any distribution over the payoffs achievable by mixed strategies is achievable by behavioral strategies as well.

To see that behavioral strategies are not always sufficient, consider the game of imperfect recall of Fig. 11.4. Upon reducing the game to strategic form, we find the matrix

$$\begin{array}{c} \\ (f,c) \\ (f,d) \\ (g,c) \\ (g,d) \end{array} \begin{array}{c} \begin{array}{cc} a & \;\; b \end{array} \\ \begin{pmatrix} 1 & -1 \\ 1 & 0 \\ 0 & 2 \\ -1 & 2 \end{pmatrix} \end{array}$$

The top and bottom rows may be removed by domination, so it is easy to see that the unique optimal mixed strategies for I and II are $(0, 2/3, 1/3, 0)$ and $(2/3, 1/3)$ respectively. The value is $2/3$. However, Player I's optimal strategy is not achievable by behavioral strategies. A behavioral strategy for I is given by two numbers, p_f, the probability of choice f in the first information set, and p_c, the probability of choice c in the second information set. This leads to the mixed strategy, $(p_f p_c, p_f(1 - p_c), (1 - p_f)p_c, (1 - p_f)(1 - p_c))$. The strategy $(0, 2/3, 1/3, 0)$ is not of this form since if the first component is zero, that is if $p_f p_c = 0$, then either $p_f = 0$ or $p_c = 0$, so that either the second or third component must also be zero.

If the rules of the game require players to use behavioral strategies, as is the case for certain models of bridge, then the game may not have a value. This means that if Player I is required to announce his behavioral strategy first, then he is at a distinct disadvantage. The game of Fig. 11.4 is an example of this. (see Exercise 11.)

11.9 Exercises

1. **The Silver Dollar.** Player II chooses one of two rooms in which to hide a silver dollar. Then, Player I, not knowing which room contains the dollar, selects one of the rooms to search. However, the search is not always successful. In fact, if the dollar is in room #1 and I searches there, then (by a chance move) he has only probability 1/2 of finding it, and if the dollar is in room #2 and I searches there, then he has only probability 1/3 of finding it. Of course, if he searches the wrong room, he certainly won't find it. If he does find the coin, he keeps it; otherwise the dollar is returned to Player II. Draw the game tree.
2. **Two Guesses for the Silver Dollar.** Draw the game tree for exercise 1, if when I is unsuccessful in his first attempt to find the dollar, he is given a second chance to choose a room and search for it with the same probabilities of success, independent of his previous search. (Player II does not get to hide the dollar again.)
3. **A Statistical Game.** Player I has two coins. One is fair (probability 1/2 of heads and 1/2 of tails) and the other is biased with probability 1/3 of heads and 2/3 of tails. Player I knows which coin is fair and which is biased. He selects one of the coins and tosses it. The outcome of the toss is announced to II. Then II must guess whether I chose the fair or biased coin. If II is correct there is no payoff. If II is incorrect, she loses 1. Draw the game tree.
4. **A Forgetful Player.** A fair coin (probability 1/2 of heads and 1/2 of tails) is tossed and the outcome is shown to Player I. On the basis of the outcome of this toss, I decides whether to bet 1 or 2. Then Player II hearing the amount bet but not knowing the outcome of the toss, must guess whether the coin was heads or tails. Finally, Player I (or, more realistically, his partner), remembering the amount bet and II's guess, but not remembering the outcome of the toss, may double or pass. II wins if her guess is correct and loses if her guess is incorrect.

The absolute value of the amount won is [the amount bet (+1 if the coin comes up heads)] ($\times 2$ if I doubled). Draw the game tree.

5. **A One-Shot Game of Incomplete Information.** Consider the two games $G_1 = \begin{pmatrix} 6 & 0 \\ 0 & 0 \end{pmatrix}$ and $G_2 = \begin{pmatrix} 3 & 0 \\ 0 & 6 \end{pmatrix}$. One of these games is chosen to be played at random with probability $1/3$ for G_1 and probability $2/3$ for G_2. The game chosen is revealed to Player I but not to Player II. Then Player I selects a row, 1 or 2, and simultaneously Player II chooses a column, 1 or 2, with payoff determined by the selected game. Draw the game tree. (If the game chosen by nature is played repeatedly with Player II learning only the pure strategy choices of Player I, this is called a *repeated game of incomplete information.* There is a large literature concerning such games; see for example, the books of Aumann and Maschler (1995) and Sorin (2002).)
6. **Basic Endgame in Poker.** Generalize Basic Endgame in poker by letting the probability of receiving a winning card be an arbitrary number p, $0 \leq p \leq 1$, and by letting the bet size be an arbitrary number $b > 0$. (In Fig. 11. 2, $1/4$ is replaced by p and $3/4$ is replaced by $1 - p$. Also 3 is replaced by $1 + b$ and -3 is replaced by $-(1 + b)$.) Find the value and optimal strategies. (Be careful. For $p \geq (2 + b)/(2 + 2b)$ there is a saddle point. When you are finished, note that for $p < (2 + b)/(2 + 2b)$, Player II's optimal strategy does not depend on p!) For other generalizations, see Ferguson and Ferguson (2007).
7. (a) Find the equivalent strategic form of the game with the game tree:

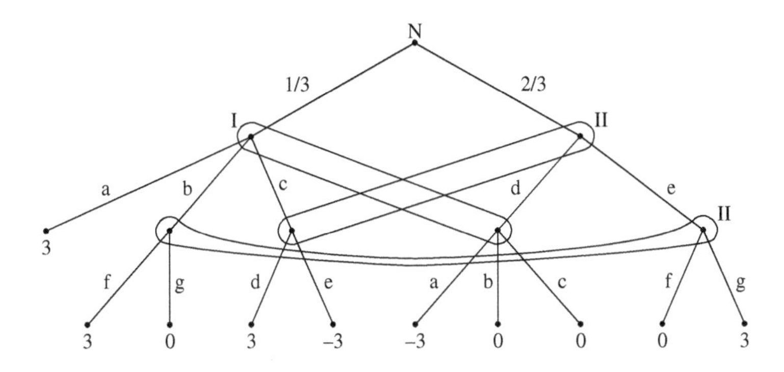

(b) Solve the game.

8. (a) Find the equivalent strategic form of the game with the game tree:

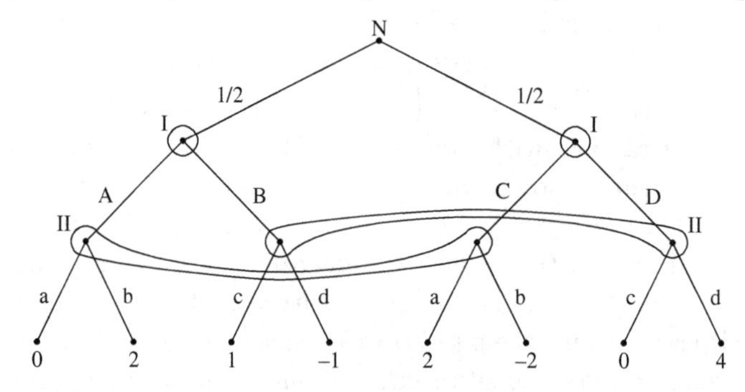

   (b) Solve the game.

9. Coin A has probability 1/2 of heads and 1/2 of tails. Coin B has probability 1/3 of heads and 2/3 of tails. Player I must predict "heads" or "tails". If he predicts heads, coin A is tossed. If he predicts tails, coin B is tossed. Player II is informed as to whether I's prediction was right or wrong (but she is not informed of the prediction or the coin that was used), and then must guess whether coin A or coin B was used. If II guesses correctly she wins $1 from I. If II guesses incorrectly and I's prediction was right, I wins $2 from II. If both are wrong there is no payoff.

   (a) Draw the game tree.
   (b) Find the equivalent strategic form of the game.
   (c) Solve.

10. Find the equivalent strategic form and solve the game of

    (a) Exercise 1.
    (b) Exercise 2.
    (c) Exercise 3.
    (d) Exercise 4.
    (e) Exercise 5.

11. Suppose, in the game of Fig. 11.4, that Player I is required to use behavioral strategies. Show that if Player I is required to announce his behavioral strategy first, he can only achieve a lower value of 1/2. Whereas, if Player II is required to announce her strategy first, Player I has a behavioral strategy reply that achieves the upper value of 2/3 at least.

12. (Beasley (1990), Chap. 6.) Player I draws a card at random from a full deck of 52 cards. After looking at the card, he bets either 1 or 5 that

the card he drew is a face card (king, queen or jack, probability 3/13). Then Player II either concedes or doubles. If she concedes, she pays I the amount bet (no matter what the card was). If she doubles, the card is shown to her, and Player I wins twice his bet if the card is a face card, and loses twice his bet otherwise.

(a) Draw the game tree. (You may argue first that Player I always bets 5 with a face card and Player II always doubles if Player I bets 1.)
(b) Find the equivalent normal form.
(c) Solve.

12 Recursive and Stochastic Games

12.1 Matrix Games with Games as Components

We consider now matrix games in which the outcome of a particular choice of pure strategies of the players may be that the players have to play another game. Let us take a simple example.

Let G_1 and G_2 denote 2×2 games with matrices

$$G_1 = \begin{pmatrix} 0 & 3 \\ 2 & -1 \end{pmatrix} \quad \text{and} \quad G_2 = \begin{pmatrix} 0 & 1 \\ 4 & 3 \end{pmatrix}$$

and let G denote the 2×2 game whose matrix is represented by

$$G = \begin{pmatrix} G_1 & 4 \\ 5 & G_2 \end{pmatrix}.$$

The game G is played in the usual manner with Player I choosing a row and Player II choosing a column. If the entry in the chosen row and column is a number, II pays I that amount and the game is over. If I chooses row 1 and II chooses column 1, then the game G_1 is played. If I chooses row 2 and II chooses column 2, then G_2 is played.

We may analyze the game G by first analyzing G_1 and G_2.

G_1: Optimal for I is (1/2,1/2)
Optimal for II is (2/3,1/3)
$\text{Val}(G_1) = 1$.

G_2: Optimal for I is (0,1)
Optimal for II is (0,1)
$\text{Val}(G_2) = 3$.

If after playing the game G the players end up playing G_1, then they can expect a payoff of the value of G_1, namely 1, on the average. If the players end up playing G_2, they can expect an average payoff of the value of G_2, namely 3. Therefore, the game G can be considered equivalent to the game with matrix

$$\begin{pmatrix} 1 & 4 \\ 5 & 3 \end{pmatrix}$$

G : Optimal for I is (2/5,3/5)
Optimal for II is (1/5,4/5)
Val(G)=17/5.

This method of solving the game G may be summarized as follows. *If the matrix of a game G has other games as components, the solution of G is the solution of the game whose matrix is obtained by replacing each game in the matrix of G by its value.*

Decomposition. This example may be written as a 4×4 matrix game. The four pure strategies of Player I may be denoted $\{(1,1),(1,2),(2,1),(1,2)\}$, where (i,j) represents: use row i in G, and if this results in G_i being played use row j. A similar notation may be used for Player II. The 4×4 game matrix becomes

$$G = \left(\begin{array}{cc|cc} 0 & 3 & 4 & 4 \\ 2 & -1 & 4 & 4 \\ \hline 5 & 5 & 0 & 1 \\ 5 & 5 & 4 & 3 \end{array}\right).$$

We can solve this game by the methods of Chap. 10.

Conversely, suppose we are given a game G and suppose after some rearrangement of the rows and of the columns the matrix may be decomposed into the form

$$G = \left(\begin{array}{c|c} \mathrm{G}_{11} & \mathrm{G}_{12} \\ \hline \mathrm{G}_{21} & \mathrm{G}_{22} \end{array}\right)$$

where G_{11} and G_{22} are arbitrary matrices and G_{12} and G_{21} are constant matrices. (A constant matrix has the same numerical value for all of its entries.) Then we can solve G by the above method, pretending that as the first move the players choose a row and column from the 2×2 decomposed matrix. See Exercise 1(b).

12.2 Multistage Games

Of course, a game that is the component of some matrix game may itself have other games as components, in which case one has to iterate the above method to obtain the solution. This works if there are a finite number of stages.

Example 1. The Inspection Game. (Dresher (1962)) Player II must try to perform some forbidden action in one of the next n time periods. Player I is allowed to inspect II secretly just once in the next n time periods. If II acts while I is inspecting, II loses 1 unit to I. If I is not inspecting when II acts, the payoff is zero.

Let G_n denote this game. We obtain the iterative equations

$$G_n = \begin{matrix} \\ \text{inspect} \\ \text{wait} \end{matrix} \begin{matrix} \text{act} & \text{wait} \\ \left(\begin{matrix} 1 \\ 0 \end{matrix}\right. & \left.\begin{matrix} 0 \\ G_{n-1} \end{matrix}\right) \end{matrix} \qquad \text{for} \quad n = 2, 3, \ldots$$

with boundary condition $G_1 = (1)$. We may solve iteratively.

$$\begin{aligned} \text{Val}(G_1) &= 1 \\ \text{Val}(G_2) &= \text{Val}\begin{pmatrix} 1 & 0 \\ 0 & 1 \end{pmatrix} = 1/2 \\ \text{Val}(G_3) &= \text{Val}\begin{pmatrix} 1 & 0 \\ 0 & 1/2 \end{pmatrix} = 1/3 \\ &\vdots \\ \text{Val}(G_n) &= \text{Val}\begin{pmatrix} 1 & 0 \\ 0 & 1/(n-1) \end{pmatrix} = 1/n \end{aligned}$$

since inductively, $\text{Val}(G_n) = \frac{1}{n-1}/(1 + \frac{1}{n-1}) = 1/n$. The optimal strategy in the game G_n for both players is $(1/n, (n-1)/n)$. For other games of this sort, see the book by Garnaev (2000).

Example 2. Guess it! (Isaacs (1955); see also Gardner (1978), p. 40) As a more complex example of a multistage game, consider the following game loosely related to the game Cluedo. From a deck with $m + n + 1$ distinct cards, m cards are dealt to Player I, n cards are dealt to Player II, and the remaining card, called the "target card", is placed face down on the table. Players know their own cards but not those of their opponent. The objective is to guess correctly the target card. Players alternate moves, with Player I starting. At each move, a player may either

(1) guess at the target card, in which case the game ends, with the winner being the player who guessed if the guess is correct, and his opponent if the guess is incorrect, or
(2) ask if the other player holds a certain card. If the other player has the card, that card must be shown and is removed from play.

With a deck of say 11 cards and each player receiving 5 cards, this is a nice playable game that illustrates need for bluffing in a clear way. If a player asks about a card that is in his own hand, he knows what the answer will be. We call such a play a *bluff*. If a player asks about a card not in his hand, we say he is *honest*. If a player is always honest and the card he asks about is the target card, the other player will know that the requested card is the target card and so will win. Thus a player must bluff occasionally. Bluffing may also lure the opponent into a wrong guess at the target card.

Let us denote this game with Player I to move by $G_{m,n}$. The game $G_{m,0}$ is easy to play. Player I can win immediately. Since his opponent has no cards, he can tell what the target card is. Similarly, the game $G_{0,n}$ is easy to solve. If Player I does not make a guess immediately, his opponent will win on the next move. However, his probability of guessing correctly is only $1/(n+1)$. Valuing 1 for a win and 0 for a loss from Player I's viewpoint, the value of the game is just the probability I wins under optimal play. We have

$$\text{Val}(G_{m,0}) = 1 \quad \text{for all } m \geq 0, \quad \text{and} \quad \text{Val}(G_{0,n}) = \frac{1}{n+1} \quad \text{for all } n \geq 0. \tag{1}$$

If Player I asks for a card that Player II has, that card is removed from play and it is Player II's turn to move, holding $n-1$ cards to her opponent's m cards. This is exactly the game $G_{n-1,m}$ but with Player II to move. We denote this game by $\bar{G}_{n-1,m}$. Since the probability that Player I wins is one minus the probability that Player II wins, we have

$$\text{Val}(\bar{G}_{n,m}) = 1 - \text{Val}(G_{n,m}) \quad \text{for all } m \text{ and } n. \tag{2}$$

Suppose Player I asks for a card that Player II does not have. Player II must immediately decide whether or not Player I was bluffing. If she decides Player I was honest, she will announce the card Player I asked for as her guess at the target card, and win if she was right and lose if she was wrong.

If she decides Player I was bluffing and she is wrong, Player I will win on his turn. If she is correct, the card Player I asked for is removed from his hand, and the game played next is $\bar{G}_{n,m-1}$.

Using such considerations, we may write the game as a multistage game in which a stage consists of three pure strategies for Player I (honest, bluff, guess) and two pure strategies for Player II (*ignore* the asked card, *call* the bluff by guessing the asked card). The game matrix becomes, for $m \geq 1$ and $n \geq 1$,

$$G_{m,n} = \begin{array}{c} \\ \text{honest} \\ \text{bluff} \\ \text{guess} \end{array} \begin{array}{c} \begin{array}{cc} \text{ignore} & \qquad\qquad \text{call} \end{array} \\ \begin{pmatrix} \frac{n}{n+1}\bar{G}_{n-1,m} + \frac{1}{n+1} & \frac{n}{n+1}\bar{G}_{n-1,m} \\ \bar{G}_{n,m-1} & 1 \\ \frac{1}{n+1} & \frac{1}{n+1} \end{pmatrix} \end{array}. \tag{3}$$

This assumes that if Player I asks honestly, he chooses among the $n+1$ unknown cards with probability $1/(n+1)$ each; also if he bluffs, he chooses among his m cards with probability $1/m$ each. That this may be done follows from the invariance considerations of Sec. 9.6.

As an example, the upper left entry of the matrix is found as follows. With probability $n/(n+1)$, Player I asks a card that is in Player II's hand and the game becomes $\bar{G}_{n-1,m}$; with probability $1/(n+1)$, Player I asks the target card, Player II ignores it and Player I wins on his next turn, i.e. gets 1. The upper right entry is similar, except this time if the asked card is the target card, Player II guesses it and Player I gets 0.

It is reasonable to assume that if $m \geq 1$ and $n \geq 1$, Player I should not guess, because the probability of winning is too small. In fact if $m \geq 1$ and $n \geq 1$, there is a strategy for Player I that dominates guessing, so that the last row of the matrix may be deleted. This strategy is: On the first move, ask any of the $m+n+1$ cards with equal probability $1/(m+n+1)$ (i.e. use row 1 with probability $(n+1)/(m+n+1)$ and row 2 with probability $m/(m+n+1)$), and if Player II doesn't guess at her turn, then guess at the next turn. We must show that Player I wins with probability at least $1/(n+1)$ whether or not Player II guesses at her next turn. If Player II guesses, her probability of win is exactly $1/(m+1)$ whether or not the asked card is one of hers. So Player I's win probability is $m/(m+1) \geq 1/2 \geq 1/(n+1)$. If Player II does not guess, then at Player I's next turn, Player II has at most n cards (she may have $n-1$) so again Player I's win probability is at least $1/(n+1)$.

So the third row may be removed in (3) and the games reduce to

$$G_{m,n} = \begin{array}{c} \\ \text{honest} \\ \\ \text{bluff} \end{array} \begin{array}{c} \text{ignore} \qquad\qquad\qquad \text{call} \\ \begin{pmatrix} \frac{n}{n+1}\bar{G}_{n-1,m} + \frac{1}{n+1} & \frac{n}{n+1}\bar{G}_{n-1,m} \\ \bar{G}_{n,m-1} & 1 \end{pmatrix} \end{array} \tag{4}$$

for $m \geq 1$ and $n \geq 1$. These 2×2 games are easily solved recursively, using the boundary conditions (1). One can find the value and optimal strategies of $G_{m,n}$ after one finds the values of $G_{n,m-1}$ and $G_{n-1,m}$ and uses (2). For example, the game $G_{1,1}$ reduces to the game with matrix $\begin{pmatrix} 3/4 & 1/4 \\ 0 & 1 \end{pmatrix}$. The value of this game is 1/2, an optimal strategy for Player I is (2/3,1/3) (i.e. bluff with probability 1/3), and the optimal strategy of player II is (1/2,1/2).

One can also show that for all $m \geq 1$ and $n \geq 1$ these games do not have saddle points. In fact, one can show more: that $\text{Val}(G_{m,n})$ is nondecreasing in m and nonincreasing in n. (The more cards you have in your hand, the better.) Let $V_{m,n} = \text{Val}(G_{m,n})$. Then using $\text{Val}\begin{pmatrix} a & b \\ c & d \end{pmatrix} = (ad - bc)/(a + d - b - c)$, we have after some algebra

$$\begin{aligned} V_{m,n} &= \text{Val}\begin{pmatrix} \frac{n}{n+1}(1 - V_{n-1,m}) + \frac{1}{n+1} & \frac{n}{n+1}(1 - V_{n-1,m}) \\ (1 - V_{n,m-1}) & 1 \end{pmatrix} \\ &= \frac{1 + n(1 - V_{n-1,m})V_{n,m-1}}{1 + (n+1)V_{n,m-1}} \end{aligned}$$

for $m \geq 1$ and $n \geq 1$. This provides a simple direct way to compute the values recursively.

The following table gives the computed values as well as the optimal strategies for the players for small values of m and n.

$m\backslash n$	1	2	3	4	5	6
1	0.5000 0.3333 0.5000	0.5000 0.2500 0.5000	0.4000 0.2000 0.4000	0.3750 0.1667 0.3750	0.3333 0.1429 0.3333	0.3125 0.1250 0.3125
2	0.6667 0.5000 0.3333	0.5556 0.3333 0.3333	0.5111 0.2667 0.2889	0.4500 0.2143 0.2500	0.4225 0.1818 0.2301	0.3871 0.1563 0.2055
3	0.6875 0.5000 0.3750	0.6250 0.3750 0.3250	0.5476 0.2857 0.2762	0.5126 0.2361 0.2466	0.4667 0.1967 0.2167	0.4411 0.1701 0.1984
4	0.7333 0.5556 0.3333	0.6469 0.3947 0.3092	0.5966 0.3134 0.2634	0.5431 0.2511 0.2342	0.5121 0.2122 0.2118	0.4749 0.1806 0.1899
5	0.7500 0.5714 0.3333	0.6809 0.4255 0.2908	0.6189 0.3278 0.2566	0.5810 0.2691 0.2284	0.5389 0.2229 0.2062	0.5112 0.1917 0.1885
6	0.7714 0.6000 0.3143	0.6972 0.4410 0.2834	0.6482 0.3488 0.2461	0.6024 0.2808 0.2236	0.5704 0.2362 0.2028	0.5353 0.2003 0.1854

Table of values and optimal strategies of $G_{m,n}$ for $1 \leq m, n \leq 6$. The top number in each box is the value, the middle number is the probability with which Player I should bluff, and the bottom number is the probability with which Player II should call the asked card.

12.3 Recursive Games. ϵ-Optimal Strategies

In some games with games as components, it may happen that the original game comes up again. Such games are called *recursive*. A simple example is

$$G = \begin{pmatrix} G & 1 \\ 1 & 0 \end{pmatrix}.$$

This is an infinite game. If the players always play the first row and first column, the game will be played forever. No matter how unlikely such a possibility is, the mathematical definition is not complete until we say what the payoff is if G is played forever. Let us say that II pays I Q units if they both choose their first pure strategy forever, and write

$$G = \begin{pmatrix} G & 1 \\ 1 & 0 \end{pmatrix}, Q.$$

We are not automatically assured the existence of a value or the existence of optimal strategies in infinite games. However, it is easy to see that the value of G exists and is equal to 1 no matter what the value of the number Q is. The analysis can be made as follows.

II can restrict her losses to at most 1 by choosing the second column. If $Q \geq 1$, I can guarantee winning at least 1 by playing his first row forever. But if $Q < 1$, this won't work. It turns out that an optimal strategy for I, guaranteeing him at least 1, does not exist in this case. However, for any $\epsilon > 0$ there is a strategy for I that guarantees him an average gain of at least $1 - \epsilon$. Such a strategy, that guarantees a player an average payoff within ϵ of the value, is called ϵ*-optimal.* In this case, the strategy that continually uses the mixed strategy $(1 - \epsilon, \epsilon)$ (top row with probability $1 - \epsilon$ and bottom row with probability ϵ) is ϵ-optimal for I. The use of such a strategy by I insures that he will eventually choose row 2, so that the payoff is bound to be 0 or 1 and never Q. The best that Player II can do against this strategy is to choose column 2 immediately, hoping that I chooses row 2. The expected payoff would then be $1 \cdot (1 - \epsilon) + 0 \cdot \epsilon = 1 - \epsilon$.

In summary, for the game G above, the value is 1; Player II has an optimal strategy, namely column 2; If $Q \geq 1$, the first row forever is optimal for I; if $Q < 1$, there is no optimal strategy for I, but the strategy $(1 - \epsilon, \epsilon)$ forever is ϵ-optimal for I.

Consider now the game

$$G_0 = \begin{pmatrix} G_0 & 5 \\ 1 & 0 \end{pmatrix}, Q.$$

For this game, the value depends on Q. If $Q \geq 1$, the first row forever is optimal for I, and the value is Q if $1 \leq Q \leq 5$, and the value is 5 if $Q \geq 5$. For $Q < 1$, the value is 1; however, in contrast to the game G above, I has an optimal strategy for the game G_0, for example $(1/2, 1/2)$ forever. II's optimal strategy is the first column forever if $Q < 5$, the second column if $Q > 5$ and anything if $Q = 5$.

In analogy to what we did for games with games as components, we might attempt to find the value v of such a game by replacing G_0 by v in the matrix and solving the equation

$$v = \text{Val}\begin{pmatrix} v & 5 \\ 1 & 0 \end{pmatrix}$$

for v. Here there are many solutions to this equation. The set of all solutions to this equation is the set of numbers v in the interval $1 \le v \le 5$. (Check this!)

This illustrates a general result that the equation, given by equating v to the value of the game obtained by replacing the game in the matrix by v, always has a solution equal to the value of the game. It may have more solutions but the value of the game is that solution that is closest to Q. For more information on these points, consult the papers of Everett (1957) and of Milnor and Shapley (1957).

Example 3. Let

$$G = \begin{pmatrix} G & 1 & 0 \\ 1 & 0 & G \\ 0 & G & 1 \end{pmatrix}, Q.$$

Then, if the value of G is v,

$$v = \text{Val}\begin{pmatrix} v & 1 & 0 \\ 1 & 0 & v \\ 0 & v & 1 \end{pmatrix} = \frac{1+v}{3}.$$

This equation has a unique solution, $v = 1/2$. This must be the value for all Q. The strategy (1/3,1/3,1/3) forever is optimal for both players.

Example 4. The basic game of Dare is played as follows. Player I, the leader, and Player II, the challenger, simultaneously "pass" or "dare". If both pass, the payoff is zero (and the game is over). If I passes and II dares, I wins 1. If I dares and II passes, I wins 3. If both dare, the basic game is played over with the roles of the players reversed (the leader becomes the challenger and vice versa). If the players keep daring forever, let the payoff be zero. We might write

$$G = \begin{matrix} \\ \text{pass} \\ \text{dare} \end{matrix} \begin{matrix} \text{pass} & \text{dare} \\ \left(\begin{matrix} 0 \\ 3 \end{matrix} \right. & \left. \begin{matrix} 1 \\ -G^T \end{matrix} \right) \end{matrix},$$

where $-G^T$ represents the game with the roles of the players reversed. (Its matrix is the negative of the transpose of the matrix G.) The value of $-G^T$ is the negative of the value of G.

If v represents the value of G, then $v \geq 0$ because of the top row. Therefore the matrix for G with $-G^T$ replaced by $-v$ does not have a saddle point, and we have

$$v = \text{Val}\begin{pmatrix} 0 & 1 \\ 3 & -v \end{pmatrix} = \frac{3}{4+v}.$$

This gives the quadratic equation, $v^2 + 4v - 3 = 0$. The only nonnegative solution is $v = \sqrt{7} - 2 = 0.64575\cdots$. The optimal strategy for I is $((5 - \sqrt{7})/3, (\sqrt{7} - 2)/3)$ and the optimal strategy for II is $(3 - \sqrt{7}, \sqrt{7} - 2)$.

Example 5. Consider the following three related games.

$$G_1 = \begin{pmatrix} G_2 & 0 \\ 0 & G_3 \end{pmatrix}, \quad G_2 = \begin{pmatrix} G_1 & 1 \\ 1 & 0 \end{pmatrix}, \quad G_3 = \begin{pmatrix} G_1 & 2 \\ 2 & 0 \end{pmatrix}$$

and suppose the payoff if the games are played forever is Q. Let us attempt to solve these games. Let $v_1 = \text{Val}(G_1)$, $v_2 = \text{Val}(G_2)$, and $v_3 = \text{Val}(G_3)$. Player I can guarantee that $v_1 > 0$, $v_2 > 0$ and $v_3 > 0$ by playing $(1/2, 1/2)$ forever. In addition, $v_2 \leq 1$ and $v_3 \leq 2$, which implies $v_1 < 1$. Therefore none of the games has a saddle point and we may write

$$v_1 = \frac{v_2 v_3}{v_2 + v_3}, \qquad v_2 = \frac{1}{2 - v_1}, \qquad v_3 = \frac{4}{4 - v_1}.$$

Substituting the latter two equations into the former, we obtain

$$\begin{aligned} \frac{v_1}{2 - v_1} + \frac{4v_1}{4 - v_1} &= \frac{4}{(2 - v_1)(4 - v_1)} \\ 5v_1^2 - 12v_1 + 4 &= 0 \\ (5v_1 - 2)(v_1 - 2) &= 0. \end{aligned}$$

Since $0 < v_1 < 1$, this implies that $v_1 = 2/5$. Hence

Game	value	opt. for I and II
G_1	2/5	(16/25, 9/25)
G_2	5/8	(5/8, 3/8)
G_3	10/9	(5/9, 4/9)

independent of the value of Q.

12.4 Stochastic Movement Among Games

We may generalize the notion of a recursive game by allowing the choice of the next game played to depend not only upon the pure strategy choices of the players, but also upon chance. Let $G_1, \ldots, G_n$ be games and let $p_1, \ldots, p_n$ be probabilities that sum to one. We use the notation, $p_1 G_1 + \cdots + p_n G_n$, to denote the situation where the game to be played next is chosen at random, with game G_i being chosen with probability p_i, $i = 1, \ldots, n$. Since, for a given number z, the 1×1 matrix (z) denotes the trivial game in which II pays I z, we may, for example, use $\frac{1}{2}G_1 + \frac{1}{2}(3)$ to represent the situation where G_1 is played if a fair coin comes up heads, and II pays I 3 otherwise.

Example 6. Let G_1 and G_2 be related as follows:

$$G_1 = \begin{pmatrix} \frac{1}{2}G_2 + \frac{1}{2}(0) & 1 \\ 2 & 0 \end{pmatrix}, \quad G_2 = \begin{pmatrix} \frac{2}{3}G_1 + \frac{1}{3}(-2) & 0 \\ 0 & -1 \end{pmatrix}.$$

The game must eventually end (with probability 1). In fact, the players could not play forever even if they wanted to. Even if they choose the first row and first column forever, eventually the game would end with a payoff of 0 or -2. Thus we do not need to specify any payoff if play continues forever. To solve, let $v_i = \text{Val}(G_i)$ for $i = 1, 2$. Then $0 \le v_1 \le 1$ and $-1 \le v_2 \le 0$, so neither game has a saddle point. Hence,

$$v_1 = \text{Val}\begin{pmatrix} \frac{1}{2}v_2 & 1 \\ 2 & 0 \end{pmatrix} = \frac{4}{6 - v_2} \qquad \text{and}$$

$$v_2 = \text{Val}\begin{pmatrix} \frac{2}{3}v_1 - \frac{2}{3} & 0 \\ 0 & -1 \end{pmatrix} = -\frac{2(1 - v_1)}{5 - 2v_1}.$$

Thus

$$v_1 = \frac{4}{6 + \frac{2(1-v_1)}{5-2v_1}} = \frac{2(5 - 2v_1)}{16 - 7v_1}.$$

This leads to the quadratic equation, $7v_1^2 - 20v_1 + 10 = 0$, with solution, $v_1 = (10 - \sqrt{30})/7 = 0.646\cdots$. We may substitute back into the equation for v_2 to find $v_2 = -(2\sqrt{30} - 10)/5 = -0.191\cdots$. From these values one can easily find the optimal strategies for the two games.

Example 7. A coin with probability 2/3 of heads is tossed. Both players must guess whether the coin will land heads or tails. If I is right and II is wrong, I wins 1 if the coin is heads and 4 if the coin is tails and the game is over. If I is wrong and II is right, there is no payoff and the game is over. If both players are right, the game is played over. But if both players are wrong, the game is played over with the roles of the players reversed. If the game never ends, the payoff is Q.

If we denote this game by G, then

$$G = \begin{array}{c} \\ \text{heads} \\ \text{tails} \end{array} \begin{array}{c} \begin{array}{cc} \text{heads} & \text{tails} \end{array} \\ \begin{pmatrix} \frac{2}{3}G + \frac{1}{3}(-G^T) & \frac{2}{3}(1) + \frac{1}{3}(0) \\ \frac{2}{3}(0) + \frac{1}{3}(4) & \frac{2}{3}(-G^T) + \frac{1}{3}G \end{pmatrix} \end{array}.$$

If we let its value be denoted by v, then

$$v = \text{Val}\begin{pmatrix} \frac{1}{3}v & \frac{2}{3} \\ \frac{4}{3} & -\frac{1}{3}v \end{pmatrix}.$$

If $v \geq 2$, then there is a saddle at the upper right corner with $v = 2/3$. This contradiction shows that $v < 2$ and there is no saddle. Therefore,

$$v = \frac{8 + v^2}{18} \qquad \text{or} \qquad v^2 - 18v + 8 = 0.$$

This has a unique solution less than two,

$$v = 9 - \sqrt{73} = 0.456\cdots$$

from which we may calculate the optimal strategy for I:

$$\left(\frac{13 - \sqrt{73}}{6}, \frac{\sqrt{73} - 7}{6}\right) = (0.743\cdots, 0.256\cdots)$$

and the optimal strategy for II:

$$\left(\frac{11 - \sqrt{73}}{6}, \frac{\sqrt{73} - 5}{6}\right) = (0.409\cdots, 0.591\cdots).$$

The value and optimal strategies are independent of Q.

12.5 Stochastic Games

If to the features of the games of the previous section is added the possibility of a payoff at each stage until the game ends, the game is called a *Stochastic Game.* This seems to be the proper level of generality for theoretical treatment of multistage games. It is an area of intense contemporary research. See for example the books of Filar and Vrieze (1997) and Maitra and Sudderth (1996). Stochastic games were introduced by Shapley in (1953) in a beautiful paper that has been reprinted in Raghavan *et al.* (1991), and in Kuhn (1997). In this section, we present Shapley's main result.

A Stochastic Game, $\mathbf{G}$, consists of a finite set of positions or states, $\{1, 2, \ldots, N\}$, one of which is specified as the starting position. We denote by $G^{(k)}$ the game in which k is the starting position. Associated with each state, k, is a matrix game, $\mathbf{A}^{(k)} = \left(a_{ij}^{(k)}\right)$. If the stochastic game is in state k, the players simultaneously choose a row and column of $\mathbf{A}^{(k)}$, say i and j. As a result, two things happen. First, Player I wins the amount $a_{ij}^{(k)}$ from Player II. Second, with probabilities that depend on i, j and k, the game either stops, or it moves to another state (possibly the same one). The probability that the game stops is denoted by $s_{ij}^{(k)}$, and the probability that the next state is ℓ is denoted by $P_{ij}^{(k)}(\ell)$, where

$$s_{ij}^{(k)} + \sum_{\ell=1}^{N} P_{ij}^{(k)}(\ell) = 1 \qquad (5)$$

for all i, j and k.

The payoffs accumulate throughout the game until it stops. To make sure the game eventually stops, we make the assumption that all the stopping probabilities are positive. Let s denote the smallest of these probabilities,

$$s = \min_{i,j,k} s_{ij}^{(k)} > 0. \qquad (6)$$

Under this assumption, the probability is one that the game ends in a finite number of moves. This assumption also makes the expected accumulated payoff finite no matter how the game is played, since if M denotes the largest of the absolute values of the payoffs, $M = \max_{i,j,k} |a_{ij}^{(k)}|$, then the total expected payoff to either player is bounded by

$$M + (1-s)M + (1-s)^2 M + \cdots = M/s. \qquad (7)$$

Player I wishes to maximize the total accumulated payoff and Player II to minimize it. We use a modification of the notation of the previous section to describe this game.

$$G^{(k)} = \left(a_{ij}^{(k)} + \sum_{\ell=1}^{N} P_{ij}^{(k)}(\ell)\, G^{(\ell)} \right). \tag{8}$$

Note that the probabilities in each component of this matrix sum to less than one. It is understood that with the remaining probability, $s_{ij}^{(k)}$, the game ends. It should be noted that in contrast to the previous section, a payoff does not end the game. After a payoff is made, it is then decided at random whether the game ends and, if not, which state should be played next.

Since no upper bound can be placed on the length of the game, this is an infinite game. A strategy for a player must specify for every n how to choose an action if the game reaches stage n. In general, theory does not guarantee a value. Moreover, the choice of what to do at stage n may depend on what happened at all previous stages, so the space of possible strategies is extremely complex.

Nevertheless, in stochastic games, the value and optimal strategies for the players exist for every starting position. Moreover, optimal strategies exist that have a very simple form. Strategies that prescribe for a player a probability distribution over his choices that depends only on the game, G_k, being played and not on the stage n or past history are called *stationary strategies.* The following theorem states that there exist stationary optimal strategies.

Theorem 1. (Shapley (1953a)) *Each game $G^{(k)}$ has a value, $v(k)$. These values are the unique solution of the set of equations,*

$$v(k) = \mathrm{Val}\left(a_{ij}^{(k)} + \sum_{\ell=1}^{N} P_{ij}^{(k)}(\ell)\, v(\ell) \right) \qquad \text{for } k = 1, \ldots, N. \tag{9}$$

Each player has a stationary optimal strategy that in state k uses the optimal mixed strategy for the game with matrix

$$\mathbf{A}^{(k)}(\mathbf{v}) = \left(a_{ij}^{(k)} + \sum_{\ell=1}^{N} P_{ij}^{(k)}(\ell)\, v(\ell) \right), \tag{10}$$

where $\mathbf{v}$ represents the vector of values, $\mathbf{v} = (v(1), \ldots, v(N))$.

In Eq. (9), we see the same principle as in the earlier sections: the value of a game is the value of the matrix game (8) with the games replaced by their values. A proof of this theorem may be found in Appendix 3.

Example 8. As a very simple example, consider the following stochastic game with one state, call it G.

$$G = \begin{pmatrix} 1+(3/5)G & 3+(1/5)G \\ 1+(4/5)G & 2+(2/5)G \end{pmatrix}.$$

From Player II's viewpoint, column 1 is better than column 2 in terms of immediate payoff, but column 2 is more likely to end the game sooner than column 1, so that it should entail smaller future payoffs. Which column should she choose?

Assume that all strategies are active, i.e. that the game does not have a saddle point. We must check when we are finished to see if the assumption was correct. Then

$$\begin{aligned} v &= \mathrm{Val}\begin{pmatrix} 1+(3/5)v & 3+(1/5)v \\ 1+(4/5)v & 2+(2/5)v \end{pmatrix} \\ &= \frac{(1+(4/5)v)(3+(1/5)v)-(1+(3/5)v)(2+(2/5)v)}{1+(4/5)v+3+(1/5)v-1-(3/5)v-2-(2/5)v} \\ &= 1+v-(2/25)v^2. \end{aligned}$$

This leads to

$$(2/25)v^2 = 1.$$

Solving this quadratic equation gives two possible solutions $v = \pm\sqrt{25/2} = \pm(5/2)\sqrt{2}$. Since the value is obviously positive, we must use the plus sign. This is $v = (5/2)\sqrt{2} = 3.535$. If we put this value into the matrix above, it becomes

$$\begin{pmatrix} 1+(3/2)\sqrt{2} & 3+(1/2)\sqrt{2} \\ 1+2\sqrt{2} & 2+\sqrt{2} \end{pmatrix}.$$

The optimal strategy for Player I in this matrix is $\mathbf{p} = (\sqrt{2}-1, 2-\sqrt{2}) = (0.414, 0.586)$, and the optimal strategy for Player II is $\mathbf{q} = (1-\sqrt{2}/2, \sqrt{2}/2) = (0.293, 0.707)$. Since these are probability vectors, our assumption is correct and these are the optimal strategies, and $v = (5/2)\sqrt{2}$ is the value of the stochastic game.

12.6 Approximating the Solution

For a general stochastic game with many states, Eq. (9) becomes a rather complex system of simultaneous nonlinear equations. We cannot hope to solve such systems in general. However, there is a simple iterative method of approximating the solution. This is based on Shapley's proof of Theorem 1, and is called *Shapley iteration.*

First we make a guess at the solution, call it $\mathbf{v}_0 = (v_0(1), \ldots, v_0(N))$. Any guess will do. We may use all zero's as the initial guess, $\mathbf{v}_0 = \mathbf{0} = (0, \ldots, 0)$. Then given $\mathbf{v}_n$, we define inductively, $\mathbf{v}_{n+1}$, by the equations,

$$v_{n+1}(k) = \mathrm{Val}\left(a_{ij}^{(k)} + \sum_{\ell=1}^{N} P_{ij}^{(k)}(\ell)\, v_n(\ell) \right) \qquad \text{for } k = 1, \ldots, N. \tag{11}$$

With $\mathbf{v}_0 = \mathbf{0}$, the $v_n(k)$ have an easily understood interpretation. $v_n(k)$ is the value of the stochastic game starting in state k if there is forced stopping if the game reaches stage n. In particular, $v_1(k) = \mathrm{Val}(\mathbf{A}_k)$ for all k.

The proof of Theorem 1 shows that $v_n(k)$ converges to the true value, $v(k)$, of the stochastic game starting at k. Two useful facts should be noted. First, the convergence is at an exponential rate: the maximum error goes down at least as fast as $(1-s)^n$. (See Corollary 1 of Appendix 3.) Second, the maximum error at stage $n+1$ is at most the maximum change from stage n to $n+1$ multiplied by $(1-s)/s$. (See Corollary 2 of Appendix 3.)

Let us take an example of a stochastic game with two positions. The corresponding games $G^{(1)}$ and $G^{(2)}$, are related as follows:

$$G^{(1)} = \begin{pmatrix} 4 + 0.3G^{(1)} & 0 + 0.4G^{(2)} \\ 1 + 0.4G^{(2)} & 3 + 0.5G^{(1)} \end{pmatrix} \quad G^{(2)} = \begin{pmatrix} 0 + 0.5G^{(1)} & -5 \\ -4 & 1 + 0.5G^{(2)} \end{pmatrix}.$$

Using $\mathbf{v}_0 = (0, 0)$ as the initial guess, we find $\mathbf{v}_1 = (2, -2)$, since

$$v_1(1) = \mathrm{Val}\begin{pmatrix} 4 & 0 \\ 1 & 3 \end{pmatrix} = 2, \quad v_1(2) = \mathrm{Val}\begin{pmatrix} 0 & -5 \\ -4 & 1 \end{pmatrix} = -2.$$

The next iteration gives

$$v_2(1) = \mathrm{Val}\begin{pmatrix} 4.6 & -0.8 \\ 0.2 & 4 \end{pmatrix} = 2.0174, \quad v_2(2) = \mathrm{Val}\begin{pmatrix} 1 & -5 \\ -4 & 0 \end{pmatrix} = -2.$$

Continuing, we find

$$\begin{aligned} v_3(1) &= 2.0210, & v_3(2) &= -1.9983,\\ v_4(1) &= 2.0220, & v_4(2) &= -1.9977,\\ v_5(1) &= 2.0224, & v_5(2) &= -1.9974,\\ v_6(1) &= 2.0225, & v_6(2) &= -1.9974. \end{aligned}$$

The smallest stopping probability is 0.5, so the rate of convergence is at least $(0.5)^n$ and the maximum error of $\mathbf{v}_6$ is at most 0.0002.

The optimal strategies using $\mathbf{v}_6$ are easily found. For game $G^{(1)}$, the optimal strategies are $\mathbf{p}^{(1)} = (0.4134, 0.5866)$ for Player I and $\mathbf{q}^{(1)} = (0.5219, 0.4718)$ for Player II. For game $G^{(2)}$, the optimal strategies are $\mathbf{p}^{(2)} = (0.3996, 0.6004)$ for Player I and $\mathbf{q}^{(2)} = (0.4995, 0.5005)$ for Player II.

12.7 Exercises

1. (a) Solve the system of games

$$G = \begin{pmatrix} 0 & G_1 \\ G_2 & G_3 \end{pmatrix}, \quad G_1 = \begin{pmatrix} 4 & 3 \\ 1 & 2 \end{pmatrix},$$

$$G_2 = \begin{pmatrix} 0 & 6 \\ 5 & 1 \end{pmatrix}, \quad G_3 = \begin{pmatrix} 0 & -2 \\ -2 & 0 \end{pmatrix}.$$

(b) Solve the games with matrices

$$(\mathrm{b}_1) \begin{pmatrix} 0 & 6 & 0 & 2 \\ 0 & 3 & 0 & 5 \\ 5 & 0 & 2 & 0 \\ 1 & 0 & 4 & 0 \end{pmatrix} \qquad (\mathrm{b}_2) \begin{pmatrix} 3 & 1 & 5 & 2 & 2 \\ 1 & 3 & 5 & 2 & 2 \\ 4 & 4 & 1 & 2 & 2 \\ 1 & 1 & 1 & 6 & 3 \\ 1 & 1 & 1 & 4 & 7 \end{pmatrix}.$$

2. **The Inspection Game.** Let $G_{m,n}$ denote the inspection game in which I is allowed m inspections in the n time periods. (Thus, for $1 \le n \le m$, $\mathrm{Val}(G_{m,n}) = 1$, while for $n \ge 1$, $\mathrm{Val}(G_{0,n}) = 0$.) Find the iterative structure of the games and solve.
3. **A Game of Endurance.** II must count from n down to zero by subtracting either one or two at each stage. I must guess at each stage whether II is going to subtract one or two. If I ever guesses incorrectly at any stage, the game is over and there is no payoff. Otherwise, if

I guesses correctly at each stage, he wins 1 from II. Let G_n denote this game, and use the initial conditions $G_0 = (1)$ and $G_1 = (1)$. Find the recursive structure of the games and solve. (In the solution, you may use the notation F_n to denote the Fibonacci sequence, 1, 1, 2, 3, 5, 8, 13,..., with definition $F_0 = 1$, $F_1 = 1$, and for $n \geq 2$, $F_n = F_{n-1} + F_{n-2}$.)

4. Solve the sequence of games, $G_0, G_1, \ldots$, where

$$G_0 = \begin{pmatrix} 3 & 2 \\ 1 & G_1 \end{pmatrix}, \ldots, G_n = \begin{pmatrix} n+3 & n+2 \\ n+1 & G_{n+1} \end{pmatrix}, \ldots.$$

Assume that if play continues forever, the payoff is zero.

5. (a) In the game "Guess it!", $G_{1,n}$, with $m = 1$ and arbitrary n, show that Player I's optimal strategy if to bluff with probability $1/(n+2)$.
   (b) Show that Player II's optimal strategy in $G_{1,n}$ is to call the asked card with probability $V_{1,n}$, the value of $G_{1,n}$.
6. **Recursive Games.**
   (a) Solve the game $G = \begin{pmatrix} G & 2 \\ 0 & 1 \end{pmatrix}, Q$.
   (b) Solve the game $G = \begin{pmatrix} G & 1 & 1 \\ 1 & 0 & G \\ 1 & G & 0 \end{pmatrix}, Q$.
7. Consider the following three related games.

$$G_1 = \begin{pmatrix} G_2 & 1 \\ 1 & 0 \end{pmatrix}, \quad G_2 = \begin{pmatrix} G_3 & 0 \\ 0 & 2 \end{pmatrix}, \quad G_3 = \begin{pmatrix} G_1 & 1 \\ 1 & 0 \end{pmatrix}$$

and suppose the payoff is Q if the games are played forever. Solve.

8. Consider the following three related games

$$G_1 = \begin{pmatrix} G_1 & G_2 & G_3 \\ G_2 & G_3 & G_1 \\ G_3 & G_1 & G_2 \end{pmatrix}, \quad G_2 = \begin{pmatrix} G_1 & 0 \\ 0 & 2 \end{pmatrix}, \quad G_3 = \begin{pmatrix} G_2 & 1 \\ 1 & 0 \end{pmatrix}$$

and suppose the payoff is Q if the games are played forever. Solve.

9. There is one point to go in the match. The player that wins the last point while serving wins the match. The server has two strategies, high and low. The receiver has two strategies, near and far. The probability

the server wins the point is given in the accompanying table.

	near	far
high	0.8	0.5
low	0.6	0.7

If the server misses the point, the roles of the players are interchanged and the win probabilities for given pure strategies are the same for the new server. Find optimal strategies for server and receiver, and find the probability the server wins the match.

10. Player I tosses a coin with probability p of heads. For each $k = 1, 2, \ldots$, if I tosses k heads in a row he may stop and challenge II to toss the same number of heads; then II tosses the coin and wins if and only if he tosses k heads in a row. If I tosses tails before challenging II, then the game is repeated with the roles of the players reversed. If neither player ever challenges, the game is a draw.
    (a) Solve when $p = 1/2$.
    (b) For arbitrary p, what are the optimal strategies of the players? Find the limit as $p \to 1$ of the probability that I wins.
11. Solve the following stochastic game.

$$G = \begin{pmatrix} 4 & 1 + (1/3)G \\ 0 & 1 + (2/3)G \end{pmatrix}.$$

12. Consider the following stochastic game with two positions.

$$G^{(1)} = \begin{pmatrix} 2 & 2 + (1/2)G^{(2)} \\ 0 & 4 + (1/2)G^{(2)} \end{pmatrix}$$

$$G^{(2)} = \begin{pmatrix} -4 & 0 \\ -2 + (1/2)G^{(1)} & -4 + (1/2)G^{(1)} \end{pmatrix}$$

    (a) Solve Eq. (9) exactly for the values $v(1)$ and $v(2)$.
    (b) Carry out Shapley iteration to find $\mathbf{v}_2$ starting with the initial guess $\mathbf{v}_0 = (0, 0)$, and compare with the exact values found in (a).

13 Infinite Games

In this chapter, we treat infinite two-person, zero-sum games. These are games (X, Y, A), in which at least one of the strategy sets, X and Y, is an infinite set. The famous example of Exercise 10.8.3, he-who-chooses-the-larger-integer-wins, shows that an infinite game may not have a value. Even worse, the example of Exercise 10.8.5 shows that the notion of a value may not even make sense in infinite games without further restrictions. This latter problem will be avoided when we assume that the function $A(x, y)$ is either bounded above or bounded below.

13.1 The Minimax Theorem for Semi-Finite Games

The minimax theorem for finite games states that every finite game has a value and both players have optimal mixed strategies. The first theorem below generalizes this result to the case where only one of the players has a finite number of pure strategies. The conclusion is that the value exists and the player with a finite number of pure strategies has an optimal mixed strategy. But first we must discuss mixed strategies and near optimal strategies for infinite games.

Mixed Strategies for Infinite Games: First note that for infinite games, the notion of a mixed strategy is somewhat open to choice. Suppose the strategy set, Y, of Player II is infinite. The simplest choice of a mixed strategy is a *finite distribution* over Y. This is a distribution that gives all its probability to a finite number of points. Such a distribution is described by a finite number of points of Y, say $y_1, y_2, \ldots, y_n$, and a set of probabilities, $q_1, q_2, \ldots, q_n$ summing to one with the understanding that point y_j is chosen with probability q_j. We will denote the set of finite distributions on Y by Y_F^*.

When Y is an interval of the real line, we may allow as a mixed strategy any distribution over Y given by its distribution function, $F(z)$. Here, $F(z)$ represents the probabiity that the randomly chosen pure strategy, y, is less than or equal to z. The advantage of enlarging the set of mixed strategies is that it then becomes more likely that an optimal mixed strategy will exist. The payoff for using such a strategy is denoted by $A(x, F)$ for $x \in X$.

Near Optimal Strategies for Infinite Games: When a game has a finite value and an optimal strategy for a player does not exist, that player must be content to choosing a strategy that comes within ϵ of achieving the value of the game for some small $\epsilon > 0$. Such a strategy is called an ϵ-optimal strategy and was discussed in Chap. 12.

In infinite games, we allow the value to be $+\infty$ or $-\infty$. For example, the value is $+\infty$ if for every number B, however large, there exists a mixed strategy $\mathbf{p}$ for Player I such that $A(\mathbf{p}, y) \geq B$ for all $y \in Y$. A simple example would be: $X = [0, \infty)$, Y arbitrary, and $A(x, y) = x$ independent of y. The value is $+\infty$, since for any B, Player I can guarantee winning at least B by choosing any $x \geq B$. Such a strategy might be called B-optimal. We will refer to both ϵ-optimal and B-optimal strategies as *near optimal strategies.*

The Semi-Finite Minimax Theorem. For finite $X = \{x_1, \ldots, x_m\}$, we denote the set of mixed strategies of Player I as usual by X^*. If Player I uses $\mathbf{p} \in X^*$ and Player II uses $\mathbf{q} \in Y_F^*$, then the average payoff is denoted by

$$A(\mathbf{p}, \mathbf{q}) = \sum_i \sum_j p_i A(x_i, y_j) q_j. \tag{1}$$

We denote the set of mixed strategies of Player II by Y^*, but we shall always assume that $Y_F^* \subset Y^*$.

Consider the semi-finite two-person zero-sum game, (X, Y, A), in which X is a finite set, Y is an arbitrary set, and $A(x, y)$ is the payoff function — the winnings of Player I if he chooses $x \in X$ and Player II chooses $y \in Y$. To avoid the possibility that the average payoff does not exist or that the value might be $-\infty$, we assume that the payoff function, $A(x, y)$, is bounded below. By bounded below, we mean that there is a number M such that $A(x, y) > M$ for all $x \in X$ and all $y \in Y$. This assumption is weak from the point of view of utility theory because, as mentioned in Appendix 1, it is customary to assume that utility is bounded.

It is remarkable that the minimax theorem still holds in this situation. Specifically the value exists and Player I has a minimax strategy. In addition, for every $\epsilon > 0$, Player II has an ϵ-minimax strategy within Y_F^*.

Theorem 13.1. *If X is finite and A is bounded below, then the game (X, Y, A) has a finite value and Player I has an optimal mixed strategy. In addition, if X has m elements, then Player II has near optimal strategies that give weight to at most m points of Y.*

This theorem is valid without the asumption that A is bounded below provided Player II is restricted to finite strategies, i.e. $Y^* = Y_F^*$. See Exercise 1. However, the value may be $-\infty$, and the notion of near optimal strategies must be extended to this case.

By symmetry, if Y is finite and A is bounded above, then the game (X, Y, A) has a value and Player II has an optimal mixed strategy.

Solving Semi-Finite Games. Here are two methods that may be used to solve semi-finite games. We take X to be the finite set, $X = \{x_1, \ldots, x_m\}$.

METHOD 1. The first method is similar to the method used to solve $2 \times n$ games presented in Sec. 8.2. For each fixed $y \in Y$, the payoff, $A(\mathbf{p}, y)$, is a linear function of $\mathbf{p}$ on the set $X^* = \{\mathbf{p} = (p_1, \ldots, p_m) : p_i \geq 0, \sum p_i = 1\}$. The optimal strategy for Player I is that value of $\mathbf{p}$ that maximizes the *lower envelope*, $f(\mathbf{p}) \equiv \inf_{y \in Y} A(\mathbf{p}, y)$. Note that $f(\mathbf{p})$, being the infimum of a collection of concave continuous (here linear) functions, is concave and continuous on X^*. Since X^* is compact, there exists a $\mathbf{p}$ at which the maximum of $f(\mathbf{p})$ is attained. General methods for solving concave maximization problems are available.

Example 1. Player I chooses $x \in \{x_1, x_2\}$, Player II chooses $y \in [0, 1]$, and the payoff is

$$A(x, y) = \begin{cases} y & \text{if } x = x_1, \\ (1 - y)^2 & \text{if } x = x_2. \end{cases}$$

Let p denote the probability that Player I chooses x_1 and let $\mathbf{p} = (p, 1 - p)$. For a given choice of $y \in Y$ by Player II, the expected payoff to Player I is $A(\mathbf{p}, y) = py + (1 - p)(1 - y)^2$. The minimum of $A(\mathbf{p}, y)$ over y occurs at $p - (1 - p)2(1 - y) = 0$, or $y = (2 - 3p)/(2 - 2p)$; except that for

$p > 2/3$, the minimum occurs at $y = 0$. So, the lower envelope is

$$f(\mathbf{p}) = \min_y A(\mathbf{p}, y) = \begin{cases} p\dfrac{4-5p}{4-4p} & \text{if } p \le 2/3 \\ 1-p & \text{if } p \ge 2/3. \end{cases}$$

The maximum of this function occurs for $p \le 2/3$, and is easily found to be $p = 1-(1/\sqrt{5}) = 0.553\ldots$. The optimal strategy for Player II occurs at that value of y for which the slope of $A(\mathbf{p}, y)$ (as a function of p) is zero. This occurs when $y = (1-y)^2$. We find $y = (3-\sqrt{5})/2 = 0.382\ldots$ is an optimal pure strategy for Player II. This is also the value of the game. See Fig. 13.1, in which the lower envelope is shown as the thick line.

METHOD 2. S-GAMES. (Blackwell and Girshick (1954)) Let $X = \{1, 2, \ldots, m\}$, and let S be a non-empty convex subset of m-dimensional Euclidean space, $\mathbb{R}^m$, and assume that S is bounded below. Player II chooses a point $\mathbf{s} = (s_1, \ldots, s_m)$ in S, and simultaneously Player I chooses a coordinate $i \in X$. Then Player II pays s_i to Player I. Such games are called S-games.

This game arises from the semi-finite game (X, Y, A) with $X = \{1, 2, \ldots, m\}$ by letting $S_0 = \{\mathbf{s} = (A(1, y), \ldots, A(m, y)) : y \in Y\}$. Choosing $y \in Y$ is equivalent to choosing $\mathbf{s} \in S_0$. Although S_0 is not necessarily convex, Player II can, by using a mixed strategy, choose a probability mixture of points in S_0. This is equivalent to choosing a point s in S, where S is the convex hull of S_0.

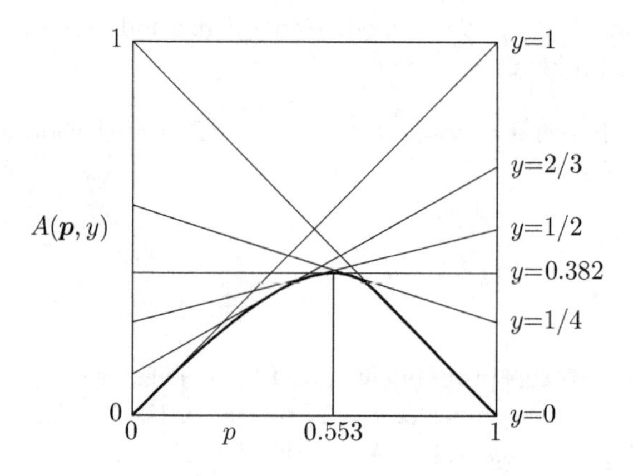

Fig. 13.1.

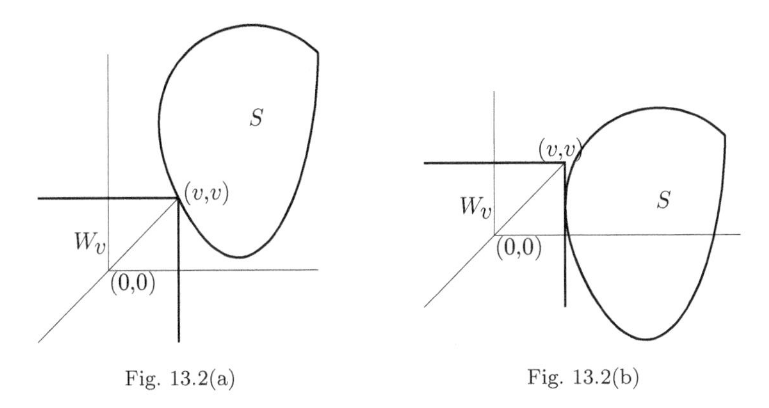

Fig. 13.2(a) Fig. 13.2(b)

To solve the game, let W_c denote the "wedge" at the point $(c, \ldots, c)$ on the diagonal in $\mathbb{R}^m$,

$$W_c = \{\mathbf{s} : s_i \leq c \quad \text{for all } i = 1, \ldots, m\}.$$

Start with some c such that the wedge W_c contains no points of s, i.e. $W_c \cap S = \emptyset$. Such a value of c exists from the assumption that S is bounded below. Now increase c and so push the wedge up to the right until it just touches S. See Fig. 13.2(a). This gives the value of the game:

$$v = \sup\{c : W_c \cap S = \emptyset\}.$$

Any point $\mathbf{s} \in W_v \cap S$ is an optimal pure strategy for Player II. It guarantees that II will lose no more than v. Such a strategy will exist if S is closed. If $W_v \cap S$ is empty, then any point $\mathbf{s} \in W_{v+\epsilon} \cap S$ is an ϵ-optimal strategy for Player II. The point $(v, \ldots, v)$ is not necessarily an optimal strategy for Player II. The optimal strategy could be on the side of the wedge as in Fig. 13.2(b).

To find an optimal strategy for Player I, first find a plane that separates W_v and S, i.e. that keeps W_v on one side and S on the other. Then find the vector perpendicular to this plane, i.e. the normal vector. The optimal strategy of Player I is the mixed strategy with components proportional to this normal vector.

Example 2. Let S be the set $S = \{(y_1, y_2) : y_1 \geq 0, y_2 \geq (1 - y_1)^2\}$. This is essentially the same as Example 1. The wedge first hits S at the vertex (v, v) when $v = (1 - v)^2$. The solution to this equation gives the value, $v = (3 - \sqrt{5})/2$. The point (v, v) is optimal for Player II. To find Player I's optimal strategy, we find the slope of the curve $y_2 = (1 - y_1)^2$ at the point (v, v). The slope of the curve is $-2(1-y_1)$, which at $y_1 = v$ is $-2(1-v) = 1-\sqrt{5}$. The slope of the normal is the negative of the reciprocal of this, namely $1/(\sqrt{5}-1)$. So $p_2/p_1 = 1/(\sqrt{5}-1)$, and since $p_1 + p_2 = 1$, we find $p_2(\sqrt{5}-1) = 1 - p_2$, or $p_2 = 1/\sqrt{5}$ and $p_1 = 1-(1/\sqrt{5})$. as found in Example 1.

13.2 Continuous Games

The simplest extension of the minimax theorem to a more general case is to assume that X and Y are compact subsets of Euclidean spaces, and that $A(x, y)$ is a continuous function of x and y. To conclude that optimal strategies for the players exist, we must allow arbitrary distribution functions on X and Y. Thus if X is a compact subset of m-dimensional space $\mathbb{R}^m$, X^* is taken to be the set of all distributions on $\mathbb{R}^m$ that give probability 0 to the complement of X. Similarly if Y is n-dimensional, Y^* is taken to be the set of all distributions on $\mathbb{R}^n$ giving weight 0 to the complement of Y. Then A is extended to be defined on $X^* \times Y^*$ by

$$A(P, Q) = \int\int A(x, y)\, dP(x)\, dQ(y).$$

Theorem 13.2. *If X and Y are compact subsets of Euclidean space and if $A(x, y)$ is a continuous function of x and y, then the game has a value, v, and there exist optimal strategies for the players; that is, there is a $P_0 \in X^*$ and a $Q_0 \in Y^*$ such that*

$$A(P, Q_0) \le v \le A(P_0, Q) \quad \textit{for all} \quad P \in X^* \quad \textit{and} \quad Q \in Y^*.$$

Example 1. Consider the game (X, Y, A), where $X = Y = [0, 1]$, the unit interval, and

$$A(x, y) = \begin{cases} g(x-y) & \text{if } 0 \le y \le x \\ g(1+x-y) & \text{if } 0 \le x < y \le 1, \end{cases}$$

where g is a continuous function defined on $[0, 1]$, with $g(0) = g(1)$. Here, both X^* and Y^* are the set of probability distributions on the unit interval.

Since X and Y are compact and $A(x, y)$ is continuous on $[0, 1]^2$, we have by Theorem 13.2, that the game has a value and the players have optimal strategies. Let us check that the optimal strategies for both players is the uniform distribution on $[0, 1]$. If Player I uses a uniform on [0,1] to choose x and Player II uses the pure strategy $y \in [0,1]$, the expected payoff to Player I is

$$\int_0^1 A(x,y)\,dx = \int_0^y g(1+x-y)\,dx + \int_y^1 g(x-y)\,dx$$

$$= \int_{1-y}^1 g(u)\,du + \int_0^{1-y} g(u)\,du = \int_0^1 g(u)\,du.$$

Since this is independent of y, Player I's strategy is an equalizer strategy, guaranteeing him an average payoff of $\int_0^1 g(u)\,du$. Clearly, the same analysis gives Player II this same amount if he chooses y at random according to a uniform distribution on [0,1]. So these strategies are optimal and the value is $v = \int_0^1 g(u)\,du$. It may be noticed that this example is a continuous version of a Latin square game. In fact the same solution holds even if g in not continuous. One only needs g to be integrable on $[0,1]$. □

A One-Sided Minimax Theorem. In the way that Theorem 13.1 generalized the finite minimax theorem by allowing Y to be an arbitrary set, Theorem 13.2 may be generalized to allow Y to be arbitrary, provided we keep the compactness condition for X. The continuity condition may be weakened to assuming only that $A(x,y)$ is a continuous function of x for every $y \in Y$. And even this can be weakened to assuming that $A(x,y)$ is only upper semicontinuous in x for every $y \in Y$.

A function $f(x)$ defined on X is *upper semicontinuous at a point* $x_0 \in X$, if for any sequence $x_1, x_2, \ldots$ of points in X such that $\lim_{n\to\infty} x_n = x_0$, we have $\limsup_{n\to\infty} f(x_n) \le f(x_0)$. It is *upper semicontinuous* (*usc*) *on* X if it is upper semicontinuous at every point of X. A function $f(x)$ is *lower semicontinuous* (*lsc*) *on* X if the above inequality is changed to $\liminf_{n\to\infty} f(x_n) \ge f(x_0)$, or equivalently, if the function, $-f(x)$, is upper semicontinuous. As an example, the function

$$f(x) = \begin{cases} 0 & \text{if } x < 0 \\ a & \text{if } x = 0 \\ 1 & \text{if } x > 0 \end{cases}$$

is usc on [0,1] if $a \ge 1$ and lsc if $a \le 0$. It is neither usc nor lsc if $0 < a < 1$.

Theorem 13.3. *If X is a compact subset of Euclidean space, and if $A(x,y)$ is an upper semicontinuous function of $x \in X$ for all $y \in Y$ and if A is bounded below (or if Y^* is the set of finite mixtures), then the game has a* *value, Player I has an optimal strategy in X^*, and for every $\epsilon > 0$ Player II has an ϵ-optimal strategy giving weight to a finite number of points.*

Similarly from Player II's viewpoint, if Y is a compact subset of Euclidean space, and if $A(x, y)$ is a lower semicontinuous function of $y \in Y$ for all $x \in X$ and if A is bounded above (or if X^* is the set of finite mixtures), then the game has a value and Player II has an optimal strategy in Y^*.

Example 2. Player I chooses x in [0,1] and Player II chooses y in [0,1] as a guess to the value of x. Player I wins $2|x-y|$ if Player II's guess is off by at least 1/2; otherwise, there is no payoff. Thus, $X = Y = [0,1]$, and

$$A(x,y) = \begin{cases} 2|x-y| & \text{if } |x-y| \geq 1/2, \\ 0 & \text{if } |x-y| < 1/2. \end{cases}$$

Although the payoff function is not continuous, it is upper semicontinuous in x for every $y \in Y$. Thus the game has a value and Player I has an optimal mixed strategy.

If we change the payoff, slightly in Player II's favor, so that Player I wins $2|x-y|$ if Player II's guess is off by more than 1/2, then

$$A(x,y) = \begin{cases} 2|x-y| & \text{if } |x-y| > 1/2, \\ 0 & \text{if } |x-y| \leq 1/2. \end{cases}$$

This is no longer upper semicontinuous in x for fixed y; instead it is lower semicontinuous in y for each $x \in X$. This time, the game has a value and Player II has an optimal mixed strategy. (See Exercise 4.)

13.3 Concave Games and Convex Games

If in Theorem 13.2, we add the assumption that the payoff function $A(x, y)$ is concave in x for all y or convex in y for all x, then we can conclude that one of the players has an optimal pure strategy, which is usually easy to find. Here is a one-sided version that complements Theorem 13.3. A good reference for these ideas is the book of Karlin (1959), Vol. 2.

Theorem 13.4. *Let (X, Y, A) be a game with Y arbitrary, X a compact, convex subset of $\mathbb{R}^m$, and $A(\mathbf{x}, y)$ bounded below. If $A(\mathbf{x}, y)$ is a concave function of $\mathbf{x} \in X$ for all $y \in Y$, then the game has a value and Player I* *has an optimal pure strategy. Moreover, Player II has an ϵ-optimal strategy that is a mixture of at most $m+1$ pure strategies.*

The dual statement for convex functions is: If Y is compact and convex in $\mathbb{R}^n$, and if A is bounded above and is convex in $\mathbf{y} \in Y$ for all $x \in X$, then the game has a value, Player II has an optimal pure strategy and Player I has ϵ-optimal strategies giving weight to at most $n+1$ points.

These games may be solved by a method similar to Method 1 of Sec. 13.1. Let's see how to find the optimal strategy of Player II in the convex functio case. Let $g(\mathbf{y}) = \sup_x A(x, \mathbf{y})$ be the upper envelope. Then $g(\mathbf{y})$ is finite since A is bounded above. It is also convex since the supremum of any set of convex functions is convex. Then since convex functions defined on a compact set attain their maximum, there exists a point $\mathbf{y}^*$ at which $g(\mathbf{y})$ takes on its maximum value, so that

$$A(x, \mathbf{y}^*) \leq \max_x A(x, \mathbf{y}^*) = g(\mathbf{y}^*) \quad \text{for all } x \in X.$$

Any such point is an optimal pure strategy for Player II. By choosing y^*, Player II will lose no more than $g(\mathbf{y}^*)$ no matter what Player I does. Player I's optimal strategy is more complex to describe in general; it gives weight only to points that play a role in the upper envelope at the point $\mathbf{y}^*$. These are points x such that $A(x, \mathbf{y})$ is tangent (or nearly tangent if only ϵ-optimal strategies exist) to the surface $g(\mathbf{y})$ at $\mathbf{y}^*$. It is best to consider examples.

Example 1. Estimation. Player I chooses a point $x \in X = [0,1]$, and Player II tries to "estimate" x by choosing a point $y \in Y = [0,1]$ close to x. Player II loses the square of the distance from x to y: $A(x,y) = (x-y)^2$. This is a convex function of $y \in [0,1]$ for all $x \in X$. For any x, $A(x,y)$ is bounded above by either $A(0,y)$ or $A(1,y)$, so the upper envelope is $g(y) = \max\{A(0,y), A(1,y)\} = \max\{y^2, (1-y)^2\}$. This is minimized at $y^* = 1/2$. If Player II uses y^*, she is guaranteed to lose no more than $g(y^*) = 1/4$.

Since $x = 0$ and $x = 1$ are the only two pure strategies influencing the upper envelope, and since y^2 and $(1-y)^2$ have slopes at y^* that are equal in absolute value but opposite in sign, Player I should mix 0 and 1 with equal probability. This mixed strategy has convex payoff $(1/2)(A(0,y) + A(1,y))$ with slope zero at y^*. Player I is guaranteed winning at least $1/4$, so $v = 1/4$ is the value of the game. The pure strategy y^* is optimal for Player II and the mixed strategy, 0 with probability 1/2 and 1 with probability 1/2, is optimal for Player I. In this example, $n = 1$, and Player I's optimal strategy mixes $2 = n + 1$ points. □

Theorem 13.4 may also be stated with the roles of the players reversed. If Y is arbitrary, and if X is a compact subset of $\mathbb{R}^m$ and if $A(\mathbf{x}, y)$ is bounded below and concave in $\mathbf{x} \in X$ for all $y \in Y$, then Player I has an optimal pure strategy, and Player II has an ϵ-optimal strategy mixing at most $m + 1$ pure strategies. It may also happen that $A(x, y)$ is concave in x for all y, and convex in y for all x. In that case, both players have optimal pure strategies as in the following example.

Example 2. A Convex–Concave Game. Suppose $X = Y = [0, 1]$, and $A(x, y) = -2x^2 + 4xy + y^2 - 2x - 3y + 1$. The payoff is convex in y for all x and concave in x for all y. Therefore, both players have pure optimal strategies, say x_0 and y_0. If Player II uses y_0, then $A(x, y_0)$ must be maximized by x_0. To find $\max_{x\in[0,1]} A(x, y_0)$ we take a derivative with respect to x: $\frac{\partial}{\partial x} A(x, y_0) = -4x + 4y_0 - 2$. So

$$x_0 = \begin{cases} y_0 - (1/2) & \text{if } y_0 > 1/2, \\ 0 & \text{if } y_0 \le 1/2. \end{cases}$$

Similarly, if Player I uses x_0, then $A(x_0, y)$ is minimized by y_0. Since $\frac{\partial}{\partial y} A(x_0, y) = 4x_0 + 2y - 3$, we have

$$y_0 = \begin{cases} 1 & \text{if } x_0 \le 1/4 \\ (1/2)(3 - 4x_0) & \text{if } 1/4 \le x_0 \le 3/4 \\ 0 & \text{if } x_0 \ge 3/4. \end{cases}$$

These two equations are satisfied only if $x_0 = y_0 - (1/2)$ and $y_0 = (1/2)(3 - 4x_0)$. It is then easily found that $x_0 = 1/3$ and $y_0 = 5/6$. The value is $A(x_0, y_0) = -7/12$.

It may be easier here to find the saddle-point of the surface, $z = -2x^2 + 4xy + y^2 - 2x - 3y + 1$, and if the saddle-point is in the unit square, then that is the solution. But the method used here shows what must be done in general. □

13.4 Solving Games

There are many interesting games that are more complex and that require a good deal of thought and ingenuity to find solutions. There is one tool for solving such games that is basic. This is the infinite game analog of the principle of indifference given in Chap. 9: Search for strategies that make the opponent indifferent among all his "good" pure strategies.

To be more specific, consider the game (X, Y, A) with $X = Y = [0,1]$ and $A(x, y)$ continuous. Let v denote the value of the game and let P denote the distribution that represents the optimal strategy for Player I. Then, $A(P, y)$ must be equal to v for all "good" y, which here means for all y in the support of Q for any Q that is optimal for Player II. (A point y is in the support of Q if Q gives positive probability to every open set containing y.) So to attempt to find the optimal P, we guess at the set, S, of "good" points y for Player II and search for a distribution P such that $A(P, y)$ is constant on S. Such a strategy, P, is called an equalizer strategy on S. The first example shows what is involved in this.

Example 1. Meeting Someone at the Train Station. A young lady is due to arrive at a train station at some random time, T, distributed uniformly between noon and 1 pm. She is to wait there until one of her two suitors arrives to pick her up. Each suitor chooses a time in [0,1] to arrive. If he finds the young lady there, he departs immediately with her; otherwise, he leaves immediately, disappointed. If either suitor is successful in meeting the young lady, he receives 1 unit from the other. If they choose the same time to arrive, there is no payoff. Also, if they both arrive before the young lady arrives, the payoff is zero. (She takes a taxi at 1 pm.)

Solution. Denote the suitors by I and II, and their strategy spaces by $X = [0,1]$ and $Y = [0,1]$. Let us find the function $A(x, y)$ that represents I's expected winnings if I chooses $x \in X$ and II chooses $y \in Y$. If $x < y$, I wins 1 if $T < x$ and loses 1 if $x < T < y$. The probability of the first is x and the probability of the second is $y - x$, so $A(x, y)$ is $x - (y - x) = 2x - y$ when $x < y$. When $y < x$, a similar analysis shows $A(x, y) = x - 2y$. Thus,

$$A(x, y) = \begin{cases} 2x - y & \text{if } x < y \\ x - 2y & \text{if } x > y \\ 0 & \text{if } x = y. \end{cases} \tag{2}$$

This payoff function is not continuous, nor is it upper semicontinuous or lower semicontinuous. It is symmetric in the players so if it has a value, the value is zero and the players have the same optimal strategy.

Let us search for an equalizer strategy for Player I and assume it has a density $f(x)$ on [0,1]. We would have

$$A(f,y) = \int_0^y (2x - y)f(x)\,dx + \int_y^1 (x - 2y)f(x)\,dx$$

$$= \int_0^y (x + y)f(x)\,dx + \int_0^1 (x - 2y)f(x)\,dx = \text{ constant.} \quad (3)$$

Taking a derivative with respect to y yields the equation

$$2yf(y) + \int_0^y f(x)\,dx - 2\int_0^1 f(x)\,dx = 0 \quad (4)$$

and taking a second derivative gives

$$2f(y) + 2yf'(y) + f(y) = 0 \quad \text{or} \quad \frac{f'(y)}{f(y)} = -\frac{3}{2y}. \quad (5)$$

This differential equation has the simple solution,

$$\log f(y) = -\frac{3}{2}\log(y) + c \quad \text{or} \quad f(y) = ky^{-3/2} \quad (6)$$

for some constants c and k. Unfortunately, $\int_0^1 y^{-3/2}\,dy = \infty$, so this cannot be used as a density on [0,1].

If we think more about the problem, we can see that it cannot be good to come in very early. There is too little chance that the young lady has arrived. So perhaps the "good" points are only those from some point $a > 0$ on. That is, we should look for a density $f(x)$ on $[a, 1]$ that is an equalizer from a on. So in (3) we replace the integrals from 0 to integrals from a and assume $y > a$. The derivative with respect to y gives (4) with the integrals starting from a rather than 0. And the second derivative is (5) exactly. We have the same solution (6) but for $y > a$. This time the resulting $f(y)$ on $[a, 1]$ is a density if

$$k^{-1} = \int_a^1 x^{-3/2}\,dx = -2\int_a^1 dx^{-1/2} = \frac{2(1 - \sqrt{a})}{\sqrt{a}}. \quad (7)$$

We now need to find a. That may be done by solving Eq. (4) with the integrals starting at a.

$$2yky^{-3/2} + \int_a^y kx^{-3/2}\,dx - 2 = 2ky^{-1/2} - 2k(y^{-1/2} - a^{-1/2}) - 2$$

$$= 2ka^{-1/2} - 2 = 0.$$

So $ka^{-1/2} = 1$, which implies $1 = 2(1 - \sqrt{a})$ or $a = 1/4$, which in turn implies $k = 1/2$. The density

$$f(x) = \begin{cases} 0 & \text{if } 0 < x < 1/4 \\ (1/2)x^{-3/2} & \text{if } 1/4 < x < 1 \end{cases} \tag{8}$$

is an equalizer for $y > 1/4$ and is therefore a good candidate for the optimal strategy. We should still check at points y less than $1/4$. For $y < 1/4$, we have from (3) and (8)

$$A(f, y) = \int_{1/4}^{1} (x - 2y)(1/2)x^{-3/2}\, dx = \int_{1/4}^{1} \frac{1}{2\sqrt{x}}\, dx - 2y = \frac{1}{2} - 2y.$$

So

$$A(f, y) = \begin{cases} (1 - 4y)/2 & \text{for } y < 1/4 \\ 0 & \text{for } y > 1/4. \end{cases} \tag{9}$$

This guarantees I at least 0 no matter what II does. Since II can use the same strategy, the value of the game is 0 and (8) is an optimal strategy for both players.

Example 2. Competing Investors. Two investors compete to see which of them, starting with the same initial fortune, can end up with the larger fortune. The rules of the competition allow them to invest in any fair game. That is, they can invest nonnegative amounts in any game whose expected return per unit invested is 1.

Suppose the investors start with 1 unit of fortune each (and we assume money is infinitely divisible). Then the players can exchange their initial fortune for the random outcome of any distribution on $[0, \infty)$ that has mean 1.

Thus the players have the same pure strategy sets. They both choose a distribution on $[0, \infty)$ with mean 1, say Player I chooses F with mean 1, and Player II chooses G with mean 1. Then Z_1 is chosen from F and Z_2 is chosen from G independently, and I wins if $Z_1 > Z_2$, II wins if $Z_2 > Z_1$ and it is a tie if $Z_1 = Z_2$. What distributions should the investors choose?

The game is symmetric in the players, so the value if it exists is zero, and both players have the same optimal strategy. Here the strategy spaces are very large, much larger than in the Euclidean case. But it turns out that the solution is easy to describe. The optimal strategy for both players is the uniform distribution on the interval (0,2):

$$F(z) = \begin{cases} z/2 & \text{for } 0 \leq z \leq 2 \\ 1 & \text{for } z > 2. \end{cases}$$

This is a distribution on $[0, \infty]$ with mean 1 and so it is an element of the strategy space of both players. Suppose Player I uses F. Then the probability that I loses is

$$\mathrm{P}(Z_1 < Z_2) = \mathrm{E}[\mathrm{P}(Z_1 < Z_2|Z_2)] \leq \mathrm{E}[Z_2/2] = (1/2)\mathrm{E}[Z_2] = 1/2.$$

The inequality follows from $\mathrm{P}(Z_1 < z_2) \leq z_2/2$ for all $z_2 \geq 0$. So the probability I wins is at least $1/2$. Since the game is symmetric, Player II by using the same strategy can keep Player I's probability of winning to at most $1/2$.

13.5 Uniform [0,1] Poker Models

The study of two-person Uniform [0,1] poker models goes back to Borel (1938) and von Neumann and Morgenstern (1944). We present these two models here. In these models, the set of possible "hands" of the players is the interval, $[0, 1]$. Players I and II are dealt hands x and y respectively in $[0, 1]$ according to a uniform distribution over the interval $[0, 1]$. Throughout the play, both players know the value of their own hand, but not that of the opponent. We assume that x and y are *independent* random variables; that is, learning the value of his own hand gives a player no information about the hand of his opponent.

There follows some rounds of betting in which the players take turns acting. After the dealing of the hands, all actions that the players take are announced. Except for the dealing of the hands at the start of the game, this would be a game of perfect information. Games of this sort, where, after an initial random move giving secret information to the players, the game is played with no further random moves of nature, are called games of *almost perfect information* (see Sorin and Ponssard (1980)).

It is convenient to study the action part of games of almost complete information by what we call the *betting tree*. This is distinct from the Kuhn tree in that it neglects the information sets that may arise from the initial distribution of hands. The examples below illustrate this concept.

The Borel Model: La Relance. Both players contribute an ante of 1 unit into the pot and receive independent uniform hands on the interval $[0, 1]$. Player I acts first either by folding and thus conceding the pot to Player II, or by betting a prescribed amount $\beta > 0$ which he adds to the pot. If Player I bets, then Player II acts either by folding and thus conceding the pot to Player I, or by calling and adding β to the pot. If Player II calls the bet of Player I, the hands are compared and the player with the higher hand wins the entire pot. That is, if $x > y$ then Player I wins the pot; if $x < y$ then Player II wins the pot. We do not have to consider the case $x = y$ since this occurs with probability 0.

The betting tree is

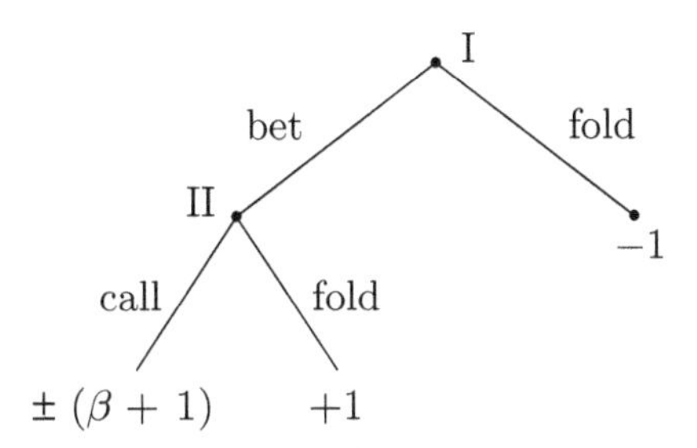

In this diagram, the plus-or-minus sign indicates that the hands are compared, and the higher hand wins the amount $\beta + 1$.

It is easy to see that the optimal strategy for Player II must be of the form for some number b in the interval [0,1]: fold if $y < b$ and call if $y > b$. The optimal value of b may be found using the principle of indifference. Player II chooses b to make I indifferent between betting and folding when I has some hand $x < b$. If I bets with such an x, he, wins 2 (the pot) if II has $y < b$ and loses β if II has $y > b$. His expected winnings are in this case, $2b - \beta(1 - b)$. On the other hand, if I folds he wins nothing. (This views the game as a constant-sum game. It views the money already put into the pot as a sunk cost, and so the sum of the payoffs of the players is 2 whatever the outcome. This is a minor point but it is the way most poker players view the pot.) He will be indifferent between betting and folding if

$$2b - \beta(1 - b) = 0$$

from which we conclude

$$b = \beta/(2 + \beta). \tag{10}$$

Player I's optimal strategy is not unique, but all of his optimal strategies are of the form: if $x > b$, bet; and if $x < b$, do anything provided the total probability that you fold is b^2. For example, I may fold with his worst hands, i.e. with $x < b^2$, or he may fold with the best of his hands less than b, i.e. with $b - b^2 < x < b$, or he may, for all $0 < x < b$, simply toss a coin with probability b of heads and fold if the coin comes up heads.

The value of the game may be computed as follows. Suppose Player I folds with any $x < b^2$ and bets otherwise and suppose Player II folds with $y < b$. Then the payoff in the unit square has the values given in the following diagram. The values in the upper right corner cancel and the rest is easy to evaluate. The value is $v(\beta) = -(\beta+1)(1-b)(b-b^2)+(1-b^2)b-b^2$, or, recalling $b = \beta/(2+\beta)$,

$$v(\beta) = -b^2 = -\frac{\beta^2}{(2+\beta)^2}. \tag{11}$$

Thus, the game is in favor of Player II.

We summarize in

Theorem 13.5. *The value of la relance is given by* (11). *An optimal strategy for Player I is to bet if* $x > b - b^2$ *and to fold otherwise, where* b *is given in* (10). *An optimal strategy for Player II is to call if* $y > b$ *and to fold otherwise.*

As an example, suppose $\beta = 2$, where the size of the bet is the size of the pot. Then $b = 1/2$. An optimal strategy for Player I is to bet if $x > 1/4$

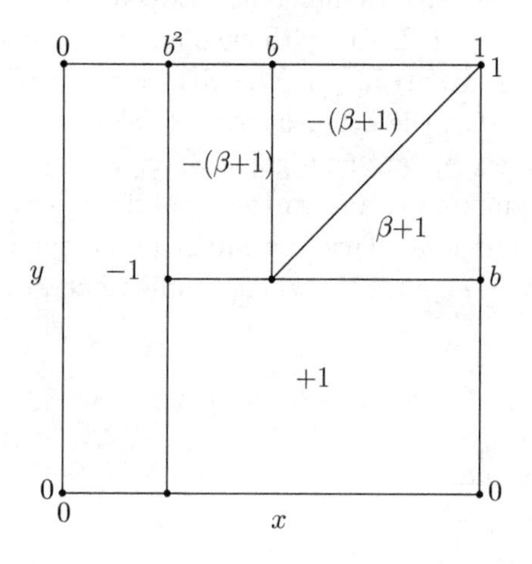

and fold otherwise; the optimal strategy of Player II is to call if $y > 1/2$. The game favors Player II, whose expected return is $1/4$ unit each time the game is played.

If I bets when $x < b$, he knows he will lose if called, assuming II is using an optimal strategy. Such a bet is called a *bluff*. In la relance, it is necessary for I to bluff with probability b^2. Which of the hands below b he chooses to bluff with is immaterial as far as the value of the game is concerned. However, there is a secondary advantage to bluffing (betting) with the hands just below b, that is, with the hands from $b - b^2$ to b. Such a strategy takes maximum advantage of a mistake the other player may make.

A given strategy σ for a player is called a *mistake* if there exists an optimal strategy for the opponent which when used against σ gives the opponent an expected payoff better than the value of the game. In la relance, it is a mistake for Player II to call with some $y < b$ or to fold with some $y > b$. If II calls with some $y < b$, then I can gain from the mistake most profitably if he bluffs only with his best hands below b.

A strategy is said to be *admissible* for a player if no other strategy for that player does better against one strategy of the opponent without doing worse against some other strategy of the opponent. The rule of betting if and only if $x > b - b^2$ is the unique admissible optimal strategy for Player I.

The von Neumann Model. The model of von Neumann differs from the model of Borel in one small but significant respect. If Player I does not bet, he does not necessarily lose the pot. Instead the hands are immediately compared and the higher hand wins the pot. We say Player I checks rather than folds. This provides a better approximation to real poker and a clearer example of the concept of "bluffing" in poker. The betting tree of von Neumann's poker is the same as Borel's except that the -1 payoff on the right branch is changed to ± 1.

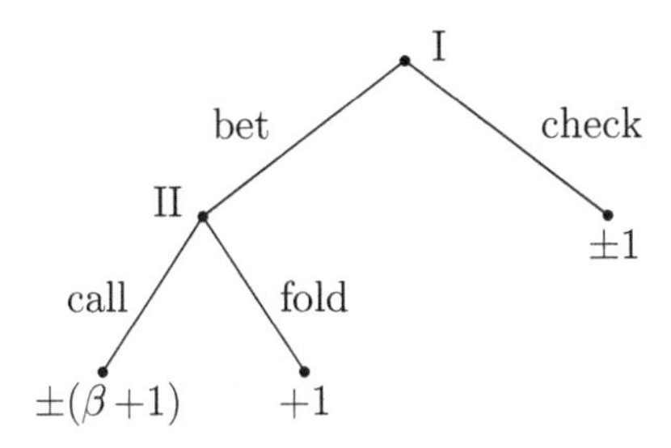

This time it is Player I that has a unique optimal strategy. It is of the form for some numbers a and b with $a < b$: bet if $x < a$ or if $x > b$, and check otherwise. Although there are many optimal strategies for Player II (and von Neumann finds all of them), one can show that there is a unique admissible one and it has the simple form: call if $y > c$ for some number c. It turns out that $0 < a < c < b < 1$.

I: | bet | check | bet |
0 a b 1

II: | fold | call |
0 c 1

The region $x < a$ is the region in which Player I bluffs. It is noteworthy that Player I must bluff with his worst hands, and not with his moderate hands. It is a mistake for Player I to do otherwise. Here is a rough explanation of this somewhat counterintuitive feature. Hands below c may be used for bluffing or checking. For bluffing it doesn't matter much which hands are used; one expects to lose them if called. For checking though it certainly matters; one is better off checking with the better hands.

Let us apply the principle of indifference to find the optimal values of a, b and c. This will lead to three equations in three unknowns, known as the *indifference equations* (not to be confused with difference equations). First, Player II should be indifferent between folding and calling with a hand $y = c$. Again we use the gambler's point of view of the game as a constant sum game, where winning what is already in the pot is considered as a bonus. If II folds, she wins zero. If she calls with $y = c$, she wins $(\beta+2)$ if $x < a$ and loses β if $x > b$. Equating her expected winnings gives the first indifference equation,

$$(\beta + 2)a - \beta(1 - b) = 0. \tag{12}$$

Second, Player I should be indifferent between checking and betting with $x = a$. If he checks with $x = a$, he wins 2 if $y < a$, and wins nothing otherwise, for an expected return of $2a$. If he bets, he wins 2 if $y < c$ and loses β if $y > c$, for an expected return of $2c - \beta(1 - c)$. Equating these gives the second indifference equation,

$$2c - \beta(1 - c) = 2a. \tag{13}$$

Third, Player I should be indifferent between checking and betting with $x = b$. If he checks, he wins 2 if $y < b$. If he bets, he wins 2 if $y < c$ and wins $\beta+2$ if $c<y<b$, and loses β if $y>b$, for an expected return of $2c+(\beta+2)(b-c)-\beta(1-b)$. This gives the third indifference equation,

$$2c+(\beta+2)(b-c)-\beta(1-b)=2b,$$

which reduces to

$$2b-c=1. \tag{14}$$

The optimal values of a, b and c can be found by solving Eqs. (12), (13) and (14) in terms of β. The solution is

$$a=\frac{\beta}{(\beta+1)(\beta+4)},\quad b=\frac{\beta^2+4\beta+2}{(\beta+1)(\beta+4)},\quad c=\frac{\beta(\beta+3)}{(\beta+1)(\beta+4)}. \tag{15}$$

The value is

$$v(\beta)=a=\beta/((\beta+1)(\beta+4)). \tag{16}$$

This game favors Player I. We summarize this in

Theorem 3.6. *The value of von Neumann's poker is given by* (16). *An optimal strategy for Player I is to check if $a<x<b$ and to bet otherwise, where a and b are given in* (15). *An optimal strategy for Player II is to call if $y>c$ and to fold otherwise, where c is given in* (15).

For pot-limit poker where $\beta=2$, we have $a=1/9$, $b=7/9$, and $c=5/9$, and the value is $v(2)=1/9$.

It is interesting to note that there is an optimal bet size for Player I. It may be found by setting the derivative of $v(\beta)$ to zero and solving the resulting equation for β. It is $\beta=2$. In other words, the optimal bet size is the size of the pot, as in pot-limit poker!

13.6 Exercises

1. Let $X=\{-1,1\}$, let $Y=\{\ldots,-2,-1,0,1,2,\ldots\}$ be the set of all integers, and let $A(x,y)=xy$.
   (a) Show that if we take $Y^*=Y^*_F$, the set of all finite distributions on Y, then the value exists, is equal to zero and both players have optimal strategies.
   (b) Show that if Y^* is taken to be the set of all distributions on Y, then we can't speak of the value, because Player II has a strategy, $\mathbf{q}$, for which the expected payoff, $A(x,\mathbf{q})$ doesn't exist for any $x\in X$.

2. Simultaneously, Player I chooses $x \in \{x_1, x_2\}$, and Player II chooses $y \in [0,1]$; then I receives

$$A(x,y) = \begin{cases} y & \text{if } x = x_1 \\ e^{-y} & \text{if } x = x_2 \end{cases}$$

from II. Find the value and optimal strategies for the players.
3. Player II chooses a point (y_1, y_2) in the ellipse $(y_1-3)^2+4(y_2-2)^2 \leq 4$. Simultaneously, Player I chooses a coordinate $k \in \{1,2\}$ and receives y_k from Player II. Find the value and optimal strategies for the players.
4. Solve the two games in Example 2 of Sec. 13.2. Hint: Use domination to remove some pure strategies.
5. Consider the game with $X = [0,1]$, $Y = [0,1]$, and

$$A(x,y) = \begin{cases} 0 & \text{if } x = y \\ -1 & \text{if } x = 0 \text{ and } y > 0 \\ +1 & \text{if } y = 0 \text{ and } x > 0 \\ -1 & \text{if } 0 < y < x \\ +1 & \text{if } 0 < x < y \end{cases}$$

Note that $A(x,y)$ is not usc in x for all y nor lsc in y for all x. Show the game does not have a value.
6. **The Greedy Game.** Each player can demand from the other as much as desired between zero and one, but there is a penalty for being too greedy. The player who demands more than his opponent must pay a fine of b to the other, where b is a fixed number, $0 \leq b \leq 1/2$. Thus we have the game (X, Y, A) where $X = Y = [0,1]$, and

$$A(x,y) = x - y + \begin{cases} +b & \text{if } x < y \\ 0 & \text{if } x = y \\ -b & \text{if } x > y. \end{cases}$$

Solve.

7. Find optimal strategies and the value of the following games.

   (a) $X = Y = [0,1]$ and $A(x,y) = \begin{cases} (x-y)^2 & \text{if } x \le y \\ 2(x-y)^2 & \text{if } x \ge y. \end{cases}$ (Underestimation is the more serious error of Player II.)

   (b) $X = Y = [0,1]$ and $A(x,y) = xe^{-y} + (1-x)y$.

8. **Hide and Seek in a Compact, Convex Set.**

   (a) Let S be the triangle in the plane with vertices $(-1,0)$, $(1,0)$, and $(0,2)$. Player I chooses a point $\mathbf{x}$ in S in which to hide, and Player II chooses a point $\mathbf{y}$ in S to seek. The payoff to Player I is the square of Euclidean distance between $\mathbf{x}$ and $\mathbf{y}$. Thus, $X = S$, $Y = S$, and $A(\mathbf{x},\mathbf{y}) = \|\mathbf{x} - \mathbf{y}\|^2$. Solve.

   (b) See if you can formulate a procedure for solving the above game of Hide and Seek if S is an arbitrary compact, convex set in $\mathbb{R}^n$.

9. **The Wallet Game.** Two players each put a random amount with mean one into their wallets. The player whose wallet contains the smaller amount wins the larger amount from the opponent.

   Carroll, Jones and Rykken (2001) show that this game does not have a value. But suppose we restrict the players to putting at most some amount b in their wallets. Here is the game:

   Player I, resp. Player II, chooses a distribution F, resp. G, on the interval $[0,b]$ with mean 1, where $b > 1$. Then independent random variables, X from F and Y from G, are chosen. If $X < Y$, Player I wins Y from Player II. If $X > Y$, Player II wins X from Player I, and if $X = Y$, there is no payoff. So the payoff function is

$$\begin{aligned} A(F,G) &= \mathrm{E}(Y\mathrm{I}(X < Y) - X\mathrm{I}(X > Y)) \\ &= \mathrm{E}((Y + X)\mathrm{I}(X < Y)) - 1 + \mathrm{E}(X\mathrm{I}(X = Y)). \end{aligned} \tag{17}$$

   The game is symmetric, so if the value exists, the value is zero, and the players have the same optimal strategies. Find an optimal strategy for the players. Hint: Search among distributions F having a density f on the interval (a,b) for some $a < 1$. Note that the last term on the right of Eq. (17) disappears for such distributions.

10. **The Multiplication Game.** (See Morrison (2010)) Players I and II simutaneously select positive numbers x and y. Player I wins $+1$ if the product xy, written in decimal form has initial significant digit 1, 2 or 3. Thus, the pure strategy spaces are $X = Y = (0,\infty)$ and the

payoff function is

$$A(x, y) = \begin{cases} +1 & \text{if the initial significant digit is 1, 2 or 3} \\ 0 & \text{otherwise.} \end{cases}$$

Solve.

Hint: (1) First note that both players may restrict their pure strategy sets to $X = Y = [1, 10)$ so that

$$A(x, y) = I\{1 \le xy < 4 \text{ or } 10 \le xy < 40\}.$$

(2) Take logs to the base 10. Let $u = \log_{10}(x)$ and $v = \log_{10}(y)$. Now, players I and II choose u and v in [0,1) with payoff

$$B(u, v) = I\{0 \le u + v < c \text{ or } 1 \le u + v < 1 + c\}$$

where $c = \log_{10}(4) = 0.60206\ldots$. Solve the game in this form and translate back to the original game.

11. Suppose, in la relance, that when Player I checks, Player II is given a choice between checking in which case there is no payoff, and calling in which case the hands are compared and the higher hand wins the antes.

    (a) Draw the betting tree.
    (b) Assume optimal strategies of the following form. I checks if and only if $0 < a < x < b < 1$ for some a and b. If I bets, then II calls if $y > c$, and if Player I checks, Player II calls if $y > d$, where $a \le c \le b$ and $a \le d \le b$. Find the indifference equations.
    (c) Solve the equations when $\beta = 2$, and find the value in this case. Which player has the advantage?

12. **Last Round Betting.** Here is a game that occurs in the last round of blackjack or baccarat tournaments, and also in the television game show, Final Jeopardy. For the general game, see Ferguson and Melolidakis (1997).

    In the last round of betting in a contest to see who can end up with the most money, Player I starts with \$70 and Player II starts with \$100. Simultaneously, Player I must choose an amount to bet between \$0 and \$70 and Player II must choose an amount between \$0 and \$100. Then the players independently play games with probability 0.6 of winning the bet and 0.4 of losing it. The player who has the most money at the end wins a big prize. If they end up with the same amount of money, they share the prize.

We may set this up as a game (X, Y, A), with $X = [0, 0.7]$, $Y = [0, 1.0]$, measured in units of \$100, and assuming money is infinitely divisible. We assume the payoff, $A(x, y)$, is the probability that Player I wins the game plus one-half the probabiity of a tie, when I bets x and II bets y. The probability that both players win their bets is $0.6 * 0.6 = 0.36$, the probability that both players lose their bets is $0.4 * 0.4 = 0.16$, and the probability that I wins his bet and II loses her bet is $0.6 * 0.4 = 0.24$. Therefore,

$$\begin{aligned} \mathrm{P}(\text{I wins}) &= 0.36\, I(0.7 + x > 1 + y) + 0.24\, I(0.7 + x > 1 - y) \\ &\quad + 0.16\, I(0.7 - x > 1 - y) = 0.36\, I(x - y > 0.3) \\ &\quad + 0.24\, I(x + y > 0.3) + 0.16\, I(y - x > 0.3) \\ \mathrm{P}(\text{a tie}) &= 0.36\, I(0.7 + x = 1 + y) + 0.24\, I(0.7 + x = 1 - y) \\ &\quad + 0.16\, I(0.7 - x = 1 - y) = 0.36\, I(x - y = 0.3) \\ &\quad + 0.24\, I(x + y = 0.3) + 0.16\, I(y - x = 0.3), \end{aligned}$$

where $I(\cdot)$ represents the indicator function. This gives

$$A(x, y) = \mathrm{P}(\text{I wins}) + \frac{1}{2}\mathrm{P}(\text{a tie}) = \begin{cases} 0.60 & \text{if } y < x - 0.3 \\ 0.40 & \text{if } y > x + 0.3 \\ 0.00 & \text{if } y + x < 0.3 \\ 0.24 & \text{if } y > x - 0.3, \\ & \qquad y < x + 0.3,\ y + x > 0.3 \\ 0.42 & \text{if } 0 < y = x - 0.3 \\ 0.32 & \text{if } 0 < x = y - 0.3 \\ 0.12 & \text{if } x + y = 0.3,\ x > 0,\ y > 0 \\ 0.30 & \text{if } x = 0.3,\ y = 0 \\ 0.20 & \text{if } x = 0,\ y = 0.3 \end{cases}$$

Find the value of the game and optimal strategies for both players. (Hint: Both players have an optimal strategy that give probability to only two points.)

Part III

Two-Person General-Sum Games

14 Bimatrix Games — Safety Levels

The simplest case to consider beyond two-person zero-sum games are the two-person non-zero-sum games. Examples abound in Economics: the struggle between labor and management, the competition between two producers of a single good, the negotiations between buyer and seller, and so on. Good reference material may be found in books of Owen and Straffin already cited. The material treated in Part III is much more oriented to economic theory. For a couple of good references with emphasis on applications in economics, consult the books, Gibbons (1992) and Bierman and Fernandez (1993).

14.1 General-Sum Strategic Form Games

Two-person general-sum games may be defined in extensive form or in strategic form. The *normal* or *strategic* form of a two-person game is given by two sets X and Y of pure strategies of the players, and two real-valued functions $u_1(x,y)$ and $u_2(x,y)$ defined on $X \times Y$, representing the payoffs to the two players. If I chooses $x \in X$ and II chooses $y \in Y$, then I receives $u_1(x,y)$ and II receives $u_2(x,y)$.

A finite two-person game in strategic form can be represented as a matrix of ordered pairs, sometimes called a bimatrix. The first component of the pair represents Player I's payoff and the second component represents Player II's payoff. The matrix has as many rows as Player I has pure strategies and as many columns as Player II has pure strategies.

For example, the bimatrix

$$\begin{pmatrix} (1,4) & (2,0) & (-1,1) & (0,0) \\ (3,1) & (5,3) & (3,-2) & (4,4) \\ (0,5) & (-2,3) & (4,1) & (2,2) \end{pmatrix} \tag{1}$$

represents the game in which Player I has three pure strategies, the rows, and Player II has four pure strategies, the columns. If Player I chooses row 3 and Player II column 2, then I receives -2 (i.e. he loses 2) and Player II receives 3.

An alternative way of describing a finite two-person game is as a pair of matrices. If m and n representing the number of pure strategies of the two players, the game may be represented by two $m \times n$ matrices $\mathbf{A}$ and $\mathbf{B}$. The interpretation here is that if Player I chooses row i and Player II chooses column j, then I wins a_{ij} and II wins b_{ij}, where a_{ij} and b_{ij} are the elements in the ith row, jth column of $\mathbf{A}$ and $\mathbf{B}$ respectively. Note that $\mathbf{B}$ represents the winnings of Player II rather than her losses as would be the case for a zero-sum game. The game of bimatrix (1) is represented as $(\mathbf{A}, \mathbf{B})$, where

$$\mathbf{A} = \begin{pmatrix} 1 & 2 & -1 & 0 \\ 3 & 5 & 3 & 4 \\ 0 & -2 & 4 & 2 \end{pmatrix} \quad \text{and} \quad \mathbf{B} = \begin{pmatrix} 4 & 0 & 1 & 0 \\ 1 & 3 & -2 & 4 \\ 5 & 3 & 1 & 2 \end{pmatrix}. \tag{2}$$

Note that the game is zero-sum if and only if the matrix $\mathbf{B}$ is the negative of the matrix $\mathbf{A}$, i.e. $\mathbf{B} = -\mathbf{A}$.

14.2 General-Sum Extensive Form Games

The extensive form of a game may be defined in the same manner as it was defined in Part II. The only difference is that since now the game is not zero-sum, the payoff cannot be expressed as a single number. We must indicate at each terminal vertex of the Kuhn tree how much is won by Player I and how much is won by Player II. We do this by writing this payoff as an ordered pair of real numbers whose first component indicates the amount that is paid to Player I, and whose second component indicates the amount paid to player II. The following Kuhn tree is an example.

If the first move by chance happens to go down to the right, if Player II chooses a, and if Player I happens to choose c, the payoff is $(2, -1)$, which means that Player I wins 2 and Player II loses 1. Note that the second component represents Player II's winnings rather than losses. In particular,

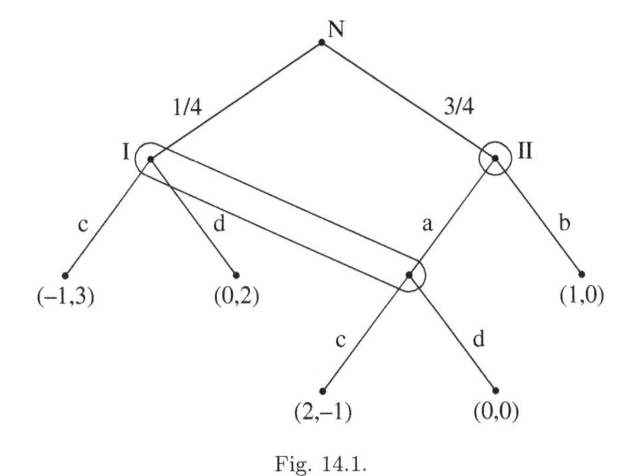

Fig. 14.1.

a game is a zero-sum game if and only if the components of each payoff vector sum to zero.

14.3 Reducing Extensive Form to Strategic Form

The problem of reducing a general sum game in extensive form to one in strategic form is solved in a completely similar manner as for the case of a zero-sum game. The only difference is that the payoffs are ordered pairs. If there are random moves, the outcome is a random distribution over these ordered pairs which is replaced by the average of the ordered pairs. This is done by taking the corresponding average over each component of the pair separately.

As an illustration, consider the game of Fig. 14.1. Player I has two pure strategies, $X = \{c, d\}$, and Player II has two pure strategies, $Y = \{a, b\}$. The corresponding strategic form of this game is given by the 2×2 bimatrix,

$$\begin{array}{c} \\ c \\ d \end{array} \begin{array}{c} \quad a \qquad\qquad b \\ \begin{pmatrix} (5/4, 0) & (2/4, 3/4) \\ (0, 2/4) & (3/4, 2/4) \end{pmatrix}. \end{array} \tag{3}$$

For example, the component of the first row, first column is computed as follows. One-fourth of the time nature goes left, and I uses c, resulting in a payoff of $(-1, 3)$. Three-fourths of the time nature goes right, Player II uses a and Player I uses c, giving a payoff of $(2, -1)$. Therefore, the average payoff is $(1/4)(-1,3)+(3/4)(2,-1)=(-1/4+6/4,3/4-3/4)=(5/4,0)$. The other components of the bimatrix are computed similarly.

14.4 Overview

The analysis of two-person games is necessarily more complex for general-sum games than for zero-sum games. When the sum of the payoffs is no longer zero (or constant), maximizing one's own payoff is no longer equivalent to minimizing the opponent's payoff. The minimax theorem does not apply to bimatrix games. One can no longer expect to play "optimally" by simply looking at one's own payoff matrix and guarding against the worst case. Clearly, one must take into account the opponent's matrix and the reasonable strategy options of the opponent. In doing so, we must remember that the opponent is doing the same. The general-sum case requires other more subtle concepts of solution.

The theory is generally divided into two branches, the *noncooperative theory* and the *cooperative theory*. In the noncooperative theory, either the players are unable to communicate before decisions are made, or if such communication is allowed, the players are forbidden or are otherwise unable to make a binding agreement on a joint choice of strategy. The main noncooperative solution concept is the *strategic equilibrium*. This theory is treated in the next two chapters. In the cooperative theory, it is assumed that the players are allowed to communicate before the decisions are made. They may make threats and counterthreats, proposals and counterproposals, and hopefully come to some compromise. They may jointly agree to use certain strategies, and it is assumed that such an agreement can be made binding.

The cooperative theory itself breaks down into two branches, depending on whether or not the players have comparable units of utility and are allowed to make monetary *side payments* in units of utility as an incentive to induce certain strategy choices. The corresponding solution concept is called the *TU cooperative value* if side payments are allowed, and the *NTU cooperative value* if side payments are forbidden or otherwise unattainable. The initials TU and NTU stand for "transferable utility" and "non-transferable utility" respectively.

14.5 Safety Levels

One concept from zero-sum games carries over and finds important use in general sum games. This is the safety level, or the amount that each player can guarantee winning on the average. In a bimatrix game with $m \times n$ matrices $\mathbf{A}$ and $\mathbf{B}$, Player I can guarantee winning on the average at least

$$v_{\mathrm{I}} = \max_{\mathbf{p}} \min_{j} \sum_{i=1}^{m} p_i a_{ij} = \mathrm{Val}(\mathbf{A}). \tag{4}$$

This is called the *safety level of Player I.* (This is by definition the lower value of $\mathbf{A}$, which by the minimax theorem is also the upper value or the value of $\mathbf{A}$. So we may write $v_{\mathrm{I}} = \mathrm{Val}(\mathbf{A})$.) Player I can achieve this payoff without considering the payoff matrix of Player II. A strategy, $\mathbf{p}$, that achieves the maximum in (4) is called a *maxmin strategy for Player I.*

Similarly, the *safety level of Player II* is

$$v_{\mathrm{II}} = \max_{\mathbf{q}} \min_{i} \sum_{j=1}^{n} b_{ij} q_j = \mathrm{Val}(\mathbf{B}^T), \tag{5}$$

since Player II can guarantee winning this amount on the average. Any strategy $\mathbf{q}$, that achieves the maximum in (5) is a *maxmin strategy for Player II.* (Note as a technical point that v_{II} is the value of $\mathbf{B}^T$, the transpose of $\mathbf{B}$. It is not the value of $\mathbf{B}$. This is because $\mathrm{Val}(\mathbf{B})$ is defined as the value of a game where the components represent winnings of the row chooser and losses of the column chooser.) An example should clarify this.

Consider the example of the following game.

$$\begin{pmatrix} (2,0) & (1,3) \\ (0,1) & (3,2) \end{pmatrix} \quad \text{or} \quad \mathbf{A} = \begin{pmatrix} 2 & 1 \\ 0 & 3 \end{pmatrix} \quad \text{and} \quad \mathbf{B} = \begin{pmatrix} 0 & 3 \\ 1 & 2 \end{pmatrix}.$$

From the matrix $\mathbf{A}$, we see that Player I's maxmin strategy is $(3/4, 1/4)$ and his safety level is $v_{\mathrm{I}} = 3/2$. From the matrix $\mathbf{B}$, we see the second column dominates the first. (Again these are II's winnings; she is trying to maximize). Player II guarantees winning at least $v_{\mathrm{II}} = 2$ by using her maxmin strategy, namely column 2. Note that this is the value of $\mathbf{B}^T$ (whereas $\mathrm{Val}(\mathbf{B}) = 1$).

Note that if both players use their maxmin strategies, then Player I only gets v_{I}, whereas Player II gets $(3/4)3 + (1/4)2 = 11/4$. This is pleasant for Player II. But if Player I looks at $\mathbf{B}$, he can see that II is very likely to choose column 2 because it strictly dominates column 1. Then Player I would get 3 which is greater than v_{I}, and Player II would get $v_{\mathrm{II}} = 2$.

The payoff (3, 2) from the second row, second column, is rather stable. If each believes the other is going to choose the second strategy, then each would choose the second strategy. This is one of the main viewpoints of noncooperative game theory, where such a strategy pair is called a strategic equilibrium.

In TU cooperative game theory, where the units used to measure I's payoff are assumed to be the same as the units used to measure Player II's payoff, the players will jointly agree on (3, 2), because it gives the largest sum, namely 5. However, in the agreement the players must also specify how the 5 is to be divided between the two players. The game is not symmetric; Player II has a threat to use column 1 and Player I has no similar threat. We will see later some of the suggestions on how to split the 5 between the players.

The NTU theory is more complex since it is assumed that the players measure their payoffs in noncomparable units. Side payments are not feasible or allowed. Any deviation from the equilibrium (3, 2) would have to be an agreed upon mixture of the other three payoffs. (The only one that makes sense to mix with (3, 2) is the payoff (1, 3).)

14.6 Exercises

1. Convert the following extensive form game to strategic form.

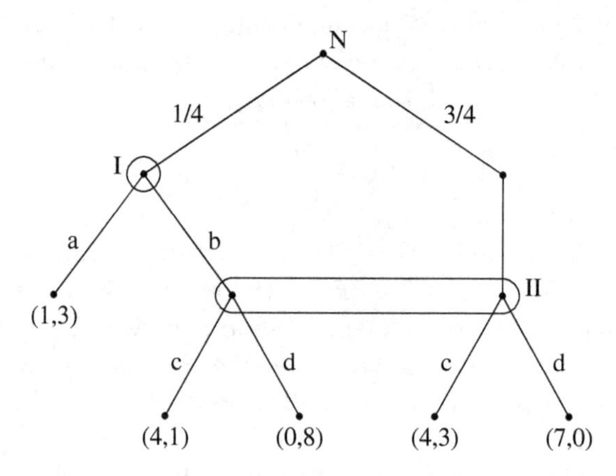

2. Find the safety levels and maxmin strategies for the players in the bimatrix games,

   (a) $\begin{pmatrix} (1,1) & (5,0) \\ (0,5) & (4,4) \end{pmatrix}$.

   (b) $\begin{pmatrix} (3,10) & (1,5) \\ (2,0) & (4,20) \end{pmatrix}$.

3. Contestants I and II in a game show start the last round with winnings of \$400 and \$500 dollars respectively. Each must decide to pass or gamble, not knowing the choice of the other. A player who passes keeps the money he/she started with. If Player I gambles, he wins \$200 with probability 1/2 or loses his entire \$400 with probability 1/2. If Player II gambles, she wins or loses \$200 with probability 1/2 each. These outcomes are independent. Then the contestant with the higher amount at the end wins a bonus of \$400.
   (a) Draw the Kuhn tree.
   (b) Put into strategic form.
   (c) Find the safety levels.
4. **A Coordination Game.** The following game is a coordination game. The safety levels and maxmin strategies for the players indicate that the first row, first column would be chosen giving both players 4. Yet if they could coordinate on the second row, second column, they would receive 6 each.

$$\begin{pmatrix} (4,4) & (4,0) \\ (0,4) & (6,6) \end{pmatrix}$$

   Suppose you, as row chooser, are playing this game once against a person chosen at random from this class. Which row would you choose? or, if you prefer, which mixed strategy would you use? Your score on this question depends on what the other students in the class do. You must try to predict what they are going to do. Do not reveal your answer to this question to the other students in the class.

15 Noncooperative Games

Two-person general-sum games and n-person games for $n > 2$ are more difficult to analyze and interpret than the zero-sum two-person games of Part II. The notion of "optimal" behavior does not extend to these more complex situations. In the noncooperative theory, it is assumed that the players cannot overtly cooperate to attain higher payoffs. If communication is allowed, no binding agreements may be formed. One possible substitute for the notion of a "solution" of a game is found in the notion of a strategic equilibrium.

15.1 Strategic Equilibria

A finite n-person game in strategic form is given by n nonempty finite sets, X_1, $X_2, \ldots, X_n$, and n real-valued functions u_1, $u_2, \ldots, u_n$, defined on $X_1 \times X_2 \times \cdots \times X_n$. The set X_i represents the pure strategy set of player i and $u_i(x_1, x_2, \ldots, x_n)$ represents the payoff to player i when the pure strategy choices of the players are x_1, $x_2, \ldots, x_n$, with $x_j \in X_j$ for $j = 1, 2, \ldots, n$.

Definition. A vector of pure strategy choices $(x_1, x_2, \ldots, x_n)$ with $x_i \in X_i$ for $i = 1, \ldots, n$ is said to be a pure strategic equilibrium, or PSE for short, if for all $i = 1, 2, \ldots, n$, and for all $x \in X_i$,

$$u_i(x_1, \ldots, x_{i-1}, x_i, x_{i+1}, \ldots, x_n) \geq u_i(x_1, \ldots, x_{i-1}, x, x_{i+1}, \ldots, x_n). \quad (1)$$

Equation (1) says that if the players other than player i use their indicated strategies, then the best player i can do is to use x_i. Such a pure strategy choice of player i is called a *best response* to the strategy choices of the other players. The notion of strategic equilibrium may be stated: a particular selection of strategy choices of the players forms a PSE if each player is using a best response to the strategy choices of the other players.

Consider the following examples with two players,

$$(a) \begin{pmatrix} (3,3) & (0,0) \\ (0,0) & (5,5) \end{pmatrix} \quad (b) \begin{pmatrix} (3,3) & (4,3) \\ (3,4) & (5,5) \end{pmatrix}.$$

In (a), the first row, first column, denoted $\langle 1,1 \rangle$, is a strategic equilibrium with equilibrium payoff $(3,3)$. If each believes the other is going to choose the first strategy, neither player will want to change to the second strategy. The second row, second column, $\langle 2,2 \rangle$, is also a strategic equilibrium. Since its equilibrium payoff is $(5,5)$, both players prefer this equilibrium. In (b), the first row, first column, $\langle 1,1 \rangle$, is still an equilibrium according to the definition. Neither player can gain by changing strategy. On the other hand, neither player can be hurt by changing, and if they both change, they both will be better off. So the equilibrium $\langle 1,1 \rangle$ is rather unstable.

Example (a) is of a game in which the players receive the same payoff, but are not allowed to communicate. If they were allowed to communicate, they would choose the joint action giving the maximum payoff. Other examples of this nature occur in the class of rendezvous games, in which two players randomly placed on a graph, each not knowing the position of the other, want to meet in minimum time. See the book of Alpern and Gal (2003).

If players in a noncooperative game are allowed to communicate and do reach some informal agreement, it may expect to be a strategic equilibrium. Since no binding agreements may be made, the only agreements that may be expected to occur are those that are *self-enforcing*, in which no player can gain by unilaterally violating the agreement. Each player is maximizing his return against the strategy the other player announced he will use.

It is useful to extend this definition to allow the players to use mixed strategies. We denote the set of probabilities over k points by $\mathbb{P}_k$:

$$\mathbb{P}_k = \left\{ \mathbf{p} = (p_1, \ldots, p_k) : \ p_i \geq 0 \quad \text{for} \quad i = 1, \ldots, k, \quad \text{and} \quad \sum_1^k p_i = 1 \right\}. \tag{2}$$

Let m_i denote the number of pure strategy choices of player i, so that the set X_i has m_i elements. Then the set of mixed strategies of player i is just $\mathbb{P}_{m_i}$. It is denoted by X_i^* where $X_i^* = \mathbb{P}_{m_i}$.

We denote the set of elements of X_i by the first m_i integers, $X_i = \{1, 2, \ldots, m_i\}$. Suppose that for $i = 1, 2, \ldots, n$, Player i uses $\mathbf{p}_i = (p_{i1}, p_{i2}, \ldots, p_{im_i}) \in X_i^*$. Then the average payoff to player j is

$$g_j(\mathbf{p}_1, \ldots, \mathbf{p}_n) = \sum_{i_1=1}^{m_1} \cdots \sum_{i_n=1}^{m_n} p_{1i_1} \cdots p_{ni_n} u_j(i_1, \ldots, i_n). \tag{3}$$

Then the analogous definition of equilibrium using mixed strategies is as follows.

Definition. A vector of mixed strategy choices $(\mathbf{p}_1, \mathbf{p}_2, \ldots, \mathbf{p}_n)$ with $\mathbf{p}_i \in X_i^*$ for $i = 1, \ldots, n$ is said to be a strategic equilibrium, or SE for short, if for all $i = 1, 2, \ldots, n$, and for all $\mathbf{p} \in X_i^*$,

$$g_i(\mathbf{p}_1, \ldots, \mathbf{p}_{i-1}, \mathbf{p}_i, \mathbf{p}_{i+1}, \ldots, \mathbf{p}_n) \geq g_i(\mathbf{p}_1, \ldots, \mathbf{p}_{i-1}, \mathbf{p}, \mathbf{p}_{i+1}, \ldots, \mathbf{p}_n). \tag{4}$$

Any mixed strategy $\mathbf{p}_i$ that satisfies (4) for all $\mathbf{p} \in X_i^*$ is a *best response* of player i to the mixed strategies of the other players. Thus, a particular selection of mixed strategy choices of the players forms an SE if and only if each player is using a best response to the strategy choices of the other players. No player can gain by unilaterally changing strategy. Note that a PSE is a special case of an SE.

This notion of best response represents a practical way of playing a game: *Make a guess at the probabilities that you think your opponents will play their various pure strategies, and choose a best response to this.* This is an example of the famous *Bayesian* approach to decision making. Of course in a game, this may be a dangerous procedure. Your opponents may be better at this type of guessing than you.

The first question that arises is “Do there always exist strategic equilibria?”. This question was resolved in 1951 by John Nash in the following theorem which generalizes von Neumann's minimax theorem. In honor of this achievement, strategic equilibria are also called *Nash equilibria.*

Theorem. *Every finite n-person game in strategic form has at least one strategic equilibrium.*

A proof of this theorem using the Brouwer Fixed Point Theorem is given in Appendix 4. This proof is an existence proof and gives no indication of how to go about finding equilibria. However, in the case of bimatrix games where $n = 2$, the Lemke–Howson algorithm (*SIAM*, 1964, **12**, 413–423) may be used to compute strategic equilibria in a finite number of steps using a simplex-like pivoting algorithm (see Parthasarathy and Raghavan (1971) for example). An interesting consequence of this method is that, under a nondegeneracy condition, the number of SE's is finite and odd!

One of the difficulties of the noncooperative theory is that there are usually many equilibria giving different payoff vectors as we shall see in the following examples. Another difficulty is that even if there is a unique strategic equilibrium, it may not be considered as a reasonable solution or a predicted outcome. In the rest of this chapter we restrict attention to $n = 2$, the two-person case.

15.2 Examples

Example 1. A Coordination Game. Consider the game with bimatrix

$$\begin{pmatrix} (3,3) & (0,2) \\ (2,1) & (5,5) \end{pmatrix}$$

and corresponding payoff matrices

$$\mathbf{A} = \begin{pmatrix} 3 & 0 \\ 2 & 5 \end{pmatrix} \quad \text{and} \quad \mathbf{B} = \begin{pmatrix} 3 & 2 \\ 1 & 5 \end{pmatrix}.$$

The corresponding maxmin (MM) strategies are (1/2, 1/2) for Player I and (3/5, 2/5) for Player II. The safety levels are $(v_I, v_{II}) = (5/2, 13/5)$.

Here there are two obvious pure strategic equilibria (PSE's) corresponding to the payoffs (3, 3) and (5, 5). Both players prefer the second SE because it gives them both 5 instead of 3. If they could coordinate their actions, this outcome would be expected. However, if they cannot communicate and if both players believe the other is going to choose the first strategy, then they are both going to choose the first strategy and receive the payoff 3. One cannot say the outcome (3, 3) is irrational. If that's the way things have always been, then one who tries to change things hurts oneself. This phenomenon occurs often, usually with many players. To try to change the structure of a language or the typewriter keyboard or the system of measurement requires a lot of people to change simultaneously before any advantage is realized.

There is a third less obvious equilibrium point that sometimes occurs in these games. If each player has an equalizing strategy for the other player's matrix, then that pair of strategies forms an equilibrium. This is because if an opponent uses a strategy that makes it not matter what you do, then anything you do is a best response, in particular the equalizing strategy on the opponent's matrix. (Recall that an equalizing strategy is one that gives the same average payoff to the opponent no matter what the opponent does.)

Let us find this *equalizing strategic equilibrium* for the above game. Note that each player uses the matrix of his opponent. Player I has the equalizing strategy $\mathbf{p} = (4/5, 1/5)$ for $\mathbf{B}$, and Player II has the equalizing strategy $\mathbf{q} = (5/6, 1/6)$ for $\mathbf{A}$. If the players use these strategies, the average payoff is $(5/2, 13/5)$, the same as the safety levels.

Is it possible that the average payoff from a strategic equilibrium is less than the safety level for one of the players? The answer is no. (See Exercise 1.) Therefore the strategic equilibrium $(\mathbf{p}, \mathbf{q})$ is as poor a strategic equilibrium as you can get. Moreover, it is extremely unstable. It is true that it does neither player any good to deviate from his/her equilibrium strategy, but on the other hand it does not harm a player to change to another strategy.

In the above example, the payoffs for the three SE's are all different. The players have the same preferences as to which of the three outcomes they would prefer. In the next example, the players have different preferences between the two pure strategic equilibria.

Example 2. The Battle of the Sexes. Suppose the matrices are

$$\begin{array}{c} \\ a \\ b \end{array}\begin{array}{c} \begin{array}{cc} a & b \end{array} \\ \begin{pmatrix} (2,1) & (0,0) \\ (0,0) & (1,2) \end{pmatrix} \end{array} \quad \text{so that} \quad \mathbf{A} = \begin{array}{c} \\ a \\ b \end{array}\begin{array}{c} \begin{array}{cc} a & b \end{array} \\ \begin{pmatrix} 2 & 0 \\ 0 & 1 \end{pmatrix} \end{array} \quad \text{and} \quad \mathbf{B} = \begin{array}{c} \\ a \\ b \end{array}\begin{array}{c} \begin{array}{cc} a & b \end{array} \\ \begin{pmatrix} 1 & 0 \\ 0 & 2 \end{pmatrix} \end{array}.$$

The name of this game arises as a description of the game played between a husband and wife in choosing which movie to see, a or b. They prefer different movies, but going together is preferable to going alone. Perhaps this should be analyzed as a cooperative game, but we analyze it here as a noncooperative game.

The pure strategy vectors (a, a) and (b, b) are both PSE's but Player I prefers the first and Player II the second.

First note that the safety levels are $v_{\text{I}} = v_{\text{II}} = 2/3$, the same for both players. Player I's MM strategy is (1/3, 2/3), while Player II's MM strategy is (2/3, 1/3). There is a third strategic equilibrium given by the equalizing strategies $\mathbf{p} = (2/3, 1/3)$ and $\mathbf{q} = (1/3, 2/3)$. The equilibrium payoff for this equilibrium point, $(v_{\text{I}}, v_{\text{II}}) = (2/3, 2/3)$, is worse for both players than either of the other two equilibrium points.

Example 3. The Prisoner's Dilemma. It may happen that there is a unique SE but that there are other outcomes that are better for both players. Consider the game with bimatrix

$$\begin{array}{c} \\ \text{cooperate} \\ \text{defect} \end{array} \begin{array}{c} \begin{array}{cc} \text{cooperate} & \text{defect} \end{array} \\ \begin{pmatrix} (3,3) & (0,4) \\ (4,0) & (1,1) \end{pmatrix} \end{array}$$

In this game, Player I can see that no matter which column Player II chooses, he will be better off if he chooses row 2. For if Player I chooses row 2 rather than row 1, he wins 4 rather than 3 if Player II chooses column 1, and he wins 1 rather than 0 if she chooses column 2. In other words, Player I's second strategy of choosing the second row strictly dominates the strategy of choosing the first. On the other hand, the game is symmetric. Player II's second column strictly dominates her first. However, if both players use their dominant strategies, each player receives 1, whereas if both players use their dominated strategies, each player receives 3.

A game that has this feature, that both players are better off if together they use strictly dominated strategies, is called the Prisoner's Dilemma. The story that leads to this bimatrix and gives the game its name is as follows. Two well-known crooks are captured and separated into different rooms. The district attorney knows he does not have enough evidence to convict on the serious charge of his choice, but offers each prisoner a deal. If just one of them will turn state's evidence (i.e. rat on his confederate), then the one who confesses will be set free, and the other sent to jail for the maximum sentence. If both confess, they are both sent to jail for the minimum sentence. If both exercise their right to remain silent, then the district attorney can still convict them both on a very minor charge. In the numerical matrix above, we take the units of measure of utility to be such that the most disagreeable outcome (the maximum sentence) has value 0, and the next most disagreeable outcome (minimum sentence) has value 1. Then we let being convicted on a minor charge to have value 3, and being set free to have value 4.

This game has abundant economic application. An example is the manufacturing by two companies of a single good. Both companies may produce either at a high or a low level. If both produce at a low level, the price stays high and they both receive 3. If they both produce at the high level the price drops and they both receive 1. If one produces at the high level while the other produces at the low level, the high producer receives 4 and the low producer receives 0. No matter what the other producer does, each will be better off by producing at a high level.

15.3 Finding All PSE's

For larger matrices, it is not difficult to find all pure strategic equilibria. This may be done using an extension of the method of finding all saddle points of a zero-sum game. With the game written in bimatrix form, put an asterisk after each of Player I's payoffs that is a maximum of its column. Then put an asterisk after each of Player II's payoffs that is a maximum of its row. Then any entry of the matrix at which both I's and II's payoffs have asterisks is a PSE, and conversely.

An example should make this clear.

$$\begin{array}{c} \\ A \\ B \\ C \\ D \end{array}\begin{array}{c} \begin{array}{cccccc} a & b & c & d & e & f \end{array} \\ \begin{pmatrix} (2,1) & (4,3) & (7^*,2) & (7^*,4) & (0,5^*) & (3,2) \\ (4^*,0) & (5^*,4) & (1,6^*) & (0,4) & (0,3) & (5^*,1) \\ (1,3^*) & (5^*,3^*) & (3,2) & (4,1) & (1^*,0) & (4,3^*) \\ (4^*,3) & (2,5^*) & (4,0) & (1,0) & (1^*,5^*) & (2,1) \end{pmatrix} \end{array}$$

In the first column, Player I's maximum payoff is 4, so both 4's are given asterisks. In the first row, Player II's maximum is 5, so the 5 receives an asterisk. And so on.

When we are finished, we see two payoff vectors with double asterisks. These are the pure strategic equilibria, (C, b) and (D, e), with payoffs $(5, 3)$ and $(1, 5)$ respectively. At all other pure strategy pairs, at least one of the players can improve his/her payoff by switching pure strategies.

In a two-person zero-sum game, a PSE is just a saddle point. Many games have no PSE's, for example, zero-sum games without a saddle point. However, just as zero-sum games of perfect information always have saddle points, non-zero-sum games of perfect information always have at least one PSE that may be found by the method of backward induction.

15.4 Iterated Elimination of Strictly Dominated Strategies

Since in general-sum games different equilibria may have different payoff vectors, it is more important than in zero-sum games to find all strategic equilibria. *We may remove any strictly dominated row or column without losing any equilibrium points* (Exercise 7).

We, being rational, would not play a strictly dominated pure strategy, because there is a (possibly mixed) strategy that guarantees us a strictly better average payoff no matter what the opponent does. Similarly, if we believe the opponent is as rational as we are, we believe that he/she will not use a dominated strategy either. Therefore we may cancel any dominated pure strategy of the opponent before we check if we have any dominated pure strategies which may now be eliminated.

This argument may be iterated. If we believe our opponent not only is rational but also believes that we are rational, then we may eliminate our dominated pure strategies, then eliminate our opponent's dominated pure strategies, and then again eliminate any of our own pure strategies that have now become dominated. The ultimate in this line of reasoning is that if it is *common knowledge* that the two players are rational, then we may iteratively remove dominated strategies as long as we like. (A statement is "common knowledge" between two players if each knows the statement, and each knows the other knows the statement, and each knows the other knows the other knows the statement, ad infinitum.)

As an example of what this sort of reasoning entails, consider a game of Prisoner's Dilemma that is to be played sequentially 100 times. The last time this is to be played it is clear that rational players will choose to defect. The other strategy is strictly dominated. But now that we know what the players will do on the last game we can apply strict domination to the next to last game to conclude the players will defect on that game too. Similarly all the way back to the first game. The players will each receive 1 at each game. If they could somehow break down their belief in the other's rationality, they might receive 3 for each game.

Here is another game, called the *Centipede Game* of Robert Rosenthal (1980), that illustrates this anomaly more vividly. This is a game of perfect information with no chance moves, so it is easy to apply the iterated removal of strictly dominated strategies. The game in extensive form is given in Fig. 15.1.

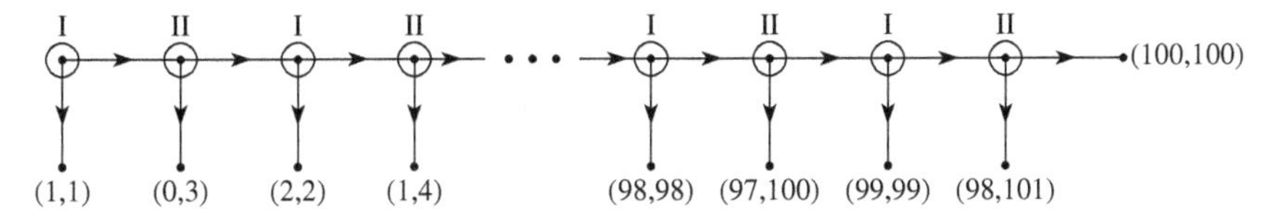

Fig. 15.1. The Centipede Game.

Since this is a game of perfect information, it may be solved by backward induction. At the last move, Player II will certainly go down instead of across since that gives her 101 instead of 100. Therefore at the next to last move, Player I will go down rather than across since that gives him 99 instead of the 98. And so forth, back to the initial position, where Player I will go down rather than across because he receives 1 instead of 0. This is the unique PSE because all eliminated strategies were strictly dominated.

Empirical evidence acquired by playing similar games shows that this gives a poor prediction of how people actually play this game. See the book of David M. Kreps (1990) *Game Theory and Economic Modeling* for a discussion.

15.5 Exercises

1. **Strategic Equilibria are Individually Rational.** A payoff vector is said to be *individually rational* if each player receives at least his safety level. Show that if $(\mathbf{p}, \mathbf{q})$ is a strategic equilibrium for the game with matrices $\mathbf{A}$ and $\mathbf{B}$, then $\mathbf{p}^T\mathbf{A}\mathbf{q} \geq v_{\mathrm{I}}$ and $\mathbf{p}^T\mathbf{B}\mathbf{q} \geq v_{\mathrm{II}}$. Thus, the payoff vector for a strategic equilibrium is individually rational.
2. Find the safety levels, the MM-strategies, and find all SE's and associated vector payoffs of the following games in strategic form.

$$\text{(a)} \begin{pmatrix} (0,0) & (2,4) \\ (2,4) & (3,3) \end{pmatrix}. \quad \text{(b)} \begin{pmatrix} (1,4) & (4,1) \\ (2,2) & (3,3) \end{pmatrix}. \quad \text{(c)} \begin{pmatrix} (0,0) & (0,-1) \\ (1,0) & (-1,3) \end{pmatrix}.$$

3. **The Game of Chicken.** Two players speed head-on toward each other and a collision is bound to occur unless one of them chickens out at the last minute. If both chicken out, everything is okay (they both win 1). If one chickens out and the other does not, then it is a great success for the player with iron nerves (payoff $= 2$) and a great disgrace for the chicken (payoff $= -1$). If both players have iron nerves, disaster strikes (both lose 2).

   (a) Set up the bimatrix of this game.
   (b) What are the safety levels, what are the MM strategies, and what is the average payoff if the players use the MM strategies?
   (c) Find all three SE's.

4. **An extensive form non-zero-sum game.** A coin with probability $2/3$ of heads and $1/3$ of tails is tossed and the outcome is shown to

player I but not to player II. Player I then makes a claim which may be true or false that the coin turned up heads or that the coin turned up tails. Then, player II, hearing the claim, must guess whether the coin came up heads or tails. Player II wins \$3 if his guess is correct, and nothing otherwise. Player I wins \$3 if I has told the truth in his claim. In addition, Player I wins an additional \$6 if player II guesses heads.

(a) Draw the Kuhn tree.
(b) Put into strategic (bimatrix) form.
(c) Find all PSE's.

5. Find all PSE's of the following games in strategic form.

$$\text{(a)} \quad \begin{pmatrix} (-3,-4) & (2,-1) & (0,6) & (1,1) \\ (2,0) & (2,2) & (-3,0) & (1,-2) \\ (2,-3) & (-5,1) & (-1,-1) & (1,-3) \\ (-4,3) & (2,-5) & (1,2) & (-3,1) \end{pmatrix}.$$

$$\text{(b)} \quad \begin{pmatrix} (0,0) & (1,-1) & (1,1) & (-1,0) \\ (-1,1) & (0,1) & (1,0) & (0,0) \\ (1,0) & (-1,-1) & (0,1) & (-1,1) \\ (1,-1) & (-1,0) & (1,-1) & (0,0) \\ (1,1) & (0,0) & (-1,-1) & (0,0) \end{pmatrix}.$$

6. Consider the bimatrix game: $\begin{pmatrix} (0,0) & (1,2) & (2,0) \\ (0,1) & (2,0) & (0,1) \end{pmatrix}$.

(a) Find the safety levels for the two players.
(b) Find all PSE's.
(c) Find all SE's given by mixed equalizing strategies.

7. **Strategic Equilibria Survive Elimination of Strictly Dominated Strategies.** Suppose row 1 is strictly dominated (by a probability mixture of rows 2 through m, i.e. $a_{1j} < \sum_{i=2}^{m} x_i a_{ij}$ for all j where $x_i \geq 0$ and $\sum_2^m x_i = 1$), and suppose $(\mathbf{p}^*, \mathbf{q}^*)$ is a strategic equilibrium. Show that $p_1^* = 0$.

8. Consider the non-cooperative bimatrix game: $\begin{pmatrix} (3,4) & (2,3) & (3,2) \\ (6,1) & (0,2) & (3,3) \\ (4,6) & (3,4) & (4,5) \end{pmatrix}$.

(a) Find the safety levels, and the maxmin strategies for both players.
(b) Find as many strategic equilibria as you can.

9. A PSE vector of strategies in a game in extensive form is said to be a **subgame perfect equilibrium** if at every vertex of the game tree, the

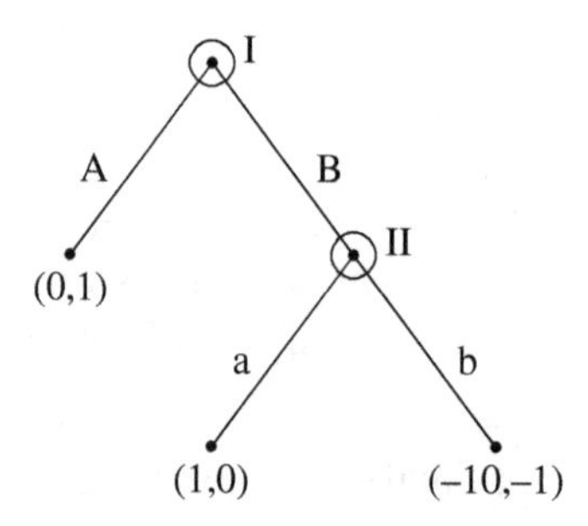

Fig. 15.2. An extensive form Game.

strategy vector restricted to the subgame beginning at that vertex is a PSE. If a game has perfect information, a subgame perfect equilibrium may be found by the method of backward induction. Figure 15.2 is an example of a game of perfect information that has a subgame perfect PSE and another PSE that is not subgame perfect.

(a) Solve the game for an equilibrium using backward induction.
(b) Put the game into strategic form.
(c) Find another PSE of the strategic form game, relate it to the extensive form game and show that it is not subgame perfect.

10. Suppose you are playing the centipede game once as Player I against a person chosen at random from this class. At what point would you choose the option to go down ending the game, assuming the opponent has not already ended the game?

    Now answer the same question assuming you are Player II.

    Your score on this question depends on what the other students in the class do. Do not reveal your answer to this question to the other students in the class.

16 Models of Duopoly

The examples given of the noncooperative theory of equilibrium have generally shown the theory to have poor predictive power. This is mainly because there may be multiple equilibria with no way to choose among them. Alternately, there may be a unique equilibrium with a poor outcome, even one found by iterated elimination of dominated strategies as in the prisoner's dilemma or the centipede game. But there are some situations, such as the prisoner's dilemma played once, in which strategic equilibria are quite reasonable predictive indicators of behavior. We begin with a model of duopoly due to Cournot (1838).

16.1 The Cournot Model of Duopoly

There are two competing firms producing a single homogeneous product. These firms must choose how much of the good to produce. The cost of producing one unit of the good is a constant c, the same for both firms. If a firm i produces the quantity q_i units of the good, then the cost to firm i is cq_i, for $i = 1, 2$. (There is no setup cost.) The price of a unit of the good is negatively related to the total amount produced. If firm 1 produces q_1 and firm 2 produces q_2 for a total of $Q = q_1 + q_2$, the price is

$$P(Q) = \begin{cases} a - Q & \text{if } 0 \le Q \le a \\ \\ 0 & \text{if } Q > a \end{cases} = (a - Q)^+ \qquad (1)$$

for some constant a. (This is not a realistic assumption, but the price will be approximately linear near the equilibrium point, and that is the main thing.) We assume the firms must choose their production quantities simultaneously; no collusion is allowed.

The pure strategy spaces for this game are the sets $X = Y = [0, \infty)$. Note that these are infinite sets, so the game is not a finite game. It would not hurt to restrict the strategy spaces to $[0, a]$; no player would like to produce more than a units because the return is zero. The payoffs for the two players are the profits,

$$u_1(q_1, q_2) = q_1 P(q_1 + q_2) - cq_1 = q_1(a - q_1 - q_2)^+ - cq_1, \tag{2}$$

$$u_2(q_1, q_2) = q_2 P(q_1 + q_2) - cq_2 = q_2(a - q_1 - q_2)^+ - cq_2. \tag{3}$$

This defines the strategic form of the game. We assume that $c < a$, since otherwise the cost of production would be at least as great as any possible return.

First, let us find out what happens in the monopolistic case when there is only one producer. That is, suppose $q_2 = 0$. Then the return to firm 1 if it produces q_1 units is $u(q_1) = q_1(a - q_1)^+ - cq_1$. The firm will choose q_1 to maximize this quantity. Certainly the maximum will occur for $0 < q_1 < a$; in this case, $u(q_1) = q_1(a - c) - q_1^2$, and we may find the point at which the maximum occurs by taking a derivative with respect to q_1, setting it to zero and solving for q_1. The resulting equation is $u'(q_1) = a - c - 2q_1 = 0$, whose solution is $q_1 = (a - c)/2$. The monopoly price is $P((a - c)/2) = (a + c)/2$, and the monopoly profit is $u((a - c)/2) = (a - c)^2/4$.

To find a duopoly PSE, we look for a pure strategy for each player that is a best response to the other's strategy. We find simultaneously the value of q_1 that maximizes (2) and the value of q_2 that maximizes (3) by setting the partial derivatives to zero.

$$\frac{\partial}{\partial q_1} u_1(q_1, q_2) = a - 2q_1 - q_2 - c = 0 \tag{4}$$

$$\frac{\partial}{\partial q_2} u_2(q_1, q_2) = a - q_1 - 2q_2 - c = 0 \tag{5}$$

(u_1 is a quadratic function of q_1 with a negative coefficient, so this root represents a point of maximum). Solving these equations simultaneously and denoting the result by q_1^* and q_2^*, we find

$$q_1^* = (a - c)/3 \quad \text{and} \quad q_2^* = (a - c)/3. \tag{6}$$

Therefore, (q_1^*, q_2^*) is a PSE for this problem.

In this SE, each firm produces less than the monopoly production, but the total produced is greater than the monopoly production. The payoff each player receives from this SE is

$$u_1(q_1^*, q_2^*) = \frac{a-c}{3}\left(a - \frac{a-c}{3} - \frac{a-c}{3}\right) - c\frac{a-c}{3} = \frac{(a-c)^2}{9}. \quad (7)$$

Note that the total amount received by the firms in this equilibrium is $(2/9)(a-c)^2$. This is less than $(1/4)(a-c)^2$, which is the amount that a monopoly would receive using the monopolistic production of $(a-c)/2$. This means that if the firms were allowed to cooperate, they could improve their profits by agreeing to share the production and profits. Thus each would produce less, $(a-c)/4$ rather than $(a-c)/3$, and receive a greater profit, $(a-c)^2/8$ rather than $(a-c)^2/9$.

On the other hand, the duopoly price is $P(q_1^* + q_2^*) = (a+2c)/3$, which is less than the monopoly price, $(a+c)/2$ (since $c < a$). Thus, the consumer is better off under a duopoly than under a monopoly.

This PSE is in fact the unique SE. This is because it can be attained by *iteratively deleting strictly dominated strategies.* To see this, consider the points at which the function u_1 has positive slope as a function of $q_1 \geq 0$ for fixed $q_2 \geq 0$. The derivative (4) is positive provided $2q_1 + q_2 < a - c$. See Fig. 16.1.

For all values of $q_2 \geq 0$, the slope is negative for all $q_1 > (a-c)/2$. Therefore, all $q_1 > (a-c)/2$ are strictly dominated by $q_1 = (a-c)/2$.

But since the game is symmetric in the players, we automatically have all $q_2 > (a-c)/2$ are strictly dominated and may be removed. When all such points are removed from consideration in the diagram, we see that for all remaining q_2, the slope is positive for all $q_1 < (a-c)/4$. Therefore, all $q_1 < (a-c)/4$ are strictly dominated by $q_1 = (a-c)/4$.

Again symmetrically eliminating all $q_2 < (a-c)/4$, we see that for all remaining q_2, the slope is negative for all $q_1 > 3(a-c)/8$. Therefore, all $q_1 > 3(a-c)/8$ are strictly dominated by $q_1 = 3(a-c)/8$. And so on, chipping a piece off from the lower end and then one from the upper end of the interval of the remaining q_1 not yet eliminated. If this is continued an infinite number of times, all q_1 are removed by iterative elimination of strictly dominated strategies except the point q_1^*, and by symmetry q_2^* for Player II.

Note that the prisoner's dilemma puts in an appearance here. Instead of using the SE obtained by removing strictly dominated strategies, both

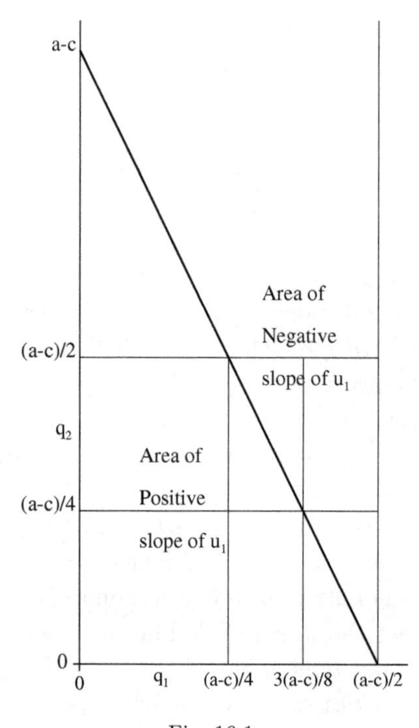

Fig. 16.1.

players would be better off if they could cooperate and produce $(a-c)/4$ each.

16.2 The Bertrand Model of Duopoly

In 1883, J. Bertrand proposed a different model of competition between two duopolists, based on allowing the firms to set prices rather than to fix production quantities. In this model, demand is a function of price rather than price a function of quantity available.

First consider the case where the two goods are identical and price information to the consumer is perfect so that the firm that sets the lower price will corner the market. We use the same price/demand function (1) solved for demand Q in terms of price P,

$$Q(P) = \begin{cases} a - P & \text{if } 0 \le P \le a \\ 0 & \text{if } P > a \end{cases} = (a-P)^+. \tag{8}$$

The actual demand is $Q(P)$ where P is the lowest price. The monopoly behavior under this model is the same as for the Cournot model of the previous section. The monopolist sets the price at $(a+c)/2$ and produces the quantity $(a-c)/2$, receiving a profit of $(a-c)^2/4$.

Suppose firms 1 and 2 choose prices p_1 and p_2 respectively. We assume that if $p_1 = p_2$ the firms share the market equally. We take the cost for a unit production again to be $c > 0$ so that the profit is $p_i - c$ times the amount sold. Then the payoff functions are

$$u_1(p_1, p_2) = \begin{cases} (p_1 - c)(a - p_1)^+ & \text{if } p_1 < p_2 \\ (p_1 - c)(a - p_1)^+/2 & \text{if } p_1 = p_2 \\ 0 & \text{if } p_1 > p_2 \end{cases} \tag{9}$$

and

$$u_2(p_1, p_2) = \begin{cases} (p_2 - c)(a - p_2)^+ & \text{if } p_2 < p_1 \\ (p_2 - c)(a - p_2)^+/2 & \text{if } p_2 = p_1 \\ 0 & \text{if } p_2 > p_1. \end{cases} \tag{10}$$

Here there is a unique PSE but it is rather disappointing. Both firms charge the production cost, $p_1^* = p_2^* = c$, and receive a payoff of zero. This is the safety level for each player. It is easy to check that this is an equilibrium. No other pair of prices can be an equilibrium because either firm could capture the entire market by slightly undercutting the other's price.

This feature of capturing the entire market by undercutting the other's price is not entirely reasonable for a number of reasons. Usually the products of two firms are not entirely interchangeable so some consumers may prefer one product to the other even if it costs somewhat more. In addition there is the problem of the consumer getting the information on the prices, and there is the feature of brand loyalty by consumers. We may modify the model in an attempt to take this into account.

The Bertrand Model with Differentiated Products. Again we assume that the firms choose prices p_1 and p_2 and that the cost of unit production is $c > 0$. Since the profits per unit produced are $p_1 - c$ and $p_2 - c$, we may assume that the prices will satisfy $p_1 \geq c$ and $p_2 \geq c$. This time we assume that the demand functions of the products of the firms for given price selections are given by

$$q_1(p_1, p_2) = (a - p_1 + bp_2)^+,$$
$$q_2(p_1, p_2) = (a - p_2 + bp_1)^+, \tag{11}$$

where $b > 0$ is a constant representing how much the product of one firm is a substitute for the product of the other. We assume $b \leq 1$ for simplicity. These demand functions are unrealistic in that one firm could conceivably charge an arbitrarily high price and still have a positive demand provided the other firm also charges a high enough price. However, this function is chosen to represent a linear approximation to the "true" demand function, appropriate near the usual price settings where the equilibrium is reached.

Under these assumptions, the strategy sets of the firms are $X = [0, \infty)$ and $Y = [0, \infty)$, and the payoff functions are

$$u_1(p_1, p_2) = q_1(p_1, p_2)(p_1 - c) = (a - p_1 + bp_2)^+(p_1 - c),$$
$$u_2(p_1, p_2) = q_2(p_1, p_2)(p_2 - c) = (a - p_2 + bp_1)^+(p_2 - c). \tag{12}$$

To find the equilibrium prices, we must find points (p_1^*, p_2^*) at which u_1 is maximized in p_1 and u_2 is maximized in p_2 simultaneously. Assuming $a - p_1 + bp_2 > 0$ and $a - p_2 + bp_1 > 0$, we find

$$\frac{\partial}{\partial p_1} u_1(p_1, p_2) = a - 2p_1 + bp_2 + c = 0,$$

$$\frac{\partial}{\partial p_2} u_2(p_1, p_2) = a - 2p_2 + bp_1 + c = 0.$$

Again the functions are quadratic in the variable of differentiation with a negative coefficient, so the resulting roots represent maxima. Solving simultaneously and denoting the result by p_1^* and p_2^*, we find

$$p_1^* = p_2^* = \frac{a + c}{2 - b}.$$

16.3 The Stackelberg Model of Duopoly

In the Cournot and Bertrand models of duopoly, the players act simultaneously. H. von Stackelberg (1934) proposed a model of duopoly in which one player, called the dominant player or leader, moves first and the outcome of that player's choice is made known to the other player before the other player's choice is made. An example might be General Motors, at times big enough and strong enough in U.S. history to play such a dominant role in the automobile industry. Let us analyze the Cournot model from this perspective.

Firm 1 chooses an amount to produce, q_1, at a cost c per unit. This amount is then told to firm 2 which then chooses an amount q_2 to produce also at a cost of c per unit. Then the price P per unit is determined by Eq. (1), $P = (a - q_1 - q_2)^+$, and the players receive $u_1(q_1, q_2)$ and $u_2(q_1, q_2)$ of Eqs. (2) and (3).

Player I's pure strategy space is $X = [0, \infty)$. From the mathematical point of view, the only difference between this model and the Cournot model is that firm 2's pure strategy space, Y, is now a set of functions mapping q_1 into q_2. However, this is now a game of perfect information that can be solved by backward induction. Since firm 2 moves last, we first find the optimal q_2 as a function of q_1. That is, we solve Eq. (5) for q_2. This gives us firm 2's strategy as

$$q_2(q_1) = (a - q_1 - c)/2. \tag{13}$$

Since firm 1 now knows that firm 2 will choose this best response, firm 1 now wishes to choose q_1 to maximize

$$\begin{aligned} u_1(q_1, q_2(q_1)) &= q_1(a - q_1 - (a - q_1 - c)/2) - cq_1 \\ &= -\frac{1}{2}q_1^2 + \frac{a-c}{2}q_1. \end{aligned} \tag{14}$$

This quadratic function is maximized by $q_1 = q_1^* = (a - c)/2$. Then firm 2's best response is $q_2^* = q_2(q_1^*) = (a - c)/4$.

Let us analyze this SE and compare its payoff to the payoff of the SE in the Cournot duopoly. Firm 1 produces the monopoly quantity and firm 2 produces less than the Cournot SE. The payoff to firm 1 is $u_1(q_1^*, q_2^*) = (a - c)^2/8$ and the payoff to firm 2 is $u_2(q_1^*, q_2^*) = (a - c)^2/16$. Therefore firm 1's profits are greater than that given by the Cournot equilibrium, and firm 2's are less. Note that the total amount produced is $(3/4)(a - c)$, which is greater than $(2/3)(a - c)$, the total amount produced under the Cournot equilibrium. This means the Stackelberg price is lower than the Cournot price, and the consumer is better off under the Stackelberg model.

The information that firm 2 received about firm 1's production has been harmful. Firm 1 by announcing its production has increased its profit. This shows that having more information may make a player worse off. More precisely, being given more information and having that fact be common knowledge may make you worse off.

16.4 Entry Deterrence

Even if a firm acts as a monopolist in a certain market, there may be reasons why it is in the best interests of the firm to charge less than the monopoly price, or equivalently, produce more than the monopoly production. One of these reasons is that the high price of the good achieved by monopoly production may attract another firm to enter the market.

We can see this in the following example. Suppose the price/demand relationship can be expressed as

$$P(Q) = \begin{cases} 17 - Q & \text{if } 0 \le Q \le 17 \\ 0 & \text{otherwise,} \end{cases} \tag{15}$$

where Q represents the total amount produced, and P represents the price per unit amount. Suppose additionally, that the cost to the firm of producing q_1 items is q_1+9. That is, there is a fixed cost of 9 and a constant marginal cost of 1 per unit quantity. The profit to the firm of producing quantity q_1 of the good is

$$u(q_1) = (17 - q_1)q_1 - (q_1 + 9) = 16q_1 - q_1^2 - 9. \tag{16}$$

The value of q_1 that maximizes the profit is found by setting the derivative of $u(q_1)$ to zero and solving for q_1:

$$u'(q_1) = 16 - 2q_1 = 0.$$

So the monopoly production is

$$q_1 = 8,$$

the monopoly price is 9, and the monopoly profit is

$$u(8) = 9 \cdot 8 - 17 = 55.$$

Suppose now a competing firm observes this market and thinks of producing a small amount, q_2, to hopefully make a small profit. Suppose also that this firm also has the same cost, q_2+9, as the monopoly firm. On producing q_2 the price will drop to $P(8+q_2) = 9 - q_2$, and the competing firm's profit will be

$$u_2 = (9 - q_2)q_2 - (q_2 + 9) = 8q_2 - q_2^2 - 9. \tag{17}$$

This is maximized at $q_2 = 4$ and the profit there is $u_2 = 7$. Since this is positive, the firm has an incentive to enter the market.

Of course, the incumbent monopolist can foresee this possibility and can calculate the negative effect it will have on the firm's profits. If the challenger enters the market with a production of 4, the price will drop to $P(8+4) = 5$, and the monopolist's profits will drop from 55 to $5 \cdot 8 - 17 = 23$. It seems reasonable that some preventative measures might be worthwhile.

If the monopolist produces a little more than the monopoly quantity, it might deter the challenger from entering the market. How much more should be produced? If the monopolist produces q_1, then the challenger's firm's profits may be computed as in (17) by

$$u_2(q_1, q_2) = (17 - q_1 - q_2)q_2 - (q_2 + 9).$$

This is maximized at $q_2 = (16 - q_1)/2$ for a profit of

$$u_2(q_1, (16 - q_1)/2) = (16 - q_1)^2/4 - 9.$$

The profit is zero if $(16 - q_1)^2 = 36$, or equivalently, if $q_1 = 10$.

This says that if the monopolist produces 10 rather than 8, then the challenger can see that it is not profitable to enter the market.

However, the monopolist's profits are reduced by producing 10 rather than 8. From (16) we see that the profits to the firm when $q_1 = 10$ are

$$u_1(10) = 7 \cdot 10 - 19 = 51$$

instead of 55. This is a relatively small amount to pay as insurance against the much bigger drop in profits from 55 to 23 the monopolist would suffer if the challenger should enter the market.

The above analysis assumes that the challenger believes that, even if the challenger should enter the market, the monopolist will continue with the monopoly production, or the pre-entry production. This would be the case if the incumbent monopolist were considered as the dominant player in a Stackelberg model. Note that the entry deterrence strategy pair, $q_1 = 10$ and $q_2 = 0$, does not form a strategic equilibrium in this Stackelberg model, since $q_1 = 10$ is not a best response to $q_2 = 0$. To analyze the situation properly, we should enlarge the model to allow the game to be played sequentially several times.

If this problem were analyzed as a Cournot duopoly, we would find that, at equilibrium, each firm would produce $5\frac{1}{3}$, the price would drop to $6\frac{1}{3}$, and each firm would realize a profit of $19\frac{4}{9}$. This low profit is another reason that the incumbent firm should make strong efforts to deter the entry of a challenger.

16.5 Exercises

1. (a) Suppose in the Cournot model that the firms have different production costs. Let c_1 and c_2 be the costs of production per unit for firms 1 and 2 respectively, where both c_1 and c_2 are assumed less than $a/2$. Find the Cournot equilibrium.
   (b) What happens, if in addition, each firm has a set up cost? Suppose Player I's cost of producing x is $x+2$, and II's cost of producing y is $3y+1$. Suppose also that the price function is $p(x,y) = 17-x-y$, where x and y are the amounts produced by I and II respectively. What is the equilibrium production, and what are the players' equilibrium payoffs?
2. Extend the Cournot model of Sec. 16.1 to three firms. Firm i chooses to produce q_i at cost cq_i where $c > 0$. The selling price is $P(Q) = (a-Q)^+$ where $Q = q_1 + q_2 + q_3$. What is the strategic equilibrium?
3. Modify the Bertrand model with differentiated products to allow sequential selection of the price as in Stackelberg's variation of Cournot's model. The dominant player announces a price first and then the subordinate player chooses a price. Solve by backward induction and compare to the SE for the simultaneous selection model.
4. Consider the Cournot duopoly model with the somewhat more realistic price function,

$$P(Q) = \begin{cases} \frac{1}{4}Q^2 - 5Q + 26 & \text{for } 0 \le Q \le 10, \\ 1 & \text{for } Q \ge 10. \end{cases}$$

   This price function starts at 26 for $Q = 0$ and decreases down to 1 at $Q = 10$ and then stays there. Assume that the cost, c, of producing one unit is $c = 1$ for both firms. No firm would produce more than 10 because the return for selling a unit would barely pay for the cost of producing it. Thus we may restrict the productions q_1, q_2, to the interval $[0, 10]$.

   (a) Find the monopoly production, and the optimal monopoly return.
   (b) Show that if $q_2 = 5/2$, then $u_1(q_1, 5/2)$ is maximized at $q_1 = 5/2$. Show that this implies that $q_1 = q_2 = 5/2$ is an equilibrium production in the duopoly.

5. (a) Suppose in the Stackelberg model that the firms have different production costs. Let c_1 and c_2 be the costs of production per unit for firms 1 and 2 respectively. Find the Stackelberg equilibrium. For simpicity, you may assume that both c_1 and c_2 are small, say less than $a/3$.
   (b) Suppose in the Stackelberg model, Player I's cost of producing x is $x + 2$, and II's cost of producing y is $3y + 1$. Suppose also that the price function is $p(x, y) = 17 - x - y$, where x and y are the amounts produced by I and II respectively. What is the equilibrium production, and what are the players' equilibrium payoffs?
6. Extend the Stackelberg model to three firms. For $i = 1, 2, 3$, firm i chooses to produce q_i at cost cq_i where $c > 0$. Firm 1 acts first in announcing the production q_1. Then firm 2 announces q_2, and finally firm 3 announces q_3. The selling price is $P(Q) = (a - Q)^+$ where $Q = q_1 + q_2 + q_3$. What is the strategic equilibrium?
7. **An Advertising Campaign.** Two firms may compete for a given market of total value, V, by investing a certain amount of effort into the project through advertising, securing outlets, etc. Each firm may allocate a certain amount for this purpose. If firm 1 allocates $x > 0$ and firm 2 allocates $y > 0$, then the proportion of the market that firm 1 corners is $x/(x + y)$. The firms have differing difficulties in allocating these resources. The cost per unit allocation to firm i is c_i, $i = 1, 2$. Thus the profits to the two firms are

$$M_1(x, y) = V \cdot \frac{x}{x + y} - c_1 x,$$

$$M_2(x, y) = V \cdot \frac{y}{x + y} - c_2 y.$$

   If both x and y are zero, the payoffs to both are zero.

   (a) Find the equilibrium allocations, and the equilibrium profits to the two firms, as a function of V, c_1 and c_2.
   (b) Specialize to the case $V = 1$, $c_1 = 1$, and $c_2 = 2$.

17 Cooperative Games

In one version of the noncooperative theory, communication among the players is allowed but there is no machanism to enforce any agreement the players may make. The only believable agreement among the players would be a Nash equilibrium because such an agreement would be *self-enforcing*: no player can gain by unilaterally breaking the agreement. In the cooperative theory, we allow communication among the players and we also allow binding agreements to be made. This requires some mechanism outside the game itself to enforce the agreements. With the extra freedom to make enforceable binding agreements in the cooperative theory, the players can generally achieve a better outcome. For example in the prisoner's dilemma, the only Nash equilibrium is for both players to defect. If they cooperate, they can reach a binding agreement that they both use the cooperate strategy, and both players will be better off.

The cooperative theory is divided into two classes of problems depending on whether or not there is a mechanism for transfer of utility from one player to the other. If there such a mechanism, we may think of the transferable commodity as "money", and assume that both players have a linear utility for money. We may take the scaling of the respective utilities to be such that the utility of no money is 0 and the utility of one unit of money is 1. In Sec. 17.2, we treat the **transferable utility** (TU) case. In Sec. 17.3, we treat the **nontransferable utility** (NTU) case.

17.1 Feasible Sets of Payoff Vectors

One of the main features of cooperative games is that the players have freedom to choose a joint strategy. This allows any probability mixture of the payoff vectors to be achieved. For example in the battle of the sexes, the players may agree to toss a coin to decide which movie to see. (They may also do this in the noncooperative theory, but after the coin is tossed, they are allowed to change their minds, whereas in the cooperative theory, the coin toss is part of the agreement.) The set of payoff vectors that the players can achieve if they cooperate is called the feasible set. The distinguishing feature of the TU case is that the players may make *side payments* of utility as part of the agreement. This feature results in a distinction between the NTU feasible set and the TU feasible set.

When players cooperate in a bimatrix game with matrices $(\mathbf{A}, \mathbf{B})$, they may agree to achieve a payoff vector of any of the mn points, (a_{ij}, b_{ij}) for $i = 1, \ldots, m$ and $j = 1 \ldots, n$. They may also agree to any probability mixture of these points. The set of all such payoff vectors is the convex hull these mn points. Without a transferable utility, this is all that can be achieved.

Definition. The NTU feasible set is the convex hull of the mn points, (a_{ij}, b_{ij}) for $i = 1, \ldots, m$ and $j = 1, \ldots, n$.

By making a side payment, the payoff vector (a_{ij}, b_{ij}) can be changed to $(a_{ij} + s, b_{ij} - s)$. If the number s is positive, this represents a payment from Player II to Player I. If s is negative, the side payment is from Player I to Player II. Thus the whole line of slope -1 through the point (a_{ij}, b_{ij}) is part of the TU feasible set. And we may take probability mixtures of these as well.

Definition. The TU feasible set is the convex hull of the set of vectors of the form $(a_{ij} + s, b_{ij} - s)$ for $i = 1, \ldots, m$ and $j = 1, \ldots, n$ and for arbitrary real numbers s.

As an example, the bimatrix game

$$\begin{pmatrix} (4,3) & (0,0) \\ (2,2) & (1,4) \end{pmatrix} \tag{1}$$

has two pure strategic equilibria, upper left and lower right. This game has the NTU feasible and TU feasible sets given in Fig. 17.1.

If an agreement is reached in a cooperative game, be it a TU or an NTU game, it may be expected to be such that no player can be made better off without making at least one other player worse off. Such an outcome is said to be Pareto optimal.

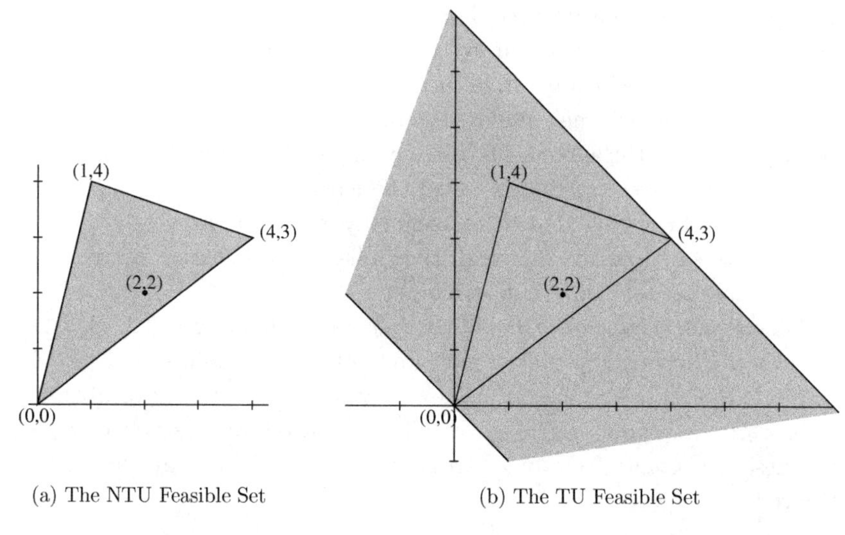

(a) The NTU Feasible Set

(b) The TU Feasible Set

Fig. 17.1.

Definition. A feasible payoff vector, (v_1, v_2), is said to be *Pareto optimal* if the only feasible payoff vector (v_1', v_2') such that $v_1' \geq v_1$ and $v_2' \geq v_2$ is the vector $(v_1', v_2') = (v_1, v_2)$.

In the example above, the Pareto feasible outcomes for the NTU game are simply the vectors on the line segment joining the points $(4, 3)$ and $(1, 4)$. The Pareto optimal outcomes for the TU game are the vectors on the line of slope -1 through the point $(4, 3)$.

For more general convex feasible sets in the plane, the set of Pareto optimal points is the set of upper right boundary points.

17.2 Cooperative Games with Transferable Utility

In this section, we restrict attention to the transferable utility case and assume that the players are "rational" in the sense that, given a choice between two possible outcomes of differing personal utility, each player will select the one with the higher utility.

The TU-Problem: In the model of the game, we assume there is a period of *preplay negotiation*, during which the players meet to discuss the possibility of choosing a joint strategy together with some possible side payment to induce cooperation. They also discuss what will happen if they cannot come to an agreement; each may threaten to use some unilateral strategy that is bad for the opponent.

If they do come to an agreement, it may be assumed that the payoff vector is Pareto optimal. This is because if the players are about to agree to some feasible vector v and there is another feasible vector, v', that is better for one of the players without making any other player worse off, that player may propose changing to the vector v', offering to transfer some of his gain in utility to the other players. The other players, being rational would agree to this proposal.

In the discussion, both players may make some threat of what strategy they will take if an agreement is not reached. However, a threat to be believable must not hurt the player who makes it to a greater degree than the opponent. Such a threat would not be credible. For example, consider the following bimatrix game

$$\begin{pmatrix} (5,3) & (0,-4) \\ (0,0) & (3,6) \end{pmatrix}. \tag{2}$$

If the players come to an agreement, it will be to use the lower right corner because it has the greatest total payoff, namely 9. Player II may argue that she should receive at least half the sum, $4\frac{1}{2}$. She may even feel generous in "giving up" as a side payment some of the 6 she would be winning. However, Player I may threaten to use row 1 unless he is given at least 5. That threat is very credible since if Player I uses row 1, Player II's cannot make a counter-threat to use column 2 because it would hurt her more than Player I. The counter-threat would not be credible.

In this model of the preplay negotiation, the threats and counter-threats may be made and remade until time to make a decision. Ultimately the players announce what threats they will carry out if agreement is not reached. It is assumed that if agreement is not reached, the players will leave the negotiation table and carry out their threats. However, being rational players, they will certainly reach agreement, since this gives a higher utility. The threats are only a formal method of arriving at a reasonable amount for the side payment, if any, from one player to the other.

The TU problem is then to choose the threats and the proposed side payment judiciously. The players use threats to influence the choice of the final payoff vector. The problem is how do the threats influence the final payoff vector, and how should the players choose their threat strategies? For two-person TU-games, there is a very convincing answer.

The TU Solution: If the players come to an agreement, then rationality implies that they will agree to play to achieve the largest possible total payoff, call it σ,

$$\sigma = \max_i \max_j (a_{ij} + b_{ij}) \tag{3}$$

as the payoff to be divided between them. That is they will jointly agree to use some row i_0 and column j_0 such that $a_{i_0 j_0} + b_{i_0 j_0} = \sigma$. Such a joint choice $\langle i_0, j_0 \rangle$, is called their *cooperative strategy.* But they must also agree on some final payoff vector (x^*, y^*), such that $x^* + y^* = \sigma$, as the appropriate division of the total payoff. Such a division may then require a *side payment* from one player to the other. If $x^* > a_{i_0 j_0}$, then Player I would receive a side payment of the difference, $x^* - a_{i_0 j_0}$, from Player II. If $x^* < a_{i_0 j_0}$, then Player II would receive a side payment of the difference, $a_{i_0 j_0} - x^*$, from Player I.

Suppose now that the players have selected their *threat strategies,* say $\mathbf{p}$ for Player I and $\mathbf{q}$ for Player II. Then if agreement is not reached, Player I receives $\mathbf{p}^T\mathbf{A}\mathbf{q}$ and Player II receives $\mathbf{p}^T\mathbf{B}\mathbf{q}$. The resulting payoff vector,

$$\mathbf{D} = \mathbf{D}(\mathbf{p}, \mathbf{q}) = (\mathbf{p}^T\mathbf{A}\mathbf{q}, \mathbf{p}^T\mathbf{B}\mathbf{q}) = (D_1, D_2) \tag{4}$$

is in the NTU feasible set and is called the *disagreement point* or *threat point.* Once the disagreement point is determined, the players must agree on the point (x, y) on the line $x + y = \sigma$ to be used as the cooperative solution. Player I will accept no less than D_1 and Player II will accept no less than D_2 since these can be achieved if no agreement is reached. But once the disagreement point has been determined, the game becomes symmetric. The players are arguing about which point on the line interval from $(D_1, \sigma - D_1)$ to $(\sigma - D_2, D_2)$ to select as the cooperative solution. No other considerations with respect to the matrices $\mathbf{A}$ and $\mathbf{B}$ play any further role. Therefore, the midpoint of the interval, namely

$$\boldsymbol{\varphi} = (\varphi_1, \varphi_2) = \left(\frac{\sigma - D_2 + D_1}{2}, \frac{\sigma - D_1 + D_2}{2} \right) \tag{5}$$

is the natural compromise. Both players suffer equally if the agreement is broken. The point, $\boldsymbol{\varphi}$, may be determined by drawing the line from $\mathbf{D}$ with 45° slope until it hits the line $x + y = \sigma$ as in Fig. 17.2.

We see from (5) what criterion the players should use to select the threat point. Player I wants to maximize $D_1 - D_2$ and Player II wants to

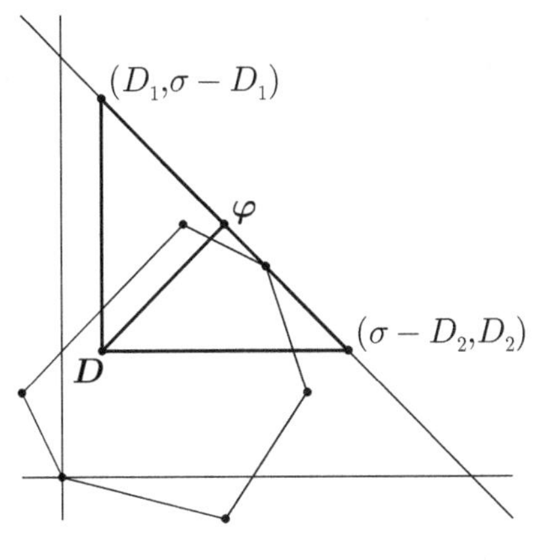

Fig. 17.2.

minimize it. This is in fact a zero-sum game with matrix $\mathbf{A} - \mathbf{B}$:

$$D_1 - D_2 = \mathbf{p}^T\mathbf{A}\mathbf{q} - \mathbf{p}^T\mathbf{B}\mathbf{q} = \mathbf{p}^T(\mathbf{A} - \mathbf{B})\mathbf{q}. \tag{6}$$

Let $\mathbf{p}^*$ and $\mathbf{q}^*$ denote optimal strategies of the game $\mathbf{A}-\mathbf{B}$ for Players I and II respectively, and let δ denote the value,

$$\delta = \mathrm{Val}(\mathbf{A} - \mathbf{B}) = \mathbf{p}^{*T}(\mathbf{A} - \mathbf{B})\mathbf{q}^*. \tag{7}$$

If Player I uses $\mathbf{p}^*$ as his threat, then the best Player II can do is to use $\mathbf{q}^*$, and conversely. When these strategies are used, the disagreement point becomes $\mathbf{D}^* = (D_1^*, D_2^*) = \mathbf{D}(\mathbf{p}^*, \mathbf{q}^*)$. Since $\delta = \mathbf{p}^{*T}\mathbf{A}\mathbf{q}^* - \mathbf{p}^{*T}\mathbf{B}\mathbf{q}^* = D_1^* - D_2^*$, we have as the TU solution:

$$\varphi^* = (\varphi_1^*, \varphi_2^*) = \left(\frac{\sigma + \delta}{2}, \frac{\sigma - \delta}{2}\right). \tag{8}$$

Suppose the players have decided on $\langle i_0, j_0 \rangle$ as the cooperative strategy to be used, where $a_{i_0 j_0} + b_{i_0 j_0} = \sigma$. To achieve the payoff (8), this requires a side payment of $(\sigma + \delta)/2 - a_{i_0 j_0}$ from Player II to Player I. If this quantity is negative, the payment of $a_{i_0 j_0} - (\sigma + \delta)/2$ goes from Player I to Player II.

Examples 1. Consider the TU game with bimatrix

$$\begin{pmatrix} (0,0) & (6,2) & (-1,2) \\ (4,-1) & (3,6) & (5,5) \end{pmatrix}.$$

There is a PSE in the first row, second column, with payoff vector (6,2). This is the matrix upon which Fig. 17.2 is based. But we shall see that the optimal disagreement point is in a somewhat different place than the one in the figure.

The maximum value of $a_{ij}+b_{ij}$ occurs in the second row third column, so the cooperative strategy is $\langle 2,3\rangle$, giving a total payoff of $\sigma=10$. If they come to an agreement, Player I will select the second row, Player II will select the third column and both players will receive a payoff of 5. They must still decide on a side payment, if any.

They consider the zero-sum game with matrix,

$$\mathbf{A}-\mathbf{B}=\begin{pmatrix}0 & 4 & -3\\ 5 & -3 & 0\end{pmatrix}.$$

The first column is strictly dominated by the last. The threat strategies are then easily determined to be

$$\mathbf{p}^*=(0.3,0.7)^T$$
$$\mathbf{q}^*=(0,0.3,0.7)^T.$$

The value of this game is: $\delta=\text{Val}\begin{pmatrix}4 & -3\\ -3 & 0\end{pmatrix}=-9/10$. Therefore from (8), the TU-value is

$$\varphi^*=((10-0.9)/2,(10+0.9)/2)=(4.55,5.45).$$

To arrive at this payoff from the agreed payoff vector, $(5,5)$, requires a side payment of 0.45 from Player I to Player II.

We may also compute the disagreement point, $\mathbf{D}^*=(D_1^*,D_2^*)$.

$$D_1^*=\mathbf{p}^{*T}\mathbf{A}\mathbf{q}^*=0.3(6\cdot 0.3-0.7)+0.7(3\cdot 0.3+5\cdot 0.7)=3.41,$$
$$D_2^*=\mathbf{p}^{*T}\mathbf{B}\mathbf{q}^*=0.3(2\cdot 0.3+2\cdot 0.7)+0.7(6\cdot 0.3+5\cdot 0.7)=4.31.$$

It is easy to see that the line from $\mathbf{D}^*$ to φ^* is $45°$, because $D_2^*-D_1^*=\varphi_2^*-\varphi_1^*=0.9$.

2. It is worthwhile to note that there may be more than one possible cooperative strategy yielding σ as the sum of the payoffs. The side payment depends on which one is used. Also there may be more than one possible disagreement point because there may be more than one pair of optimal strategies for the game $\mathbf{A}-\mathbf{B}$. However, all such disagreement points must be on the same $45°$ line, since the point φ depends on the disagreement

point only through the value, δ, and all disagreement points have the same TU-value.

Here is an example containing both possibilities:

$$\begin{pmatrix} (1,5) & (2,2) & (0,1) \\ (4,2) & (1,0) & (2,1) \\ (5,0) & (2,3) & (0,0) \end{pmatrix}.$$

There are two cooperative strategies giving total payoff $\sigma = 6$, namely $\langle 1,1\rangle$ and $\langle 2,1\rangle$. The matrix $\mathbf{A} - \mathbf{B}$ is

$$\begin{pmatrix} -4 & 0 & -1 \\ 2 & 1 & 1 \\ 5 & -1 & 0 \end{pmatrix}$$

which has a saddle-point at $\langle 2,3\rangle$. Thus $\mathbf{D} = (2,1)$ is the disagreement point, and the value is $\delta = 1$. Thus the TU cooperative value is $\varphi = (7/2, 5/2)$.

However, there is another saddle-point at $\langle 2,2\rangle$ that, of course, has the same value $\delta = 1$. But this time the disagreement point is $\varphi = (1,0)$. All such disagreement points must be on the $45°$ line through φ.

If $\langle 2,1\rangle$ is used as the cooperative strategy, the resulting vector payoff of $(4,2)$ requires that Player I pay $1/2$ to Player I. If $\langle 1,1\rangle$ is used as the cooperative strategy, the resulting vector payoff of $(5,1)$ requires that Player I pay $3/2$ to Player II.

17.3 Cooperative Games with Non-Transferable Utility

We now consider games in which side payments are forbidden. It may be assumed that the utility scales of the players are measured in noncomparable units. The players may argue, threaten, and come to a binding agreement as before, but there is no monetary unit with which the players can agree to make side payments. The players may barter goods that they own, but this must be done within the game and reflected in the bimatrix of the game.

We approach NTU games through the *Nash Bargaining Model.* This model is based on two elements assumed to be given and known to the players. One element is a compact (i.e. bounded and closed), convex set, S, in the plane. One is to think of S as the set of vector payoffs achievable by the players if they agree to cooperate. It is the analogue of the NTU-feasible set, although it is somewhat more general in that it does not have to be a polyhedral set. It could be a circle or an ellipse, for example. We refer to S as the *NTU-feasible set.*

The second element of the Nash Bargaining Model is a point, $(u^*, v^*) \in S$, called the *threat point* or *status-quo point.* Nash viewed the bargaining model as a game between two players who come to a market to barter goods. For him, the players have the option of not entering into any trade agreement at all, and it was natural for him to take the status-quo point as $(u^*, v^*) = (0, 0) \in S$. The subsequent theory allows (u^*, v^*) to be an arbitrary point of S.

Given an NTU-feasible set, S, and a threat point, $(u^*, v^*) \in S$, the problem is to decide on a feasible outcome vector for this game that will somehow reflect the value of the game to the players. That is, we want to find a point, $(\bar{u}, \bar{v}) = \mathbf{f}(S, u^*, v^*)$, to be considered a "fair and reasonable outcome" or "solution" of the game for an arbitrary compact convex set S and point $(u^*, v^*) \in S$. In the approach of Nash, "fair and reasonable" is defined by a few axioms. Then it is shown that these axioms lead to a *unique* solution, $\mathbf{f}(S, u^*, v^*)$. Here are the axioms.

Nash Axioms for $\mathbf{f}(S, u^*, v^*) = (\bar{u}, \bar{v})$.

(1) **Feasibility.** $(\bar{u}, \bar{v}) \in S$.
(2) **Pareto Optimality.** There is no point $(u, v) \in S$ such that $u \geq \bar{u}$ and $v \geq \bar{v}$ except $(\bar{u}, \bar{v})$ itself.
(3) **Symmetry.** If S is symmetric about the line $u = v$, and if $u^* = v^*$, then $\bar{u} = \bar{v}$.
(4) **Independence of irrelevant alternatives.** If T is a closed convex subset of S, and if $(u^*, v^*) \in T$ and $(\bar{u}, \bar{v}) \in T$, then $\mathbf{f}(T, u^*, v^*) = (\bar{u}, \bar{v})$.
(5) **Invariance under change of location and scale.** If $T = \{(u', v') : u' = \alpha_1 u + \beta_1, v' = \alpha_2 v + \beta_2 \text{ for } (u, v) \in S\}$, where $\alpha_1 > 0$, $\alpha_2 > 0$, β_1, and β_2 are given numbers, then

$$\mathbf{f}(T, \alpha_1 u^* + \beta_1, \alpha_2 v^* + \beta_2) = (\alpha_1 \bar{u} + \beta_1, \alpha_2 \bar{v} + \beta_2).$$

Analysis of the Axioms. It is useful to review the axioms to see which might be weakened or changed to allow other "solutions".

The first axiom is incontrovertible. The agreed outcome must be feasible.

The second axiom reflects the rationality of the players. If the players work together and reach agreement, they would not accept (u, v) as the outcome if they could also achieve $(\hat{u}, \hat{v})$ with $\hat{u} > u$ and $\hat{v} > v$. However, the second axiom is slightly stronger than this. It says that they would not accept (u, v) if they could achieve $(\hat{u}, \hat{v})$ with $\hat{u} \geq u$ and $\hat{v} > v$ (or $\hat{u} > u$ and $\hat{v} \geq v$). This plays no role in the main case of the theorem (when there is a $(u, v) \in S$ such that $u > u^*$ and $v > v^*$). But suppose S consists of the line from (0,0) to (0,1), inclusive, and $(u^*, v^*) = (0, 0)$. Player I can achieve a payoff of 0 without entering into any agreement. So to agree to the point (0,1) requires a weak kind of altruistic behavior on his part. It is true that this agreement would not hurt him, but still this weak altruism does not follow from the assumed rationality of the players.

The third axiom is a fairness axiom. If the game is symmetric in the players, there is nothing in the game itself to distinguish the players so the outcome should be symmetric.

The fourth axiom is perhaps the most controversial. It says that if two players agree that $(\bar{u}, \bar{v})$ is a fair and reasonable solution when S is the feasible set, then points in S far away from $(\bar{u}, \bar{v})$ and (u^*, v^*) are irrelevant. If S is reduced to a convex subset $T \subset S$, then as long as T still contains $(\bar{u}, \bar{v})$ and (u^*, v^*), the players would still agree on $(\bar{u}, \bar{v})$. But let S be the triangle with vertices (0,0), (0,4) and (2,0), and let the threat point be (0,0). Suppose the players agree on (1,2) as the outcome. Would they still agree on (1,2) if the feasible set were T, the quadralateral with vertices (0,0), (0,2), (1,2) and (2,0)? Conversely, if they agree on (1,2) for T, would they agree on (1,2) for S? The extra points in S cannot be used as threats because it is assumed that all threats have been accounted for in the assumed threat point. These extra points then represent unattainable hopes or ideals, which Player II admits by agreeing to the outcome (1,2).

The fifth axiom, just reflects the understanding that the utilities of the players are separately determined only up to change of location and scale. Thus, if one of the players decides to change the location and scale of his utility, this changes the numbers in the bimatrix, but does not change the game. The agreed solution should undergo the same change.

Theorem. *There exists a unique function* $\mathbf{f}$ *satisfying the Nash axioms. Moreover, if there exists a point* $(u, v) \in S$ *such that* $u > u^*$ *and* $v > v^*$*, then* $\mathbf{f}(S, u^*, v^*)$ *is that point* $(\bar{u}, \bar{v})$ *of* S *that maximizes* $(u - u^*)(v - v^*)$ *among points of* S *such that* $u \geq u^*$ *and* $v \geq v^*$*.*

Below we sketch the proof in the interesting case where there exists a point $(u, v) \in S$ such that $u > u^*$ and $v > v^*$. The uninteresting case is left to the exercises.

First we check that the point $(\bar{u}, \bar{v})$ in $S^+ = \{(u,v) \in S : u \geq u^*, v \geq v^*\}$ indeed satisfies the Nash axioms. The first four axioms are very easy to verify. To check the fifth axiom, note that

if $(u - u^*)(v - v^*)$ is maximized over S^+ at $(\bar{u}, \bar{v})$,

then $(\alpha_1 u - \alpha_1 u^*)(\alpha_2 v - \alpha_2 v^*)$ is maximized over S^+ at $(\bar{u}, \bar{v})$,

so $(\alpha_1 u + \beta_1 - \alpha_1 u^* - \beta_1)(\alpha_2 v + \beta_2 - \alpha_2 v^* - \beta_2)$ is maximized over S^+ at $(\bar{u}, \bar{v})$,

hence $(u' - \alpha_1 u^* - \beta_1)(v' - \alpha_2 v^* - \beta_2)$ is maximized over T^+ at $(\alpha_1 \bar{u} + \beta_1, \alpha_2 \bar{v} + \beta_2)$, where $T^+ = \{(u', v') \in S^+ : u' = \alpha_1 u + \beta_1, v' = \alpha_2 v + \beta_2\}$.

To see that the axioms define the point uniquely, we find what $(\bar{u}, \bar{v})$ must be for certain special sets S, and extend step by step to all closed convex sets. First note that if S is symmetric about the line $u = v$ and $(0,0) \in S$, then axioms (1), (2), and (3) imply that $\mathbf{f}(S,0,0)$ is that point $(z,z) \in S$ farthest up the line $u = v$. Axiom 4 then implies that if T is any closed bounded convex subset of the half plane $H_z = \{(u,v) : u + v \leq 2z\}$ where $z > 0$, and if $(z,z) \in T$ and $(0,0) \in T$, then $\mathbf{f}(T,0,0) = (z,z)$, since such a set T is a subset of a symmetric set with the same properties.

Now for an arbitrary closed convex set S and $(u^*, v^*) \in S$, let $(\hat{u}, \hat{v})$ be the point of S^+ that maximizes $(u - u^*)(v - v^*)$. Define α_1, β_1, α_2, and β_2 so that $\begin{cases} \alpha_1 u^* + \beta_1 = 0 \\ \alpha_1 \hat{u} + \beta_1 = 1 \end{cases}$ and $\begin{cases} \alpha_2 v^* + \beta_2 = 0 \\ \alpha_2 \hat{v} + \beta_2 = 1 \end{cases}$, and let T be as in axiom 5. According to the invariance of $(\hat{u}, \hat{v})$ under change of location and scale, the point $(1,1) = (\alpha_1 \hat{u} + \beta_1, \alpha_2 \hat{v} + \beta_2)$ maximizes $u \cdot v$ over T^*. Since the slope of the curve $uv = 1$ at the point $(1,1)$ is -1, the set T, being convex, is a subset of H_1, and so by the last sentence of the previous paragraph, $\mathbf{f}(T,0,0) = (1,1)$. By axiom 5, $\mathbf{f}(T,0,0) = (\alpha_1 \bar{u} + \beta_1, \alpha_2 \bar{v} + \beta_2)$ where $\mathbf{f}(S, u^*, v^*) = (\bar{u}, \bar{v})$. Since $(\alpha_1 \bar{u} + \beta_1, \alpha_2 \bar{v} + \beta_2) = (1,1) = (\alpha_1 \hat{u} + \beta_1, \alpha_2 \hat{v} + \beta_2)$, we have $\bar{u} = \hat{u}$ and $\bar{v} = \hat{v}$, so that $(\hat{u}, \hat{v}) = \mathbf{f}(S, u^*, v^*)$. □

Here is a geometric interpretation. (See Fig. 17.3.) Consider the curves (hyperbolas) $(u - u^*)(v - v^*) = c$ for a constant c. For large enough c, this curve will not intersect S. Now bring c down until the curve just osculates S. The NTU-solution is the point of osculation.

Moreover, at the point $(\bar{u}, \bar{v})$ of osculation, the slope of the curve is the negative of the slope of the line from (u^*, v^*) to $(\bar{u}, \bar{v})$. (Check this.)

Examples.

1. Let S be the triangle with vertices $(0,0)$, $(0,1)$ and $(3,0)$, and let the threat point be $(0,0)$ as in Fig. 17.4. The Pareto optimal boundary is the

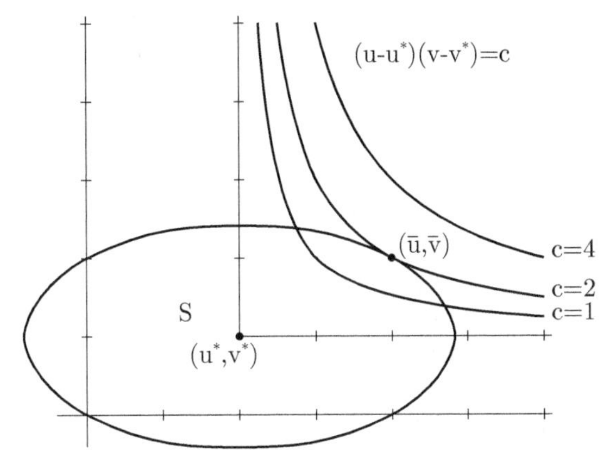

Fig. 17.3.

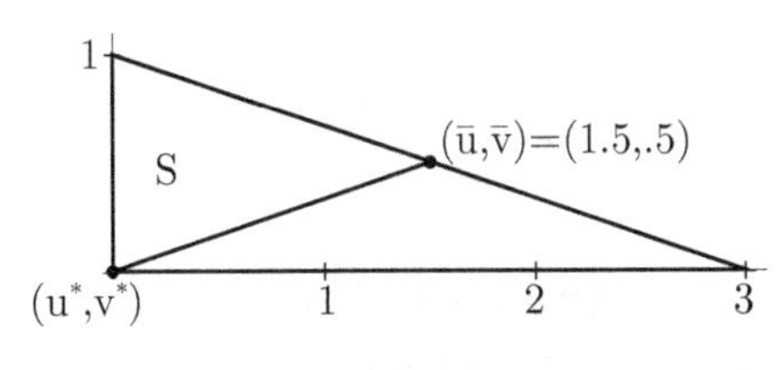

Fig. 17.4.

line from $(0,1)$ to $(3,0)$ of slope $-1/3$. The curve of the form $u \cdot v = c$ that osculates this line must have slope $-1/3$ at the point of osculation. So the slope of the line from (0,0) to $(\bar{u}, \bar{v})$ must be $1/3$. This intersects the Pareto boundary at the midpoint, $(3/2, 1/2)$. This is therefore the NTU-solution.

2. Let the NTU-feasible set be the ellipse, $S = \{(x,y) : (x-2)^2+4(y-1)^2 \leq 8\}$. Let the threat point be $(u^*, v^*) = (2,1)$. The situation looks much like that of Fig. 17.3. The point $(x,y) \in S$ that maximizes the product, $(x-2)(y-1)$, is the NTU-solution. This point must be on the Pareto optimal boundary consisting of the arc of the ellipse from $(2, 1+\sqrt{2})$ to $(2+2\sqrt{2}, 1)$. On this arc, $y - 1 = \sqrt{2-(x-2)^2/4}$, so we seek $x \in [2, 2+2\sqrt{2}]$ to maximize $(x-2)(y-1) = (x-2)\sqrt{2-(x-2)^2/4}$. The derivative of this is $\sqrt{2-(x-2)^2/4} - (x-2)^2/4\sqrt{2-(x-2)^2/4}$. Setting this to zero reduces to $(x-2)^2 = 4$, whose roots are 2 ± 2. Since $x \in [2, 2+2\sqrt{2}]$, we must have $x = 4$, and $y = 2$. Therefore $(\bar{u}, \bar{v}) = (4,2)$ is the NTU-solution of the game.

3. Consider the game with bimatrix (1), whose NTU-feasible set is given in Fig. 17.1(a). What should be taken as the threat point? If we take the view of Nash, that either player may refuse to enter into agreement, thus leaving the players at the status-quo point $(0, 0)$, then we should add this strategy of non-cooperation to the player's pure strategy sets. The resulting bimatrix of the game is really

$$\begin{pmatrix} (4,3) & (0,0) & (0,0) \\ (2,2) & (1,4) & (0,0) \\ (0,0) & (0,0) & (0,0) \end{pmatrix}. \tag{9}$$

This has the same NTU-feasible set as Fig. 17.1(a). We may take the threat point to be $(u^*, v^*) = (0, 0)$. The Pareto optimal boundary is the line segment from (1, 4) to (4, 3). This line has slope $-1/3$. However, the line of slope $1/3$ from the origin intersects the extension of this line segment at a point to the right of (4, 3). This means that $x \cdot y$ increases as (x, y) travels along this line segment from (1, 4) to (4, 3). The NTU-solution is therefore (4, 3).

The Lambda-Transfer Approach. There is another approach, due to Lloyd Shapley, for solving the NTU-problem that has certain advantages over the Nash approach. First, it relates the solution to the corresponding solution to the TU-problem. Second, it avoids the difficult-to-justify fourth axiom. Third, the threat point arises naturally as a function of the bimatrix and does not have to be specified *a priori*. Fourth, it extends to more general problems, but when specialized to the problems with status-quo point (0,0), it gives the same answer as the Nash solution.

The main difficulty with the NTU-problems is the lack of comparability of the utilities. If we pretend the utilities are measured in the same units and apply the TU-theory to arrive at a solution, it may happen that the TU-solution is in the NTU-feasible set. If it happens to be in the NTU-feasible set, the players can use it as the NTU-solution since it can be achieved without any transfer of utility. But what can be done if the TU-solution is not in the NTU-feasible set?

Recall that the utilities are not measured in the same units. Someone might suggest that an increase of one unit in Player 1's utility is worth an increase of λ units in Player 2's utility, where $\lambda > 0$. If that were so, we could analyze the game as follows. If the original bimatrix is $(\mathbf{A}, \mathbf{B})$, we first consider the game with bimatrix $(\lambda\mathbf{A}, \mathbf{B})$, solve it for the TU-solution, and then divide Player 1's payoff by λ to put it back into Player 1's original units. This is called the λ*-transfer game*. By the methods of Sec. 17.2, the TU-solution to the game with bimatrix $(\lambda A, B)$ is the vector $((\sigma(\lambda)+\delta(\lambda))/2, (\sigma(\lambda)-\delta(\lambda))/2)$, where

$$\sigma(\lambda) = \max_{ij}\{\lambda a_{ij} + b_{ij}\} \quad \text{and} \quad \delta(\lambda) = \mathrm{Val}(\lambda\mathbf{A} - \mathbf{B}).$$

The solution to the λ-transfer game is then found by dividing the first coordinate by λ, giving

$$\varphi(\lambda) = (\varphi_1(\lambda), \varphi_2(\lambda)) = \left(\frac{\sigma(\lambda)+\delta(\lambda)}{2\lambda}, \frac{\sigma(\lambda)-\delta(\lambda)}{2}\right). \tag{10}$$

If the point $\varphi(\lambda)$ is in the NTU-feasible set, it could, with the justification given earlier, be used as the NTU-solution.

It turns out that there generally exists a *unique* value of λ, call it λ^*, such that $\varphi(\lambda^*)$ is in the NTU-feasible set. This $\varphi(\lambda^*)$ can be used as the NTU-solution. The value of λ^* is called the *equilibrium exchange rate.*

The problem now is to find λ so that $\varphi(\lambda)$ is in the NTU-feasible set. In general, this may not be easy without the assistance of a computer because $\mathrm{Val}(\lambda\mathbf{A} - \mathbf{B})$ is not a simple function of λ.

However, in one case the problem becomes easy. This occurs for bimatrix games, $(\mathbf{A}, \mathbf{B})$, when the matrices $\mathbf{A}$ and $-\mathbf{B}$ have saddle points in the same position in the matrix. Such bimatrix games, when played as NTU games, are said to be *fixed threat point games.* Whatever be the value of λ in such games, the matrix game, $\lambda\mathbf{A} - \mathbf{B}$, which is used to determine the threat point, has a saddle point at that same position in the matrix. Thus, the threat strategy of the λ-transfer game will not depend on λ and the threat point is easy to determine. For example, in the game with bimatrix (9), the fixed threat strategy is the lower right corner, because both $\mathbf{A}$ and $-\mathbf{B}$ have saddle points there.

For NTU-games with a fixed threat point, the λ-transfer solution, $\varphi(\lambda^*)$, turns out to be the same as the Nash solution. So in this case we may find the λ-transfer solution by using the method already described to find the Nash solution.

To see why this is so, consider as an example the game with bimatrix,

$$\begin{pmatrix} (-1,1) & (1,3) \\ (0,0) & (3,-1) \end{pmatrix}. \tag{11}$$

Both $\mathbf{A}$ and $-\mathbf{B}$ have saddlepoints in the lower left corner. Therefore, $\lambda\mathbf{A} - \mathbf{B} = \begin{pmatrix} -\lambda - 1 & \lambda - 3 \\ 0 & 3\lambda + 1 \end{pmatrix}$ has a saddle point in the lower left corner, whatever be the value of $\lambda \geq 0$. So, (0,0) is a fixed threat point of the game.

The Nash solution is a point, $(\bar{u}, \bar{v})$, on the NTU-feasible set, S, such that the line through $(\bar{u}, \bar{v})$ with slope equal to the negative of the slope of the line from (0,0) to $(\bar{u}, \bar{v})$ is a tangent line to S at the point $(\bar{u}, \bar{v})$. Now if we change scale on the x-axis by $\lambda^* = \bar{v}/\bar{u}$, the point $(\bar{u}, \bar{v})$ goes into $(\lambda^*\bar{u}, \bar{v}) = (\bar{v}, \bar{v})$. The line from (0,0) to $(\bar{v}, \bar{v})$ is now the diagonal line and the slope of the tangent line at $(\bar{v}, \bar{v})$ is now -1. Therefore, $(\bar{v}, \bar{v})$ is the TU-solution of the λ^*-transfer game, and so $(\bar{u}, \bar{v})$ is also the λ-transfer solution! In addition, the equilibrium exchange rate is just λ^*.

In this example, $\lambda^* = 2$, and the NTU-solution is $(1.25, 2.5)$.

This argument is perfectly general and works even if the fixed threat point is not the origin. (Just change location along both axes so that the fixed threat point is the origin.) The λ-transfer solution is just the Nash solution, and the equilibrium exchange rate is just the slope of the line from the threat point to the Nash solution.

When there is no fixed threat point, one must use a direct method of finding that value of λ such that the point (10) is in the NTU-feasible set. Exercise 6 gives an idea of what is involved.

17.4 End-Game with an All-In Player

Poker is usually played with the "table-stakes" rule. This rule states that each player risks only the amount of money before him at the beginning of the hand. This means that a player can lose no more than what is in front of him on the table. It also means that a player cannot add to this amount during the play of a hand.

When a player puts all the money before him into the pot, he is said to be "all-in". When a player goes all-in, his money and an equal amount matched by each other player still contesting the pot is set aside and called the main pot. All further betting, if any, occurs only among the remaining players who are not all-in. These bets are placed in a side pot. Since the all-in player places no money in the side pot, he/she is not allowed to win it. The winner of the main pot is the player with the best hand and the winner of the side pot is the non-all-in player with the best hand. Betting in the side pot may cause a player to fold. When this happens, he/she relinquishes all rights to winning the main pot as well as the side pot. This gives the all-in player a subtle advantage. Occasionally, a player in the side pot will fold with a winning hand which allows the all-in player to win the main pot, which he would not otherwise have done. This possibility leads to some interesting problems.

As an example, consider End-Game in which there is an all-in player. Let's call the all-in player Player III. He can do nothing. He can only watch and hope to win the main pot. Let's assume he has four kings showing in a game of 5-card stud poker. Unfortunately for him, Player II has a better hand showing, say four aces. However, Player I has a possible straight flush. If he has the straight flush, he beats both players; otherwise he loses to both players. How should this hand be played? Will the all-in player with the four kings ever win the main pot?

We set up an idealized version of the problem mathematically. Let A denote the size of the main pot, and let p denote the probability that Player I has the winning hand. (Assume that the value of p is common knowledge to the players.) As in End-Game, Player I acts first by either checking or betting a fixed amount B. If Player I checks, he wins A if he has the winning hand, and Player II wins A otherwise. Player III wins nothing. If Player I bets, Player II may call or fold. If Player II calls and Player I has a winning hand, Player I wins $A+B$, Player II loses B, and Player III wins nothing. If Player II calls and Player I does not have the winning hand, Player I loses B and Player II wins $A+B$ and Player III wins nothing. If Player II folds and Player I has the winning hand, Player I wins A and the others win nothing. But if Player II folds and Player I does not have the winning hand, Player III wins A and the others win nothing.

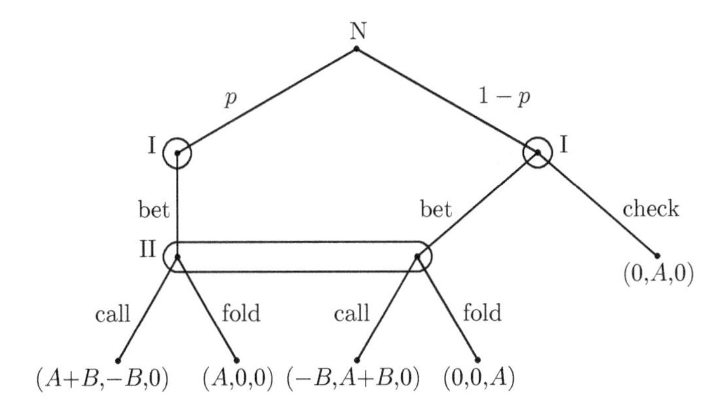

As in Basic Endgame, Player I never gains by checking with a winning hand, so Player I has two pure strategies, the *honest* strategy (bet with a winning hand, fold with a losing hand) and the *bluff* strategy (bet with a winning or losing hand). Player II also has two pure strategies, *call*, and *fold*. Player III has no strategies so the payoffs of the three players may be written in a 2×2 table as follows.

$$\begin{array}{c} \\ \textit{honest} \\ \textit{bluff} \end{array} \begin{array}{c} \begin{array}{cc} \qquad\qquad\qquad \textit{call} \qquad\qquad\qquad & \qquad\qquad \textit{fold} \end{array} \\ \begin{pmatrix} (p(A+B), -pB + (1-p)A, 0) & (pA, (1-p)A, 0) \\ (p(A+B) - (1-p)B,\ -pB + (1-p)(A+B),\ 0) & (pA, 0, (1-p)A) \end{pmatrix} \end{array}$$

Ordinarily, A will be somewhat larger than B, and p will be rather small. The analysis does not depend very much on the actual values of these three numbers, so it will be easier to understand if we take specific values for them. Let's take $A = 100$, $B = 10$ and $p = 1/10$. Then the payoff matrix becomes

$$\begin{array}{c} \\ \textit{honest} \\ \textit{bluff} \end{array} \begin{array}{c} \begin{array}{cc} \quad \textit{call} \quad & \quad \textit{fold} \end{array} \\ \begin{pmatrix} (11, 89, 0) & (10, 90, 0) \\ (2, 98, 0) & (10, 0, 90) \end{pmatrix} \end{array}$$

This is a constant-sum game for three players, but since Player III has no strategy choices and since coalitions among players are strictly forbidden by the rules of poker, it is best to consider this as non-constant-sum game between Players I and II. Removing the payoff for Player III from consideration, the matrix becomes

$$\begin{array}{c} \\ \textit{honest} \\ \textit{bluff} \end{array} \begin{array}{c} \begin{array}{cc} \quad \textit{call} \quad & \quad \textit{fold} \end{array} \\ \begin{pmatrix} (11, 89) & (10, 90) \\ (2, 98) & (10, 0) \end{pmatrix} \end{array} \tag{12}$$

First note that row 1 weakly dominates row 2, and if row 2 is removed, column 2 dominates column 1. This gives us an equilibrium at (row 1, column 2), with payoff (10,90). Player I cannot gain by betting with a losing hand. Even if the bluff is successful and Player II folds, Player III will still beat him. Worse, Player II may call his bluff and he will lose the bet.

So it seems that Player I might as well be honest. The result is that Player I never bluffs and Player II always folds when Player I bets, and the payoff is $(10, 90, 0)$. This is the accepted and time-honored course of action in real games in poker rooms around the world. But let's look at it more closely.

As long as Player II uses column 2, it doesn't hurt Player I to bluff. In fact, there are more equilibria. The strategy pair $(1-p, p)$ for Player I and column 2 for Player II is a equilibrium provided $p \le 1/99$ (Exercise 7(a)). The equilibrium with $p = 1/99$ has payoff $(10, 89\frac{1}{11}, \frac{10}{11})$. This takes payoff from Player II and gives it to Player III, something Player II wants to avoid. She may be willing to concede a small fracion of her winnings to avoid this.

Player I can play this equilibrium without loss. In fact, he can try for more by bluffing with probability greater than $p = 1/99$. If he does this, Player II's best reply is column 1. If Player I uses $p = 1/9$ and Player II uses column 2, Player I's average payoff is still $11(8/9) + 2(1/9) = 10$. A value of p between $1/99$ and $1/9$ encourages Player II to play column 1, and then Player I's average payoff will be greater than 10. So Player I wants more than 10, and Player II wants Player I not to use row 2.

Noncooperative games that are repeated in time in which side payments are not allowed may be considered as NTU cooperative games, in which the results of play over time take the place of preplay negotiations. If this game is analyzed as an NTU game by the methods of this section, the resulting payoff is $(10\frac{5}{6}, 89\frac{1}{6}, 0)$ (Exercise 7(b)). This may be achieved by Player I always playing row 1, and Player II playing column 1 with probability $5/6$. In other words, Player II calls 5 times out of 6, even though she knows that Player I never bluffs. (Another way of achieving this payoff is to have Player II call all the time, and to have Player I bluff with probability $1/54$.)

17.5 Exercises

1. For the following bimatrix games, draw the NTU and TU feasible sets. What are the Pareto optimal outcomes?

$$\text{(a)} \begin{pmatrix} (0,4) & (3,2) \\ (4,0) & (2,3) \end{pmatrix} \qquad \text{(b)} \begin{pmatrix} (3,1) & (0,2) \\ (1,2) & (3,0) \end{pmatrix}$$

(c) There is also a feasible set for non-cooperative games. For a non-cooperative bimatrix game, (A, B), the players cannot coordinate their mixed strategies, so the non-cooperative feasible set is just

$$\{(u, v) = (\mathbf{p}^T A\mathbf{q}, \mathbf{p}^T B\mathbf{q}) : \mathbf{p} \in X^*, \mathbf{q} \in Y^*\}.$$

This set may not be polyhedral. It may not even be convex. Graph the non-cooperative feasible set for the Battle of the Sexes with bimatrix $\begin{pmatrix} (2,1) & (0,0) \\ (0,0) & (1,2) \end{pmatrix}$. Note that the players cannot achieve a payoff of $(3/2, 3/2)$ using independent mixed strategies.

2. Find the cooperative strategy, the TU solution, the side payment, the optimal threat strategies, and the disagreement point for the two matrices (1) and (2) of Secs. 17.1 and 17.2.
3. Find the cooperative strategy, the TU solution, the side payment, the optimal threat strategies, and the disagreement point for the following matrices of Exercise 15.5.5. For (a), you may want to use the Matrix Game Solver on the web at http://www.tomsferguson.com/gamesolve.html.

$$\text{(a)} \quad \begin{pmatrix} (-3,-4) & (2,-1) & (0,\ 6) & (1,\ 1) \\ (2,\ 0) & (2,\ 2) & (-3,\ 0) & (1,-2) \\ (2,-3) & (-5,\ 1) & (-1,-1) & (1,-3) \\ (-4,\ 3) & (2,-5) & (1,\ 2) & (-3,\ 1) \end{pmatrix}.$$

$$\text{(b)} \quad \begin{pmatrix} (0,\ 0) & (1,-1) & (1,\ 1) & (-1,\ 0) \\ (-1,\ 1) & (0,\ 1) & (1,\ 0) & (0,\ 0) \\ (1,\ 0) & (-1,-1) & (0,\ 1) & (-1,\ 1) \\ (1,-1) & (-1,\ 0) & (1,-1) & (0,\ 0) \\ (1,\ 1) & (0,\ 0) & (-1,-1) & (0,\ 0) \end{pmatrix}.$$

4. Let $S = \{(x, y) : y \geq 0 \text{ and } y \leq 4 - x^2\}$ be the NTU-feasible set.
   (a) Find the NTU-solution if $(u^*, v^*) = (0, 0)$.
   (b) Find the NTU-solution if $(u^*, v^*) = (0, 1)$.
5. Find the NTU-solution and the equilibrium exchange rate for the following fixed threat point games.

$$\text{(a)} \quad \begin{pmatrix} (6,3) & (0,0) & (0,0) \\ (1,8) & (4,6) & (0,0) \\ (0,0) & (0,0) & (0,0) \end{pmatrix}, \quad \text{(b)} \quad \begin{pmatrix} (1,0) & (-1,1) & (0,0) \\ (3,3) & (-2,9) & (2,7) \end{pmatrix}.$$

6. Find the NTU-solution and the equilibrium exchange rates of the following games without a fixed threat point.

$$\text{(a)} \quad \begin{pmatrix} (5,2) & (0,0) \\ (0,0) & (1,4) \end{pmatrix}. \qquad \text{(b)} \quad \begin{pmatrix} (3,2) & (0,5) \\ (2,1) & (1,0) \end{pmatrix}.$$

7. (a) In Endgame with an all-in player with the values $A = 100$, and $B = 10$, show that the strategy pair $(1-p, p)$ for Player I and column 2 for Player II is an equilibrium pair with payoff $(10, 90(1-p), 90p)$ provided $0 \leq p \leq 1/99$. Show that these are the only equilibria. Show that if $p = 1/99$, Players I and II only get their safety levels.
   (b) Find the TU solution of this game. Show that the TU solution is in the NTU feasible set and so is also the NTU solution.

Part IV

Games in Coalitional Form

18 Many-Person TU Games

We now consider many-person **cooperative games**. In such games there are no restrictions on the agreements that may be reached among the players. In addition, we assume that all payoffs are measured in the same units and that there is a *transferrable utility* which allows *side payments* to be made among the players. Side payments may be used as inducements for some players to use certain mutually beneficial strategies. Thus, there will be a tendency for players, whose objectives in the game are close, to form alliances or coalitions. The structure given to the game by coalition formation is conveniently studied by reducing the game to a form in which coalitions play a central role. After defining the coalitional form of a many-person TU game, we shall learn how to transform games from strategic form to coalitional form and vice versa.

18.1 Coalitional Form. Characteristic Functions

Let $n \geq 2$ denote the number of players in the game, numbered from 1 to n, and let N denote the set of players, $N = \{1, 2, \ldots, n\}$. A *coalition*, S, is defined to be a subset of N, $S \subset N$, and the set of all coalitions is denoted by 2^N. By convention, we also speak of the empty set, $\emptyset$, as a coalition, the *empty coalition*. The set N is also a coalition, called the *grand coalition*.

If there are just two players, $n=2$, then there are four coalition, $\{\emptyset, \{1\}, \{2\}, N\}$. If there are 3 players, there are 8 coalitions, $\{\emptyset, \{1\}, \{2\}, \{3\}, \{1,2\}, \{1,3\}, \{2,3\}, N\}$. For n players, the set of coalitions, 2^N, has 2^n elements.

Definition. The *coalitional form* of an n-person game is given by the pair (N, v), where $N = \{1, 2, \ldots, n\}$ is the set of players and v is a real-valued function, called the *characteristic function* of the game, defined on the set, 2^N, of all coalitions (subsets of N), and satisfying

(i) $v(\emptyset) = 0$, and
(ii) (superadditivity) if S and T are disjoint coalitions ($S \cap T = \emptyset$), then $v(S) + v(T) \le v(S \cup T)$.

Compared to the strategic or extensive forms of n-person games, this is a very simple definition. Naturally, much detail is lost. The quantity $v(S)$ is a real number for each coalition $S \subset N$, which may be considered as the value, or worth, or power, of coalition S when its members act together as a unit. Condition (i) says that the empty set has value zero, and (ii) says that the value of two disjoint coalitions is at least as great when they work together as when they work apart. The assumption of superadditivity is not needed for some of the theory of coalitional games, but as it seems to be a natural condition, we include it in the definition.

18.2 Relation to Strategic Form

Recall that the *strategic form* of an n-person game is given by the $2n$-tuple, $(X_1, X_2, \ldots, X_n, u_1, u_2, \ldots, u_n)$, where

(1) for $i = 1, \ldots, n$, X_i is the set of pure strategies of Player i, and
(2) for $i = 1, \ldots, n$, $u_i(x_1, \ldots, x_n)$ is the payoff function to Player i, if Player 1 uses $x_1 \in X_1$, Player 2 uses $x_2 \in X_2, \ldots$, and Player n uses $x_n \in X_n$.

Transforming a game from strategic form to coalitional form entails specifying the value, $v(S)$, for each coalition $S \in 2^N$. The usual way to assign a characteristic function to a strategic form game is to define $v(S)$ for each $S \in 2^N$ as the *value* of the 2-person zero-sum game obtained when the coalition S acts as one player and the complementary coalition, $\overline{S} = N - S$, acts as the other player, and where the payoff to S is the sum of the payoffs to the players in S: $\sum_{i \in S} u_i(x_1, \ldots, x_n)$. Thus,

$$v(S) = \mathrm{Val}\left(\sum_{i \in S} u_i(x_1, \ldots, x_n) \right), \tag{1}$$

where the players in S jointly choose the x_i for $i \in S$, and the players in $\overline{S}$ choose the x_i for $i \notin S$. The value, $v(S)$, is the analogue of the safety level for coalition S. It represents the total amount that coalition S can guarantee for itself, even if the members of $\overline{S}$ gang up against it, and have as their only object to keep the sum of the payoffs to members of S as small as possible. This is a lower bound to the payoff S should receive because it assumes that the members of $\overline{S}$ ignore what possible payoffs they might receive as a result of their actions. An example of the computations involved in a three-person game is given below.

To see that v of Eq. (1) is a characteristic function, note that Condition (i) holds, since the empty sum is zero. To see that (ii) holds, note that if $\mathbf{s}$ is a set of strategies for S that guarantees them $v(S)$, and $\mathbf{t}$ is a set of strategies for T that guarantees them $v(T)$, then the set of strategies $(\mathbf{s},\mathbf{t})$ guarantees $S \cup T$ at least $v(S) + v(T)$. Perhaps other joint strategies can guarantee even more, so certainly, $v(S \cup T) \geq v(S) + v(T)$.

Every finite n-person game in strategic form can be reduced to coalitional form in this way. Often, *such a reduction to coalitional form loses important features in the game*, such as threats. So for a given characteristic function v, there are usually many games in strategic form whose reduction by the above method has characteristic function v. See Exercise 3 for a two-person game that favors one of the players, yet the reduction in coalitional form is symmetric in the players.

One way of constructing a strategic form game whose reduction to coalitional form has a given characteristic function, v, is as follows. The strategy space X_i for Player i is taken to be the set of all coalitions that contain i: $X_i = \{S \in 2^N : i \in S\}$. Then the payoff to Player i is the minimum amount, $v(\{i\})$, unless all members of the coalition, S_i, chosen by Player i, choose the same coalition as player i has, in which case the coalition S_i is given its value $v(S_i)$ which it then splits among its members. Thus the payoff function u_i is

$$u_i(S_1, \ldots, S_n) = \begin{cases} v(S_i)/|S_i| & \text{if } S_j = S_i, \text{ for all } j \in S_i \\ v(\{i\}) & \text{otherwise,} \end{cases} \tag{2}$$

where $|S_i|$ represents the number of members of the coalition S_i. Clearly, a coalition S can guarantee itself $v(S)$ simply by having each member of S select S as his coalition of choice. Moreover, since v is superadditive, the coalition S cannot guarantee more for itself by having its members form subcoalitions.

18.3 Constant-Sum Games

A game in strategic form is said to be *zero-sum* if $\sum_{i\in N} u_i(x_1,\ldots,x_n) = 0$ for all strategy choices $x_1,\ldots,x_n$ of the players. In such a game, we have $\sum_{i\in S} u_i(x_1,\ldots,x_n) = -\sum_{i\in \overline{S}} u_i(x_1,\ldots,x_n)$ for any coalition S, where $\overline{S} = N - S$ is the complement of S. This implies that in the reduction of such a game to coalitional form, the value of the game coalition S plays against $\overline{S}$ is the negative of the value of the game $\overline{S}$ plays against S, so that $v(S)+v(\overline{S}) = 0$ for all coalitions S. We may take this as the definition of a zero-sum game in coalitional form. Similarly, a strategic form game is *constant-sum* if $\sum_{i\in N} u_i(z_1,\ldots,z_n) = c$ for some constant c. By a similar reasoning, the reduction of such a game leads to $v(S) + v(\overline{S}) = c = v(N)$ for all coalitions S in a constant sum game. This may be taken as the definition of a constant-sum game in coalitional form.

Definition. A game in coalitional form is said to be *constant-sum*, if $v(S)+ v(\overline{S}) = v(N)$ for all coalitions $S \in 2^N$. It is said to be *zero-sum* if, in addition, $v(N) = 0$.

18.4 Example

Consider the three-person game with players I, II, and III with two pure strategies each and with payoff vectors:

If I chooses 1:

	III: 1	III: 2
II: 1	$(0,3,1)$	$(2,1,1)$
2	$(4,2,3)$	$(1,0,0)$

If I chooses 2:

	III: 1	III: 2
II: 1	$(1,0,0)$	$(1,1,1)$
2	$(0,0,1)$	$(0,1,1)$

Let us find the associated game in coalitional form by finding the characteristic function, v. We automatically have $v(\emptyset) = 0$. It is easy to find $v(N)$. It is the largest sum in the eight cells. This occurs for the cell $(1,2,1)$ and gives total payoff $v(N) = 9$. To find $v(\{1\})$, compute the payoff matrix for the winnings of I against (II, III):

	(II, III) 1,1	1,2	2,1	2,2
I: 1	0	2	4	1
2	1	1	0	0

The second and third columns are dominated, so $v(\{1\}) = \mathrm{Val}\begin{pmatrix} 0 & 1 \\ 1 & 0 \end{pmatrix} = 1/2$.

To find $v(\{2\})$ and $v(\{3\})$, we make similar constructions of the matrices of II's winnings vs. I and III, and III's winnings vs I and II and find the values of the resulting games. In the matrix of II's winnings, the choice 2 by II and (2, 1) by (I, III) is a saddlepoint with value $v(\{2\}) = 0$. In the matrix of III's winnings, the value is $v(\{3\}) = 3/4$.

To find $v(\{1,3\})$ say, we first construct the matrix of the sum of the winnings of I and III playing against II. This is

(I, III):	II: 1	2
1, 1	1	7
1, 2	3	1
2, 1	1	1
2, 2	2	1

The lower two rows are dominated by the second row, so that the value is $v(\{1,3\}) = \mathrm{Val}\begin{pmatrix} 1 & 7 \\ 3 & 1 \end{pmatrix} = 5/2$. Similarly, we may compute the matrix of I and II playing against III, and the matrix of II and III playing against I. Both these matrices have saddle points. We find $v(\{1,2\}) = 3$ and $v\{2,3\}) = 2$. This completes the specification of the characteristic function.

18.5 Exercises

1. Find the characteristic function of the 3-person game with players I, II, and III with two pure strategies each and with the following payoff vectors. Note that this is a zero-sum game. Hence, $v(\{1,3\}) = -v(\{2\})$, etc.

   If I chooses 1:

II \ III:	1	2
1	$(-2,1,1)$	$(1,-4,3)$
2	$(1,3,-4)$	$(10,-5,-5)$

If I chooses 2:

	III: 1	2
II: 1	$(-1,-2,3)$	$(-4,2,2)$
2	$(12,-6,-6)$	$(-1,3,-2)$

2. Find the characteristic function of the 3-person game in strategic form when the payoff vectors are:

If I chooses 1:

	III: 1	2
II: 1	$(1,2,1)$	$(3,0,1)$
2	$(-1,6,-3)$	$(3,2,1)$

If I chooses 2:

	III: 1	2
II: 1	$(-1,2,4)$	$(1,0,3)$
2	$(7,5,4)$	$(3,2,1)$

3. Consider the two-person game with bimatrix

$$\begin{pmatrix} (0,2) & (4,1) \\ (2,4) & (5,4) \end{pmatrix}$$

(a) Find the associated game in coalitional form. Note that in coalitional form the game is symmetric in the two players.
(b) Argue that the above game is actually favorable to Player 2.
(c) Find the TU-value as a game in strategic form. Note that this value gives more to Player 2 than to Player 1.
(d) If the game in coalitional form found in (a) is transformed to strategic form by the method of Eq. (2), what is the bimatrix that arises?

4. Consider the following 3-person game of perfect information. Let S denote the set $\{1,2,\ldots,10\}$. First Player 1 chooses $i \in S$. Then Player 2, knowing i, chooses $j \in S$, $j \neq i$. Finally Player 3, knowing i and j, chooses $k \in S$, $k \neq i$, $k \neq j$. The payoff given these three choices is $(|i-j|, |j-k|, |k-i|)$. Find the coalitional form of the game.

19 Imputations and the Core

In cooperative games, it is to the joint benefit of the players to form the grand coalition, N, since by superadditivity the amount received, $v(N)$, is as large as the total amount received by any disjoint set of coalitions they could form. As in the study of 2-person TU games, it is reasonable to suppose that "rational" players will agree to form the grand coalition and receive $v(N)$. The problem is then to agree on how this amount should be split among the players. In this section, we discuss one of the possible properties of an agreement on a fair division, that it be stable in the sense that no coalition should have the desire and power to upset the agreement. Such divisions of the total return are called points of the core, a central notion of game theory in economics.

19.1 Imputations

A payoff vector $\mathbf{x} = (x_1, x_2, \ldots, x_n)$ of proposed amounts to be received by the players, with the understanding that Player i is to receive x_i, is sometimes called an *imputation*. The first desirable property of an imputation is that the total amount received by the players should be $v(N)$.

Definition. A payoff vector, $\mathbf{x} = (x_1, x_2, \ldots, x_n)$, is said to be *group rational* or *efficient* if $\sum_{i=1}^{n} x_i = v(N)$.

No player could be expected to agree to receive less than that player could obtain acting alone. Therefore, a second natural condition to expect of an imputation, $\mathbf{x} = (x_1, x_2, \ldots, x_n)$, is that $x_i \geq v(\{i\})$ for all players, i.

Definition. A payoff vector, $\mathbf{x}$, is said to be *individually rational* if $x_i \geq v(\{i\})$ for all $i = 1, \ldots, n$.

Imputations are defined to be those payoff vectors that satisfy both these conditions.

Definition. An *imputation* is a payoff vector that is group rational and individually rational. The set of imputations may be written

$$\left\{\mathbf{x} = (x_1, \ldots, x_n) : \sum_{i \in N} x_i = v(N), \text{ and } x_i \geq v(\{i\}) \text{ for all } i \in N\right\}. \tag{1}$$

Thus, an imputation is an n-vector, $\mathbf{x} = (x_1, \ldots, x_n)$, such that $x_i \geq v(\{i\})$ for all i and $\sum_{i=1}^{n} x_i = v(N)$. The set of imputations is never empty, since from the superadditivity of v, we have $\sum_{i=1}^{n} v(\{i\}) \leq v(N)$. For example, one imputation is given by $\mathbf{x} = (x_1, x_2, \ldots, x_n)$, where $x_i = v(\{i\})$ for $i = 1, \ldots, n-1$, and $x_n = v(N) - \sum_{1}^{n-1} v(\{i\})$. This is the imputation most preferred by Player n. In fact the set of imputations is exactly the simplex consisting of the convex hull of the n points obtained by letting $x_i = v(\{i\})$ for all x_i except one, which is then chosen to satisfy $\sum_{1}^{n} x_i = v(N)$.

In Example 18.4, $v(\{1\}) = 1/2$, $v(\{2\}) = 0$, $v(\{3\}) = 3/4$, and $v(N) = 9$. The set of imputations is

$$\{(x_1, x_2, x_3) : x_1 + x_2 + x_3 = 9,\ x_1 \geq 1/2,\ x_2 \geq 0,\ x_3 \geq 3/4\}.$$

This is a triangle (in fact, an equilateral triangle) each of whose vertices satisfy two of the three inequalities with equality, namely, $(8\frac{1}{4}, 0, \frac{3}{4})$, $(\frac{1}{2}, 7\frac{3}{4}, \frac{3}{4})$, and $(\frac{1}{2}, 0, 8\frac{1}{2})$. These are the imputations most preferred by players 1, 2, and 3 respectively.

19.2 Essential Games

There is one trivial case in which the set of imputations consists of one point. Such a game is called inessential.

Definition. A game in coalitional form is said to be *inessential* if $\sum_{i=1}^{n} v(\{i\}) = v(N)$, and *essential* if $\sum_{i=1}^{n} v(\{i\}) < v(N)$.

If a game is inessential, then the unique imputation is $\mathbf{x} = (v(\{1\}), \ldots, v(\{n\}))$, which may be considered the "solution" of the game. Every player can expect to receive his safety level. Two-person zero-sum games are all inessential. (Exercise 1.)

From the game-theoretic viewpoint, inessential games are trivial. For every coalition S, $v(S)$ is determined by $v(S) = \sum_{i \in S} v(\{i\})$. There is no tendency for the players to form coalitions.

In Example 18.4, $v(\{1\})+v(\{2\})+v(\{3\}) = 1/2+0+3/4 < 9 = v(N)$, so the game is essential, i.e, non-trivial.

19.3 The Core

Suppose some imputation, $\mathbf{x}$, is being proposed as a division of $v(N)$ among the players. If there exists a coalition, S, whose total return from $\mathbf{x}$ is less than what that coalition can achieve acting by itself, that is, if $\sum_{i \in S} x_i < v(S)$, then there will be a tendency for coalition S to form and upset the proposed $\mathbf{x}$ because such a coalition could guarantee each of its members more than they would receive from $\mathbf{x}$. Such an imputation has an inherent instability.

Definition. An imputation $\mathbf{x}$ is said to be *unstable through a coalition* S if $v(S) > \sum_{i \in S} x_i$. We say $\mathbf{x}$ is *unstable* if there is a coalition S such that $\mathbf{x}$ is unstable through S, and we say $\mathbf{x}$ is *stable* otherwise.

Definition. The set, C, of stable imputations is called the *core*,

$$C = \left\{ \mathbf{x} = (x_1, \ldots, x_n) : \sum_{i \in N} x_i = v(N) \quad \text{and} \right.$$

$$\left. \sum_{i \in S} x_i \geq v(S), \quad \text{for all } S \subset N \right\}. \tag{2}$$

The core can consist of many points as in the examples below; but the core can also be empty. It may be impossible to satisfy all the coalitions at the same time. One may take the size of the core as a measure of stability, or of how likely it is that a negotiated agreement is prone to be upset. One class of games with empty cores are the essential constant-sum games.

Theorem 19.1. *The core of an essential constant-sum game is empty.*

Proof. Let $\mathbf{x}$ be an imputation. Since the game is essential, we have $\sum_{i \in N} v(\{i\}) < v(N)$. Then there must be a player, k, such that $x_k > v(\{k\})$, for otherwise $v(N) = \sum_{i \in N} x_i \leq \sum_{i \in N} v(\{i\}) < v(N)$. Since the game is constant-sum, we have $v(N - \{k\}) + v(\{k\}) = v(N)$. But then, $\mathbf{x}$ must be unstable through the coalition $N - \{k\}$, because $\sum_{i \neq k} x_i = \sum_{i \in N} x_i - x_k < V(N) - v(\{k\}) = v(N - \{k\})$. □

19.4 Examples

Example 1. Consider the game with characteristic function v given by

$$
\begin{array}{llll}
 & v(\{1\})=1 & v(\{1,2\})=4 & \\
v(\emptyset)=0 & v(\{2\})=0 & v(\{1,3\})=3 & v(\{1,2,3\})=8 \\
 & v(\{3\})=1 & v(\{2,3\})=5 &
\end{array}
$$

The imputations are the points (x_1, x_2, x_3) such that $x_1 + x_2 + x_3 = 8$ and $x_1 \geq 1$, $x_2 \geq 0$, $x_3 \geq 1$. This set is the triangle with vertices $(7,0,1)$, $(1,6,1)$ and $(1,0,7)$.

It is useful to plot this triangle in *barycentric coordinates.* This is done by pretending that the plane of the plot is the plane $x_1 + x_2 + x_3 = 8$, and giving each point on the plane three coordinates which add to 8. Then it is easy to draw the lines $x_1 = 1$ or the line $x_1 + x_3 = 3$ (which is the same as the line $x_2 = 5$), etc. It then becomes apparent that the set of imputations is an equilateral triangle.

On the plane $x_1 + x_2 + x_3 = 8$:

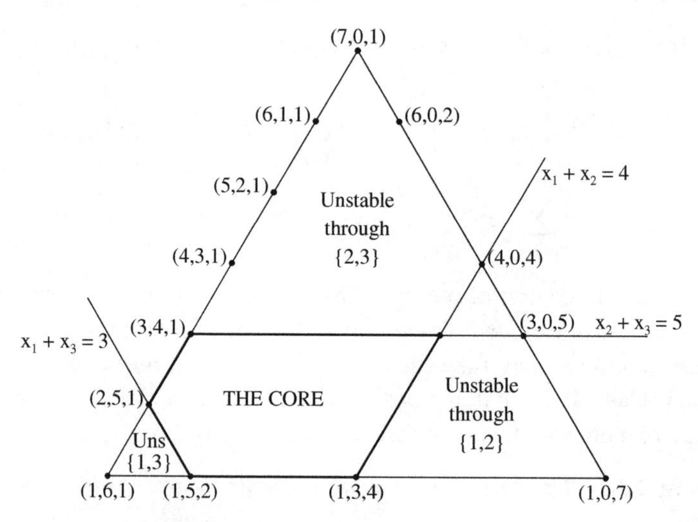

Let us find which imputations are unstable. The coalition $\{2,3\}$ can guarantee itself $v(\{2,3\}) = 5$, so all points (x_1, x_2, x_3) with $x_2 + x_3 < 5$ are unstable through $\{2,3\}$. These are the points above the line $x_2 + x_3 = 5$ in the diagram. Since $\{1,2\}$ can guarantee itself $v(\{1,2\}) = 4$, all points below and to the right of the line $x_1 + x_2 = 4$ are unstable. Finally, since $\{1,3\}$ can guarantee itself $v(\{1,3\}) = 3$, all points below the line $x_1 + x_3 = 3$ are unstable. The core is the remaining set of points in the set of imputations given by the 5-sided figure in the diagram, including the boundary.

Example 2. A certain *objet d'art* is worth a_i dollars to Player i for $i = 1, 2, 3$. We assume $0 < a_1 < a_2 < a_3$, so Player 3 values the object most. But Player 1 owns this object so $v(\{1\}) = a_1$. Players 2 and 3 by themselves can do nothing, so $v(\{2\}) = 0$, $v(\{3\}) = 0$, and $v(\{2,3\}) = 0$. If Players 1 and 2 come together, the joint worth is a_2, so $v(\{1,2\}) = a_2$. Similarly, $v(\{1,3\}) = a_3$. If all three get together, the object is still only worth a_3, so $v(N) = a_3$. Let us find the core of this game.

The core consists of all vectors (x_1, x_2, x_3) satisfying

$$\begin{array}{lll} x_1 \geq a_1 & x_1 + x_2 \geq a_2 & \\ x_2 \geq 0 & x_1 + x_3 \geq a_3 & x_1 + x_2 + x_3 = a_3 \\ x_3 \geq 0 & x_2 + x_3 \geq 0 & \end{array}$$

It follows from $x_2 = a_3 - x_1 - x_3 \leq 0$ and $x_2 \geq 0$ that $x_2 = 0$ for all points of the core. Then we find that $x_1 \geq a_2$ and $x_3 = a_3 - x_1$. Hence the core is $C = \{(x, 0, a_3 - x) : a_2 \leq x \leq a_3\}$.

This indicates that the object will be purchased by Player 3 at some purchase price x between a_2 and a_3. Player 1 ends up with x dollars and Player 3 ends up with the object minus x dollars. Player 2 plays no active role in this, but without her around Player 3 might hope to get the object for less than a_2.

19.5 Exercises

1. Show that every 2-person constant-sum game is inessential.
2. Find the set of imputations and the core of the battle of the sexes with bimatrix:

$$\begin{pmatrix} (4,2) & (1,1) \\ (0,0) & (2,4) \end{pmatrix}.$$

3. Graph the core for the 3-person game with characteristic function: $v(\emptyset) = 0$, $v(\{1\}) = 0$, $v(\{2\}) = -1$, $v(\{3\}) = 1$, $v(\{1,2\}) = 3$, $v(\{1,3\}) = 2$, $v(\{2,3\}) = 4$, and $v(N) = 5$.
4. **Definition.** A game with characteristic function v is said to be *symmetric* if $v(S)$ depends only on the number of elements of S, say $v(S) = f(|S|)$ for some function f.

   (a) In a symmetric 3-player game with $v(\{i\}) = 0$, $v(\{i,j\}) = a$ and $v(\{1,2,3\}) = 3$, for what values of a is the core non-empty?

(b) In a symmetric 4-player game with $v(\{i\})=0$, $v(\{i,j\})=a$, $v(\{i,j,k\}) = b$, and $v(N) = 4$, for what values of a and b is the core non-empty?

(c) Generalize. Find necessary and sufficient conditions on the values of $f(|S|) = v(S)$ for a symmetric game to have a non-empty core.

5. Let $\delta_i = v(N) - v(N - \{i\})$ for $i = 1, \ldots, n$. Show that the core is empty if $\sum_1^n \delta_i < v(N)$.

6. We say that Player i is a *dummy* in a game (N, v), if $v(\{i\} \cup S) = v(S)$ for all coalitions, S. In particular, $v(\{i\}) = 0$. Thus, a dummy cannot help (or harm) any coalition. Show that if Player 1 is a dummy and if $(x_1, x_2, \ldots, x_n)$ is in the core, then $x_1 = 0$.

7. **The Glove Market.** Let N consist of two types of players, $N = P \cup Q$, where $P \cap Q = \emptyset$. Let the characteristic function be defined by

$$v(S) = \min\{|S \cap P|, |S \cap Q|\}.$$

The game (N, v) is called the glove market because of the following interpretation. Each player of P owns a right-hand glove and each player of Q owns a left-hand glove. If j members of P and k members of Q form a coalition, they have $\min\{j, k\}$ complete pairs of gloves, each being worth 1. Unmatched gloves are worth nothing.

(a) Suppose $|P| = 2$ and $|Q| = 2$. Find the core.

(b) Suppose $|P| = 2$ and $|Q| = 3$. Show the core consists of a single point.

(c) Generalize to arbitrary $|P|$ and $|Q|$.

8. There are two machine owners (Players 1 and 2) and three workers (Players 3, 4 and 5). Each machine owner owns two machines. Each worker can produce 1 unit on any machine. Thus for example,
$v(\{i, k\}) = 1$ for $i = 1, 2$ and $k = 3, 4, 5$,
$v(\{i, j, k\}) = 2$ for $i = 1, 2$ and $j, k = 3, 4, 5$.
$v(\{1, 2, 3, 4, 5\}) = 3$.
Find the core.

20 The Shapley Value

We now treat another approach to n-person games in characteristic function form. The concept of the core is useful as a measure of stability. As a solution concept, it presents a set of imputations without distinguishing one point of the set as preferable to another. Indeed, the core may be empty.

Here we deal with the concept of a value. In this approach, one tries to assign to each game in coalitional form a unique vector of payoffs, called the value. The ith entry of the value vector may be considered as a measure of the value or power of the ith player in the game. Alternatively, the value vector may be thought of as an arbitration outcome of the game decided upon by some fair and impartial arbiter. The central "value concept" in game theory is the one proposed by Shapley (1953b). We define the Shapley value in this section and discuss its application to measuring power in voting systems where it is called the Shapley–Shubik power index. In Sec. 21.4, we treat another value concept, the nucleolus.

20.1 Value Functions — The Shapley Axioms

As an example of the type of reasoning involved in arbitrating a game, consider Example 1 of Sec. 19.4. Certainly the arbiter should require the players to form the grand coalition to receive 8, but how should this be split among the players? Player 2 can get nothing by himself, yet he is more valuable than 1 or 3 in forming coalitions. Which is more important? Shapley approached this problem by axiomatizing the concept of fairness.

A *value function*, ϕ, is function that assigns to each possible characteristic function of an n-person game, v, an n-tuple, $\phi(v) = (\phi_1(v), \phi_2(v), \ldots, \phi_n(v))$ of real numbers. Here $\phi_i(v)$ represents the worth or value of player i in the game with characteristic function v. The axioms of fairness are placed on the function, ϕ.

The Shapley Axioms for $\phi(v)$:

1. **Efficiency.** $\sum_{i \in N} \phi_i(v) = v(N)$.
2. **Symmetry.** If i and j are such that $v(S \cup \{i\}) = v(S \cup \{j\})$ for every coalition S not containing i and j, then $\phi_i(v) = \phi_j(v)$.
3. **Dummy Axiom.** If i is such that $v(S) = v(S \cup \{i\})$ for every coalition S not containing i, then $\phi_i(v) = 0$.
4. **Additivity.** If u and v are characteristic functions, then $\phi(u + v) = \phi(u) + \phi(v)$.

Axiom 1 is group rationality, that the total value of the players is the value of the grand coalition. The second axiom says that if the characteristic function is symmetric in Players i and j, then the values assigned to i and j should be equal. The third axiom says that if Player i is a dummy in the sense that he neither helps nor harms any coalition he may join, then his value should be zero. The strongest axiom is number 4. It reflects the feeling that the arbitrated value of two games played at the same time should be the sum of the arbitrated values of the games if they are played at different times. It should be noted that if u and v are characteristic functions, then so is $u + v$.

Theorem 20.1. *There exists a unique function ϕ satisfying the Shapley axioms.*

Proof. For a given nonempty set $S \subset N$, let w_S represent the special characteristic function, defined for all $T \subset N$ to be

$$w_S(T) = \begin{cases} 1 & \text{if } S \subset T \\ 0 & \text{otherwise.} \end{cases} \tag{1}$$

From axiom 3, $\phi_i(w_S) = 0$ if $i \notin S$. From axiom 2, if both i and j are in S, then $\phi_i(w_S) = \phi_j(w_S)$. From axiom 1, $\sum_{i \in N} \phi_i(w_S) = w_S(N) = 1$, so that $\phi_i(w_S) = 1/|S|$ for all $i \in S$. Applying similar analysis to the characteristic function cw_S for an arbitrary number, c, we find

$$\phi_i(cw_S) = \begin{cases} c/|S| & \text{for } i \in S \\ 0 & \text{for } i \notin S. \end{cases} \tag{2}$$

In the next paragraph, we show that any characteristic function, v, is representable uniquely as a weighted sum of characteristic functions of the form (1), $v = \sum_{S \subset N} c_S w_S$, for some appropriate, easily computable, constants c_S. Then axiom 4 may be applied to show that if a value function exists, it must be

$$\phi_i(v) = \sum_{\substack{S \subset N \\ i \in S}} \frac{c_S}{|S|}, \tag{3}$$

where this sum is taken over all coalitions S containing i. This works even if some of the c_S are negative, since axiom 4 also implies that $\phi(u - v) = \phi(u) - \phi(v)$, provided u, v, and $u - v$ are characteristic functions. (Just write $u = (u - v) + v$.) To complete the proof, one must show existence, namely that (3) with the c_S defined below, satisfies the Shapley axioms. This is not difficult but we defer the proof to Theorem 20.2, where we show existence by showing another value function satisfies the Shapley axioms (and therefore must be the same as (3)).

Now let us show that any v may be written as $v = \sum_{S \subset N} c_S w_S$ by finding the constants c_S. Let $c_\emptyset = 0$, and define inductively on the number of elements in T, for all $T \subset N$,

$$c_T = v(T) - \sum_{\substack{S \subset T \\ S \neq T}} c_S. \tag{4}$$

Note that each c_T is defined in terms of c_S for *all* $S \subset T$, $S \neq T$. Then,

$$\sum_{S \subset N} c_S w_S(T) = \sum_{S \subset T} c_S = c_T + \sum_{\substack{S \subset T \\ S \neq T}} c_S = v(T). \tag{5}$$

Hence, $v = \sum_{S \subset N} c_S w_S$ as was to be shown.

To show unicity, suppose there are two sets of constants, c_S and c'_S, such that

$$V(T) = \sum_{S \subset N} c_S w_S(T) = \sum_{S \subset N} c'_S w_S(T) \tag{6}$$

for all $T \subset N$. We use induction to show that $c_S = c'_S$ for all S. First if T is the singleton set $\{i\}$ in (6), then all $w_s(T)$ vanish except for $S = \{i\}$, and we have $c_{\{i\}} = c'_{\{i\}}$ for all i. Now let R be an arbitrary coalition, and suppose that $c_S = c'_S$ for all $S \subset R$. Then in (6) with $T = R$, all terms cancel except the term with $S = R$ leaving $C_r = c'_R$. □

It is interesting to note that the superadditivity of v is not needed in this proof.

20.2 Computation of the Shapley Value

The proof of Theorem 20.1 provides a method of computing the Shapley value: First find the numbers c_S inductively using Eq. (4). Then form the Shapley value using Eq. (3).

As an example of this method, consider the characteristic function of Example 1 of Sec. 19.4.

$$\begin{array}{llll} & v(\{1\})=1 & v(\{1,2\})=4 & \\ v(\emptyset)=0 & v(\{2\})=0 & v(\{1,3\})=3 & v(\{1,2,3\})=8. \\ & v(\{3\})=1 & v(\{2,3\})=5 & \end{array}$$

We find inductively, $c_{\{1\}}=v(\{1\})=1$, $c_{\{2\}}=0$ and $c_{\{3\}}=1$. Then, $c_{\{1,2\}}=v(\{1,2\})-c_{\{1\}}-c_{\{2\}}=4-1-0=3$, $c_{\{1,3\}}=3-1-1=1$, and $c_{\{2,3\}}=5-0-1=4$. Finally,

$$\begin{aligned} c_N &= v(N)-c_{\{1,2\}}-c_{\{1,3\}}-c_{\{2,3\}}-c_{\{1\}}-c_{\{2\}}-c_{\{3\}} \\ &= 8-3-1-4-1-0-1=-2. \end{aligned}$$

Hence, we know we can write v as

$$v = w_{\{1\}}+w_{\{3\}}+3w_{\{1,2\}}+w_{\{1,3\}}+4w_{\{2,3\}}-2w_{\{1,2,3\}}.$$

From this we find

$$\begin{aligned} \phi_1(v) &= 1+\frac{3}{2}+\frac{1}{2}-\frac{2}{3}=2+\frac{1}{3}, \\ \phi_2(v) &= \frac{4}{2}+\frac{3}{2}-\frac{2}{3}=2+\frac{5}{6}, \\ \phi_3(v) &= 1+\frac{1}{2}+\frac{4}{2}-\frac{2}{3}=2+\frac{5}{6}. \end{aligned}$$

The Shapley value is $\phi=(14/6,17/6,17/6)$. This point is in the core (see the diagram in Sec. 19.4). The Shapley value is not always in the core. For one thing, it could happen that the core is empty. But even if the core is not empty, the Shapley value is not necessarily in the core. (See Exercise 1, for example.)

20.3 An Alternative Form of the Shapley Value

There is an alternate way of arriving at the Shapley value that gives additional insight into its properties. Suppose we form the grand coalition by entering the players into this coalition one at a time. As each player enters the coalition, he receives the amount by which his entry increases the value of the coalition he enters. The amount a player receives by this scheme depends on the order in which the players are entered. The Shapley value is just the average payoff to the players if the players are entered in completely random order.

Theorem 20.2. *The Shapley value is given by* $\phi = (\phi_1, \ldots, \phi_n)$, *where for* $i = 1, \ldots, n$,

$$\phi_i(v) = \sum_{\substack{S \subset N \\ i \in S}} \frac{(|S|-1)!(n-|S|)!}{n!}[v(S) - v(S - \{i\})]. \tag{7}$$

The summation in this formula is the summation over all coalitions S that contain i. The quantity, $v(S) - v(S - \{i\})$, is the amount by which the value of coalition $S - \{i\}$ increases when player i joins it. Thus to find $\phi_i(v)$, merely list all coalitions containing i, compute the value of player i's contribution to that coalition, multiply this by $(|S|-1)!(n-|S|)!/n!$, and take the sum.

The interpretation of this formula is as follows. Suppose we choose a random order of the players with all $n!$ orders (permutations) of the players equally likely. Then we enter the players according to this order. If, when player i is enters, he forms coalition S (that is, if he finds $S - \{i\}$ there already), he receives the amount $[v(S) - v(S - \{i\})]$.

The probability that when i enters he will find coalition $S - \{i\}$ already there is $(|S|-1)!(n-|S|)!/n!$. The denominator is the total number of permutations of the n players. The numerator is number of these permutations in which the $|S|-1$ members of $S - \{i\}$ come first ($(|S|-1)!$ ways), then player i, and then the remaining $n - |S|$ players ($(n-|S|)!$ ways). So this formula shows that $\phi_i(v)$ is just the average amount player i contributes to the grand coalition if the players sequentially form this coalition in a random order.

As an illustration of the use of this formula, let us compute $\phi_1(v)$ again for Example 1 of Sec. 19.4. The probability that Player 1 enters first is $2!0!/3! = 1/3$, and then his payoff is $v(\{1\}) = 1$. The probability that 1 enters second and finds 2 there is 1/6, and his payoff is $v(\{1,2\}) - v(\{2\}) = 4 - 0 = 4$. The probability that 1 enters second and finds 3 there is 1/6, and the expected payoff is $v(\{1,3\}) - v(\{3\}) = 3 - 1 = 2$. The probability that 1 enters last is 1/3, and then his payoff is $v(\{1,2,3\}) - v(\{2,3\}) = 8 - 5 = 3$.

Player 1's average payoff is therefore

$$\phi_1(v) = \frac{1}{3} \cdot 1 + \frac{1}{6} \cdot 4 + \frac{1}{6} \cdot 2 + \frac{1}{3} \cdot 3 = 14/6$$

as found earlier.

The following table shows the computations for all three players simultaneously. The 6 different orders of the players are listed along with the payoffs to the players. In the first row, the players enter in the order 1, 2, 3. Player 1 receives $v(1) = 1$ upon entry; then Player 2 receives $v(1,2) - v(1) = 4 - 1 = 3$; finally Player 3 receives $v(N) - v(1,2) = 8 - 4 = 4$. Each of the six rows is equally likely, probability 1/6 each. The Shapley value is the average of the six numbers in each column.

	Player			
Order of Entry	1	2	3	Total
1 2 3	1	3	4	8
1 3 2	1	5	2	8
2 1 3	4	0	4	8
2 3 1	3	0	5	8
3 1 2	2	5	1	8
3 2 1	3	4	1	8
Average	14/6	17/6	17/6	8

Proof of Theorem 20.2. To see that formula (7) gives the Shapley value, we have only to see that it satisfies axioms 1 through 4, since we have already shown that there is at most one such function, ϕ. Axioms 2, 3, and 4 are easy to check directly from the formula of Theorem 20.2. Axiom 1 follows from the above interpretation of the formula, since in each realization of forming the grand coalition, exactly $v(N)$ is given to the players. Hence, the average amount given to the players is also $v(N)$. □

20.4 Simple Games. The Shapley–Shubik Power Index

The Shapley value has an important application in modeling the power of members of voting games. This application was developed by Shapley and Shubik in 1954 and the measure is now known as the Shapley–Shubik Power Index.

Players are members of legislature or members of the board of directors of a corporation, etc. In such games, a proposed bill or decision is either passed or rejected. Those subsets of the players that can pass bills without outside help are called winning coalitions while those that cannot are called losing coalitions. In all such games, we may take the value of a winning coalition to be 1 and the value of a losing coalition to be 0. Such games are called simple games.

Definition. A game (N, v) is *simple* if for every coalition $S \subset N$, either $v(S) = 0$ or $v(S) = 1$.

In a simple game, a coalition S is said to be a *winning* coalition if $v(S) = 1$ and a *losing* coalition if $v(S) = 0$. So in a simple game every coalition is either winning or losing. It follows from superadditivity of v that in simple games every subset of a losing coalition is losing, and every superset of a winning coalition is winning.

Typical examples of simple games are

(1) the *majority rule game* where $v(S) = 1$ if $|S| > n/2$, and $v(S) = 0$ otherwise;
(2) the *unanimity game* where $v(S) = 1$ if $S = N$ and $v(S) = 0$ otherwise; and
(3) the *dictator game* where $v(S) = 1$ if $1 \in S$ and $v(S) = 0$ otherwise.

For simple games, formula (7) for the Shapley value simplifies because the difference $[v(S) - v(S - \{i\})]$ is always zero or one. It is zero if $v(S)$ and $v(S - \{i\})$ are both zero or both one, and it is one otherwise. Therefore we may remove $[v(S) - v(S - \{i\})]$ from formula (7) provided we sum only over those coalitions S that are winning with i and losing without i. Formula (7) for the Shapley value (the Shapley–Shubik Index) becomes

$$\phi_i(v) = \sum_{\substack{S \text{ winning} \\ S-\{i\} \text{ losing}}} \frac{(|S|-1)!(n-|S|)!}{n!}. \tag{8}$$

There is a large class of simple games called *weighted voting games.* These games are defined by a characteristic function of the form

$$v(S) = \begin{cases} 1 & \text{if } \sum_{i \in S} w_i > q \\ 0 & \text{if } \sum_{i \in S} w_i \leq q \end{cases},$$

for some non-negative numbers w_i, called the *weights*, and some positive number q, called the *quota*. If $q = (1/2)\sum_{i\in N} w_i$, this is called a *weighted majority game.*

Example. Consider the game with Players 1, 2, 3, and 4, having 10, 20, 30, and 40 shares of stock respectively, in a corporation. Decisions require approval by a majority (more than 50%) of the shares. This is a weighted majority game with weights $w_1 = 10$, $w_2 = 20$, $w_3 = 30$ and $w_4 = 40$ and with quota $q = 50$.

Let us find the Shapley value of this game. The winning coalitions are $\{2,4\}$, $\{3,4\}$, $\{1,2,3\}$, and all supersets (sets containing one of these). For $i = 1$, $v(S) - v(S - \{1\}) = 0$ unless $S = \{1,2,3\}$. So

$$\phi_1(v) = \frac{2!1!}{4!} = \frac{1}{12}.$$

For $i = 2$, $v(S) - v(S - \{2\}) = 0$ unless $S = \{2,4\}$, $\{1,2,3\}$, or $\{1,2,4\}$, so that

$$\phi_2(v) = \frac{1!2!}{4!} + 2\frac{2!1!}{4!} = \frac{1}{4}.$$

Similarly, $\phi_3(v) = 1/4$. For $i = 4$, $v(S) - v(S - \{4\}) = 0$ unless $S = \{2,4\}$, $\{3,4\}$, $\{1,2,4\}$, $\{1,3,4\}$ or $\{2,3,4\}$. So

$$\phi_4(v) = 2\frac{1!2!}{4!} + 3\frac{2!1!}{4!} = \frac{5}{12}.$$

The Shapley value is $\phi = (1/12, 3/12, 3/12, 5/12)$. Note that the value is the same for players 2 and 3 although player 3 has more shares.

20.5 Exercises

1. (Market with one seller and two buyers) Player 1 owns an art object of no intrinsic worth to him. Therefore he wishes to sell it. The object is worth \$30 to player 2 and \$40 to player 3. Set this up as a game in characteristic function form. Find the Shapley value. Is the Shapley value in the core? (Refer to Example 2 of Sec. 19.4.)
2. Find the Shapley value of the game with characteristic function

$$\begin{array}{llll} & v(\{1\}) = 1 & v(\{1,2\}) = 2 & \\ v(\emptyset) = 0 & v(\{2\}) = 0 & v(\{1,3\}) = -1 & v(\{1,2,3\}) = 6. \\ & v(\{3\}) = -4 & v(\{2,3\}) = 3 & \end{array}$$

3. (a) Using the superadditivity of v, show that the Shapley value is an imputation.
   (b) Let T be a fixed coalition in N. Show that $\phi_x(v)$ is a strictly increasing function of $v(T)$ (with the other $v(S)$, $S \neq T$, held fixed) if $x \in T$, and a strictly decreasing function of $V(T)$ if $x \notin T$.
4. Find the Shapley value of the n-person game, for $n > 2$, with characteristic function,

(a) $v(S) = \begin{cases} |S| & \text{if } 1 \in S \\ 0 & \text{otherwise.} \end{cases}$ (b) $v(S) = \begin{cases} |S| & \text{if } 1 \in S \text{ or } 2 \in S \\ 0 & \text{otherwise.} \end{cases}$

(c) $v(S) = \begin{cases} |S| & \text{if } 1 \in S \text{ and } 2 \in S \\ 0 & \text{otherwise.} \end{cases}$

5. Is every simple game a weighted voting game? Prove or give a counterexample.
6. Find the Shapley value of the weighted majority game with 4 players having 10, 30, 30, and 40 shares.
7. **A Near Dictator.** Find the Shapley value of the weighted majority game with $n \geq 3$ Players in which player 1 has $2n - 3$ shares and Players 2 to n have 2 shares each.
8. Modify the example of Sec. 20.4 so that the chairman of the board may decide tie votes. (The chairman of the board is a fifth player who has no shares.) Find the Shapley value.
9. (a) (One large political party and three smaller ones.) Consider the weighted majority game with one large party consisting of $1/3$ of the votes and three equal sized smaller parties with $2/9$ of the vote each. Find the Shapley value. Is the power of the large party greater or less than its proportional size?
   (b) (Two large political parties and three smaller ones.) Consider the weighted majority game with two large parties with $1/3$ of the votes each and three smaller parties with $1/9$ of the votes each. Find the Shapley value. Is the combined power of the two larger parties greater or less than their proportional size?
10. (L. S. Shapley (1981) "Measurement of Power in Political Systems" in *Game Theory and its Applications* Proceedings in Applied Mathematics vol. 24, Amer. Math. Soc.) "County governments in New York are headed by Boards of Supervisors. Typically each municipality in a county has one seat, though a larger city may have two or more. But the supervisorial districts are usually quite unequal in population, and an

Table 20.1.

District	Population	%	No. of votes	%
Hempstead 1	728,625	57.1	31	27.0
Hempstead 2			31	27.0
Oyster Bay	285,545	22.4	28	24.3
North Hempstead	213,335	16.7	21	18.3
Long Beach	25,654	2.0	2	1.7
Glen Cove	22,752	1.8	2	1.7
Total			115	

effort is made to equalize citizen representation throughout the county by giving individual supervisors different numbers of votes in council. Table 20.1 shows the situation in Nassau County in 1964.

Under this system, a majority of 58 out of 115 votes is needed to pass a measure. But an inspection of the numerical possibilities reveals that the three weakest members of the board actually have no voting power at all. Indeed, their combined total of 25 votes is never enough to tip the scales. The assigned voting weights might just as well have been (31, 31, 28, 0, 0, 0) or (1, 1, 1, 0, 0, 0) for that matter."

The Shapley value is obviously (1/3,1/3,1/3,0,0,0). This is just as obviously unsatisfactory. In 1971, the law was changed, setting the threshold required to pass legislation to 63 votes rather than 58. Find the Shapley value under the changed rules, and compare to the above table.

11. In the United Nations Security Council, there are 15 voting nations, including the "big five". To pass a resolution, 9 out of the 15 votes are needed, but each of the big five has veto power. One way of viewing this situation is as a weighted voting game in which each of the big five gets 7 votes and each of the other 10 nations gets 1 vote, and 39 votes are required to pass a resolution. Find the Shapley value.
12. **Cost Allocation.** A scientist has been invited for consultation at three distant cities. In addition to her consultation fees, she expects travel compensation. But since these three cities are relatively close, travel expenses can be greatly reduced if she accommodates them all in one trip. The problem is to decide how the travel expenses should be split among her hosts in the three cities. The one-way travel expenses among these three cities, A, B, and C, and her home base, H, are given in the accompanying table (measured in some unspecified units).

Between H and A, cost $= 7$. Between A and B, cost $= 2$.
Between H and B, cost $= 8$. Between A and C, cost $= 4$.
Between H and C, cost $= 6$. Between B and C, cost $= 4$.

Assume that the value of the visit is the same for each of the hosts, say 20 units each. Set the problem up as a three-person game in coalitional form by specifying the characteristic function. Find the Shapley value. How much is the trip going to cost and how much should each host contribute to travel expenses?

13. **A one-product balanced market.** (Vorob'ev) Consider a market with one completely divisible commodity where the set N of players is divided into two disjoint sets, the buyers B and the sellers C, $N = B \cup C$. Each seller owns a certain amount of the commodity, say seller $k \in C$ owns y_k, and each buyer demands a certain amount, say buyer $j \in B$ demands x_j. We assume that the market is balanced, that is that the supply is equal to demand, $\sum_{k \in C} y_k = \sum_{j \in B} x_j$. We may set up such a game in characteristic function form by letting

$$v(S) = \min\left\{ \sum_{j \in S \cap B} x_j \,,\, \sum_{k \in S \cap C} y_k \right\}.$$

Thus, the value of a coalition is the total amount of trading that can be done among members of the coalition. Find the Shapley value of the game. (Hint: for each permutation of the players, consider also the reverse permutation in which the players enter the grand coalition in reverse order.)

14. (a) Consider the n-person game with players $1, 2, \ldots, n$, whose characteristic function satisfies

$$v(S) = k \quad \text{if } \{1, \ldots, k\} \subset S \quad \text{but } k+1 \notin S.$$

For example, $v(\{2,3,5\}) = 0$, $v(\{1,3,4,6\}) = 1$ and $v(\{1,2,3,5,6,7\}) = 3$. Find the Shapley value. (Use the method of Theorem 20.1.)

(b) Generalize to the case

$$v(S) = a_k \quad \text{if } \{1, \ldots, k\} \subset S \quad \text{but } k+1 \notin S,$$

where $0 \le a_1 \le a_2 \le \cdots \le a_n$.

15. **The Airport Game.** (Littlechild and Owen (1973)) Consider the following cost allocation problem. Building an airfield will benefit

n players. Player j requires an airfield that costs c_j to build, so to accommodate all the players, the field will be built at a cost of $\max_{1 \le j \le n} c_j$. How should this cost be split among the players? Suppose all the costs are distinct and let $c_1 < c_2 < \cdots < c_n$. Take the characteristic function of the game to be

$$v(S) = -\max_{j \in S} c_j.$$

(a) Let $R_k = \{k, k+1, \ldots, n\}$ for $k = 1, 2, \ldots, n$, and define the characteristic function v_k through the equation

$$v_k(S) = \begin{cases} -(c_k - c_{k-1}) & \text{if } S \cap R_k \neq \emptyset \\ 0 & \text{if } S \cap R_k = \emptyset. \end{cases}$$

Show that $v = \sum_{k=1}^{n} v_k$.

(b) Find the Shapley value.

16. **A Market with 1 Seller and m Buyers.** Player 0 owns an object of no intrinsic worth to himself. Buyer j values the object at a_j dollars, $j = 1, \ldots, m$. Suppose $a_1 > a_2 > \cdots > a_m > 0$. Set up the characteristic function and find the Shapley value.
(Answer: For the seller, $\phi_0(v) = \sum_{k=1}^{m} a_k/(k(k+1))$, and for the buyers,

$$\phi_j(v) = \frac{a_j}{j(j+1)} - 2\sum_{k=j+1}^{m} \frac{a_k}{(k-1)k(k+1)}.)$$

17. **The Core of a Simple Game.** In a simple game, (N, v), a player, i, is said to be a veto player, if $v(N - \{i\}) = 0$.
    (a) Show that the core is empty if there are no veto players.
    (b) Show, conversely, that the core is not empty if there is at least one veto player.
    (c) Characterize the core.

21 The Nucleolus

Another interesting value function for n-person cooperative games may be found in the nucleolus, a concept introduced by Schmeidler (1969). Instead of applying a general axiomatization of fairness to a value function defined on the set of all characteristic functions, we look at a fixed characteristic function, v, and try to find an imputation $\mathbf{x} = (x_1, \ldots, x_n)$ that minimizes the worst inequity. That is, we ask each coalition S how dissatisfied it is with the proposed imputation $\mathbf{x}$ and we try to minimize the maximum dissatisfaction.

21.1 Definition of the Nucleolus

As a measure of the inequity of an imputation $\mathbf{x}$ for a coalition S, the *excess* is defined to be,

$$e(\mathbf{x}, S) = v(S) - \sum_{j \in S} x_j,$$

which measures the amount (the size of the inequity) by which coalition S falls short of its potential $v(S)$ in the allocation $\mathbf{x}$. Since the core is defined as the set of imputations such that $\sum_{i \in S} x_i \geq v(S)$ for all coalitions S, we immediately have that *an imputation* $\mathbf{x}$ *is in the core if and only if all its excesses are negative or zero.*

On the principle that the one who yells loudest gets served first, we look first at those coalitions S whose excess, for a fixed allocation $\mathbf{x}$, is the largest. Then we adjust $\mathbf{x}$, if possible, to make this largest excess smaller. When the largest excess has been made as small as possible, we concentrate on the next largest excess, and adjust $\mathbf{x}$ to make it as small as possible, and so on. An example should clarify this procedure.

Example 1. The Bankruptcy Game. (O'Niell (1982)) A small company goes bankrupt owing money to three creditors. The company owes creditor A \$10,000, creditor B \$20,000, and creditor C \$30,000. If the company has only \$36,000 to cover these debts, how should the money be divided among the creditors? A *pro rata* split of the money would lead to the allocation of \$6000 for A, \$12,000 for B, and \$18,000 for C, denoted by $\mathbf{x} = (6, 12, 18)$ in thousands of dollars. We shall compare this allocation with those suggested by the Shapley value and the nucleolus.

First, we must decide on a characteristic function to represent this game. Of course we will have $v(\emptyset) = 0$ from the definition of characteristic function, and $v(ABC) = 36$ measured in thousands of dollars, By himself, A is not guaranteed to receive anything since the other two could receive the whole amount; thus we take $v(A) = 0$. Similarly, $v(B) = 0$. Creditor C is assured of receiving at least \$6000, since even if A and B receive the total amount of their claim, namely \$30,000, that will leave \$36,000 − \$30,000 = \$6000 for C. Thus we take $v(C) = 6$. Similarly, we find $v(AB) = 6$, $v(AC) = 16$, and $v(BC) = 26$.

To find the nucleolus of this game, let $\mathbf{x} = (x_1, x_2, x_3)$ be an efficient allocation (that is, let $x_1 + x_2 + x_3 = 36$), and look at the excesses as found in the table below. We may omit the empty set and the grand coalition from consideration since their excesses are always zero. To get an idea of how to proceed, consider first an arbitrary point, say the pro rata point $(6, 12, 18)$. As seen in the table, the vector of excesses is $\mathbf{e} = (-6, -12, -12, -12, -8, -4)$. The largest of these numbers is -4 corresponding to the coalition BC. This coalition will claim that every other coalition is doing better than it is. So we try to improve on things for this coalition by making $x_2 + x_3$ larger, or, equivalently, x_1 smaller (since $x_1 = 36 - x_2 - x_3$). But as we decrease the excess for BC, the excess for A will increase at the same rate and so these excesses will meet at -5, when $x_1 = 5$. It is clear that no choice of $\mathbf{x}$ can make the maximum excess smaller than -5 since at least one of the coalitions A or BC will have excess at least -5. Hence, $x_1 = 5$ is the first component of the nucleolus.

S	$v(S)$	$e(\mathbf{x}, S)$	$(6, 12, 18)$	$(5, 12, 19)$	$(5, 10.5, 20.5)$	$(6, 11, 19)$
A	0	$-x_1$	-6	-5	-5	-6
B	0	$-x_2$	-12	-12	-10.5	-11
C	6	$6 - x_3$	-12	-13	-14.5	-13
AB	6	$6 - x_1 - x_2$	-12	-11	-9.5	-11
AC	16	$16 - x_1 - x_3$	-8	-8	-9.5	-9
BC	26	$26 - x_2 - x_3$	-4	-5	-5	-4

Though x_1 is fixed, we still have x_2 and x_3 to vary subject to $x_2 + x_3 = 36 - 5 = 31$, and we choose them to make the next largest excess smaller. If we choose the point $\mathbf{x} = (5, 12, 19)$ as the next guess, we see that the next largest excess after the -5's is the -8 corresponding to coalition AC. To make this smaller, we must increase x_3 (decrease x_2). But as we do so, the excesses for coalitions B and AB increase at the same rate. Since the excess for coalition AB starts closer to -8 we find x_2 and x_3 so that $e(\mathbf{x}, AB) = e(\mathbf{x}, AC)$. This occurs at $x_2 = 10.5$ and $x_3 = 20.5$. We denote the nucleolus by ν. The nucleolus is therefore

$$\nu = (5, 10.5, 20.5).$$

It is of interest to compare this solution to the Shapley value. We may compute the Shapley value by any of the methods given in Sec. 21.3. Using the formula, we find

$$\phi_A = (1/3)(0) + (1/6)(6) + (1/6)(10) + (1/3)(10) = 6,$$
$$\phi_B = (1/3)(0) + (1/6)(6) + (1/6)(20) + (1/3)(20) = 11,$$
$$\phi_C = (1/3)(6) + (1/6)(16) + (1/6)(26) + (1/3)(30) = 19.$$

The last column in the table shows the excesses for the Shapley value. □

It is time to define more precisely the concept of the nucleolus of a game with characteristic function v. First we define an ordering on vectors that reflects the notion of smaller maximum excess as given in the above example.

Define $\mathbf{O}(\mathbf{x})$ as the vector of excesses arranged in decreasing (nonincreasing) order. In the example, if $\mathbf{x} = (6, 12, 18)$ then $\mathbf{O}(\mathbf{x}) = (-4, -6, -8, -12, -12, -12)$. On the vectors $\mathbf{O}(\mathbf{x})$ we use the lexographic order. We say a vector $\mathbf{y} = (y_1, \ldots, y_k)$ is lexographically less than a vector $\mathbf{z} = (z_1, \ldots, z_k)$, and write $\mathbf{y} <_L \mathbf{z}$, if $y_1 < z_1$, or if $y_1 = z_1$ and $y_2 < z_2$, or if $y_1 = z_1$, $y_2 = z_2$ and $y_3 < z_3$, or $\ldots$, or if $y_1 = z_1, \ldots, y_{k-1} = z_{k-1}$ and $y_k < z_k$.

That is, $\mathbf{y} <_L \mathbf{z}$ if in the first component in which $\mathbf{y}$ and $\mathbf{z}$ differ, that component of $\mathbf{y}$ is less than the corresponding component of $\mathbf{z}$. Similarly, we write $\mathbf{y} \le_L \mathbf{z}$ if either $\mathbf{y} <_L \mathbf{z}$ or $\mathbf{y} = \mathbf{z}$. The nucleolus is an efficient allocation that minimizes $\mathbf{O}(\mathbf{x})$ in the lexographic ordering.

Definition. Let $X = \{\mathbf{x} : \sum_{j=1}^{n} x_j = v(N)\}$ be the set of efficient allocations. We say that a vector $\nu \in X$ is a nucleolus if for every $\mathbf{x} \in X$ we have $\mathbf{O}(\nu) \leq_L \mathbf{O}(\mathbf{x})$.

21.2 Properties of the Nucleolus

The main properties of the nucleous are stated without proof in the following theorem.

Theorem 21.1. *The nucleolus of a game in coalitional form exists and is unique. The nucleolus is group rational, individually rational, and satisfies the symmetry axiom and the dummy axiom. If the core is not empty, the nucleolus is in the core.*

The only difficult part to prove is the uniqueness of the nucleolus. See the book of Owen for a discussion. Since the nucleolus always exists and is unique, we may speak of *the* nucleolus of a game. Like the Shapley value, the nucleolus will satisfy individual rationality if the characteristic function is super-additive or, more generally, if it is monotone in the sense that for all players i and for all coalitions S not containing i, we have $v(S) + v(\{i\}) \leq v(S \cup \{i\})$. In contrast to the Shapley value, the nucleolus will be in the core provided the core is not empty. (Exercise 1.)

Since the nucleolus satisfies the first three axioms of the Shapley value, it does not satisfy the linearity axiom.

It is interesting to see how the nucleolus and the Shapley value change in the bankrupt company example as the total remaining assets of the company change from \$0 to \$60,000, that is, as $v(N)$ changes from 0 to 60. Consider the nucleolus. If $v(N)$ is between 0 and 15, the nucleolus divides this amount equally among the players. For $v(N)$ between 15 and 25, the nucleolus splits the excess above 15 equally between B and C, while for $v(N)$ between 25 and 35, all the excess above 25 goes to C. For $v(N)$ between 35 and 45, the excess above 35 is split between B and C, and for $v(N)$ between 45 and 60, the excess above 45 is divided equally among the three players.

Nucleolus

Amount of $v(N)$ between	0 and 15	share equally
	15 and 25	B and C share
	25 and 35	C gets it all
	35 and 45	B and C share
	45 and 60	share equally

Shapley Value

Amount of $v(N)$ between		
	0 and 10	share equally
	10 and 20	B and C share
	20 and 40	C gets 2/3rd and A and B get 1/6th
	40 and 50	B and C share
	50 and 60	share equally

One notes that at $v(N) = 30$, the nucleolus and the Shapley value coincide with the pro rata point. Compared to the pro rata point, both the Shapley value and the nucleolus favor the weaker players if $v(N)$ is small, and favor the stronger players if $v(N)$ is large, more so for the Shapley value than the nucleolus.

21.3 Computation of the Nucleolus

The nucleolus is more difficult to compute than the Shapley value. In fact, the first step of finding the nucleolus is to find a vector $\mathbf{x} = (x_1, \ldots, x_n)$ that minimizes the maximum of the excesses $e(\mathbf{x}, S)$ over all S subject to $\sum x_j = v(N)$. This problem of minimizing the maximum of a collection of linear functions subject to a linear constraint is easily converted to a linear programming problem and can thus be solved by the simplex method, for example. After this is done, one may have to solve a second linear programming problem to minimize the next largest excess, and so on.

For $n = 3$, these problems are not hard, but they may be more difficult than the example of the bankrupt company. It may be useful to work out another example. Suppose

$$\begin{array}{llll} & v(\{A\}) = -1 & v(\{AB\}) = 3 & \\ v(\emptyset) = 0 & v(\{B\}) = 0 & v(\{AC\}) = 4 & v(\{ABC\}) = 5 . \\ & v(\{C\}) = 1 & v(\{BC\}) = 2 & \end{array}$$

Alone, A is in the worst position, but in forming coalitions he is more valuable. The Shapley value is $\phi = (10/6, 7/6, 13/6)$. Let us find the nucleolus.

As an initial guess, try $(1, 1, 3)$. In the table below, we see that the maximum excess occurs at the coalition AB. To improve on this, we must decrease x_3. Since the next largest excess is for coalition AC, we keep x_2 fixed (increase x_1) and choose $x_3 = 2$ to make the excess for AB equal to the excess for AC. This leads to the point $(2, 1, 2)$ whose largest excess is 0, occurring at coalitions AB and AC. To make this smaller, we must decrease both x_2 and x_3. This involves increasing x_1, and will increase the excess for BC. We can see that the best we can do will occur when the excesses for AB and AC and BC are all equal. Solving the three equations,

$$x_3 - 2 = x_2 - 1 = x_1 - 3 \quad \text{and} \quad x_1 + x_2 + x_3 = 5,$$

we find $x_3 = x_2 + 1$ and $x_1 = x_2 + 2$ so that the solution is $\mathbf{x} = (8/3, 2/3, 5/3)$. This is the nucleolus.

S	$v(S)$	$e(\mathbf{x}, S)$	$(1,1,3)$	$(2,1,2)$	$(8/3,2/3,5/3)$
A	-1	$-1 - x_1$	-2	-3	$-11/3$
B	0	$-x_2$	-1	-1	$-2/3$
C	1	$1 - x_3$	-2	-1	$-2/3$
AB	3	$3 - x_1 - x_2 = x_3 - 2$	1	0	$-1/3$
AC	4	$4 - x_1 - x_3 = x_2 - 1$	0	0	$-1/3$
BC	2	$2 - x_2 - x_3 = x_1 - 3$	-2	-1	$-1/3$

Compared to the Shapley value, the nucleolus is very generous to Player 1. The Shapley value gives Player 1 one-third of $v(N)$, whereas the nucleolus gives Player 1 more than half $v(N)$.

21.4 Exercises

1. Show that if the core is not empty, then the nucleolus is in the core.
2. Show that for a constant-sum three-person game, the nucleolus is the same as the Shapley value.
3. **The Cattle Drive.** Rancher A has some cattle ready for market, and he foresees a profit of \$1200 on the sale. But two other ranchers lie between his ranch and the market town. The owners of these ranches, B and C, can deny passage through their land or require payment of a suitable fee. The question is: What constitutes a suitable fee? The characteristic function may be taken to be: $v(A) = v(B) = v(C) = v(BC) = 0$ and $v(AB) = v(AC) = v(ABC) = 1200$.

   (a) Find the core, and note that it consists of one point. This point must then be the nucleolus. (Why?)
   (b) Find the Shapley value.
   (c) Which do you think is more suitable for settling the question of a fee, the nucleolus or the Shapley value, and why?

4. Find the nucleolus for Exercise 20.5.1. Compare to the Shapley value. How could you tell before computing it that the nucleolus was not the same as the Shapley value?
5. Find the nucleolus for Exercise 20.5.2.
6. Find the nucleolus for Exercise 20.5.4(a). You may assume that the nucleolus satisfies the symmetry axiom.
7. Find the nucleolus for Exercise 20.5.6. You may assume that the nucleolus satisfies the dummy axiom.
8. **Cost Allocation.** Three farms are connected to each other and to the main highway by a series of rough trails as shown in the figure. Each farmer would benefit by having a paved road connecting his farm to the highway. The amounts of these benefits are indicated in square brackets [...]. The costs of paving each section of the trails are also indicated on the diagram.

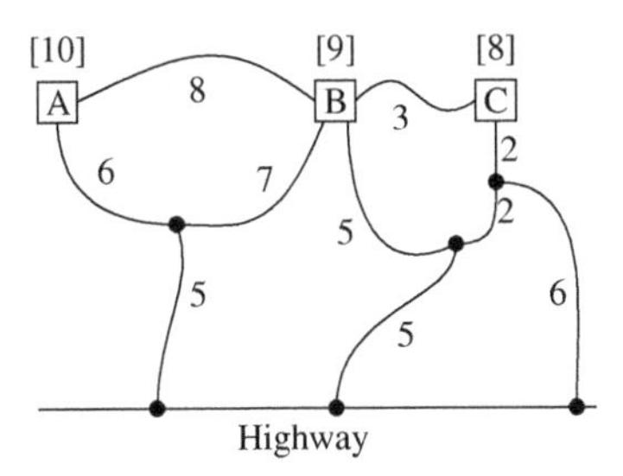

It is clear that no single farmer would find it profitable to build his own road, but a cooperative project would obviously be worthwhile.

(a) Determine the characteristic function.
(b) Find the Shapley value.
(c) Find the nucleolus.

9. **The Landowner and the Peasants.** Here is a generalization of symmetric games allowing one special player. The game is played with one landowner and m peasants, $n = m + 1$ players. The peasants can produce nothing by themselves, but neither can the landowner. All peasants are interchangeable. If k peasants and the landowner cooperate, they can jointly receive the amount $f(k)$, where $0 = f(0) < f(1) < f(2) < \cdots < f(m)$. We denote the landowner as player number 1 and the peasants as players 2 through n. Thus,

$$v(S) = \begin{cases} f(|S| - 1) & \text{if } 1 \in S, \\ 0 & \text{if } 1 \notin S. \end{cases}$$

(a) Suppose $m = 3$ and $f(x) = x$. Find the Shapley value.
(b) Suppose $m = 3$ and $f(x) = x$. Find the nucleolus.
(c) Find a simple formula for the Shapley value for general m and $f(x)$.
(d) Find a general formula for the nucleolus of this game.

10. **An Assignment Game.** Two house owners, A and B, are expecting to sell their houses to two potential buyers, C and D, each wanting to buy one house at most. Players A and B value their houses at 10 and 20 respectively, in some unspecified units. In the same units, Player C values A's house at 14 and B's house at 23, while Player D values A's house at 18 and B's house at 25.

    (a) Determine a characteristic function for the game.
    (b) Find the Shapley value.
    (c) Find the nucleolus.

11. Find the nucleolus for the n-person game of Exercise 20.5.14(a). (Answer: $\boldsymbol{\nu} = (n/2, 1, 1/2, \ldots, 1/2)$.)
12. Find the nucleolus of the Near Dictator Game of Exercise 20.5.7.

Appendix 1 Utility Theory

Much of the theory presented is based on utility theory at a fundamental level. This theory gives a justification for our assumptions (1) that the payoff functions are numerical valued and (2) that a randomized payoff may be replaced by its expectation. There are many expositions on this subject at various levels of sophistication. The basic theory was developed in the book of Von Neumann and Morgenstern (1944). Further developments are given in Savage (1954), Blackwell and Girshick (1954) and Luce and Raiffa (1957). More recent descriptions may be found in Owen (1982) and Shubik (1984), and a more complete exposition of the theory may be found in Fishburn (1988). Here is a brief description of the basics of linear utility theory.

The method a "rational" person uses in choosing between two alternative actions, a_1 and a_2, is quite complex. In general situations, the payoff for choosing an action is not necessarily numerical, but may instead represent complex entities such as "you receive a ticket to a ball game tomorrow when there is a good chance of rain and your raincoat is torn" or "you lose five dollars on a bet to someone you dislike and the chances are that he is going to rub it in". Such entities we refer to as *payoffs* or *prizes*. The "rational" person in choosing between two actions evaluates the value of the various payoffs and balances it with the probabilities with which he thinks the payoffs will occur. He may do, and usually does, such an evaluation subconsciously. We give here a mathematical model by which such choices among actions are made. This model is based on the notion that a "rational" person can express his preferences among payoffs in a method consistent with certain axioms. The basic conclusion is that the "value" to him of a payoff may be expressed as a numerical function, called a *utility*, defined on the set of payoffs, and that the preference between lotteries giving him a probability distribution over the payoffs is based only on the expected value of the utility of the lottery.

Let $\mathcal{P}$ denote the set of payoffs of the game. We use P, P_1, P_2, and so on to denote payoffs (that is, elements of $\mathcal{P}$).

Definition. A *preference relation on* $\mathcal{P}$, or simply preference on $\mathcal{P}$, is a (weak) linear ordering, $\preceq$, on $\mathcal{P}$; that is,

(a) (linearity) if P_1 and P_2 are in $\mathcal{P}$, then either $P_1 \preceq P_2$ or $P_2 \preceq P_1$ (or both), and
(b) (transitivity) if P_1, P_2 and P_3 are in $\mathcal{P}$, and if $P_1 \preceq P_2$ and $P_2 \preceq P_3$, then $P_1 \preceq P_3$.

If $P_1 \preceq P_2$ and $P_2 \preceq P_1$, then we say P_1 and P_2 are *equivalent* and write $P_1 \simeq P_2$.

We assume that our "rational" being can express his preferences over the set $\mathcal{P}$ in a way that is consistent with some preference relation. If $P_1 \preceq P_2$ and $P_1 \not\simeq P_2$, we say that our rational person *prefers* P_2 to P_1 and write $P_1 \prec P_2$. If $P_1 \simeq P_2$, we say that he is *indifferent* between P_1 and P_2. The statement $P_1 \preceq P_2$ means he either prefers P_2 to P_1 or he is indifferent between them.

Unfortunately, just knowing that a person prefers P_2 to P_1, gives us no indication of how much more he prefers P_2 to P_1. In fact, the question does not make sense until a third point of comparison is introduced. We could, for example, ask him to compare P_2 with the payoff of P_1 plus \$100 in order to get some comparison of how much more he prefers P_2 to P_1 in terms of money. We would like to go farther and express all his preferences in some numerical form. To do this however requires that we ask him to express his preferences on the space of all lotteries over the payoffs.

Definition. A *lottery* is a finite probability distribution over the set $\mathcal{P}$ of payoffs. We denote the set of lotteries by $\mathcal{P}^*$.

(A finite probability distribution is one that gives positive probability to only a finite number of points.)

If P_1, P_2 and P_3 are payoffs, the probability distribution, p, that chooses P_1 with probability 1/2, P_2 with probability 1/4, and P_3 with probability 1/4 is a lottery. We use lower case letters, p, p_1, p_2 to denote elements of $\mathcal{P}^*$. Note that the lottery p that gives probability 1 to a fixed payoff P may be identified with P, since receiving payoff P is the same as receiving payoff P with probability 1. With this identification, we may consider $\mathcal{P}$ to be a subset of $\mathcal{P}^*$.

We note that if p_1 and p_2 are lotteries and $0 \le \lambda \le 1$, then $\lambda p_1 + (1-\lambda)p_2$ is also a lottery. It is that lottery that first tosses a coin with probability λ of heads; if heads comes up, then it uses p_1 to choose an element of $\mathcal{P}$ and if tails comes up, it uses p_2. Thus $\lambda p_1 + (1-\lambda)p_2$ is an element of $\mathcal{P}^*$. Mathematically, a lottery of lotteries is just another lottery.

We assume now that our "rational" person has a preference relation not only over $\mathcal{P}$ but over $\mathcal{P}^*$ as well. One very simple way of setting up a preference over $\mathcal{P}^*$ is through a utility function.

Definition. A *utility function* is a real-valued function defined over $\mathcal{P}$.

Given a utility function, $u(P)$, we extend the domain of u to the set $\mathcal{P}^*$ of all lotteries by defining $u(p)$ for $p \in \mathcal{P}^*$ to be the expected utility: i.e. if $p \in \mathcal{P}^*$ is the lottery that chooses $P_1, P_2, \ldots, P_k$ with respective probabilities $\lambda_1, \lambda_2, \ldots, \lambda_k$, where $\lambda_i \ge 0$ and $\sum \lambda_i = 1$, then

$$u(p) = \sum_{i=1}^{k} \lambda_i u(P_i) \tag{1}$$

is the expected utility of the payoff for lottery p. Thus given a utility u, a simple preference over $\mathcal{P}^*$ is given by

$$p_1 \preceq p_2 \quad \text{if and only if} \quad u(p_1) \le u(p_2), \tag{2}$$

i.e. that lottery with the higher expected utility is preferred.

The basic question is, can we go the other way around? Given an arbitrary preference, $\preceq$ on $\mathcal{P}^*$, does there exist a utility u defined on $\mathcal{P}$ such that (2) holds? The answer is no in general, but under the following two axioms on the preference relation, the answer is yes!

A1. *If p_1, p_2 and q are in $\mathcal{P}^*$, and $0 < \lambda \le 1$, then*

$$p_1 \preceq p_2 \quad \textit{if and only if} \quad \lambda p_1 + (1-\lambda)q \preceq \lambda p_2 + (1-\lambda)q. \tag{3}$$

A2. *For arbitrary p_1, p_2 and q in $\mathcal{P}^*$,*

$$p_1 \prec p_2 \quad \textit{implies there exists a } \lambda > 0 \textit{ such that } p_1 \prec \lambda q + (1-\lambda)p_2 \tag{4}$$

and similarly,

$$p_1 \prec p_2 \quad \textit{implies there exists a } \lambda > 0 \textit{ such that } \lambda q + (1-\lambda)p_1 \prec p_2. \tag{5}$$

Axiom A1 is easy to justify. Consider a coin with probability $\lambda > 0$ of coming up heads. If the coin comes up tails you receive q. If it comes up heads you are asked to choose between p_1 and p_2. If you prefer p_2, you would naturally choose p_2. This axiom states that if you had to decide between p_1 and p_2 before learning the outcome of the toss, you would make the same decision. A minor objection to this axiom is that we might be indifferent between $\lambda p_1 + (1-\lambda)q$ and $\lambda p_2 + (1-\lambda)q$ if λ is sufficiently small, say $\lambda = 10^{-100}$, even though we prefer p_1 to p_2. Another objection comes from the person who dislikes gambles with random payoffs. He might prefer a p_2 that gives him \$2 outright to a gamble, p_1, giving him \$1 with probability 1/2 and \$3.10 with probability 1/2. But if q is \$5 for sure and $\lambda = 1/2$, he might prefer $\lambda p_1 + (1-\lambda)q$ to $\lambda p_2 + (1-\lambda)q$ on the basis of larger expected monetary reward, because the payoff is random in either case.

Axiom A2 is more debatable. It is called the continuity axiom. Condition (4) says that if $p_1 \prec \lambda q + (1-\lambda)p_2$ when $\lambda = 0$, then it holds for λ sufficiently close to 0. It might not be true if q is some really horrible event like death. It is safe to assume that for most people, $p_2 = \$100$ is strictly preferred to $p_1 = \$1$, which is strictly preferred to $q =$ death. Yet, would you ever prefer a gamble giving you death with probability λ and \$100 with probability $1-\lambda$, for some positive λ, to receiving \$1 outright? If not, then condition (4) is violated. However, people do not behave as if avoiding death is an overriding concern. They will drive on the freeway to get to the theater or the ballpark for entertainment, even though they have increased the probability of death (by a very small amount) by doing so. At any rate, Axiom A2 implies that there is no payoff infinitely less desirable or infinitely more desirable than any other payoff.

Theorem 1. *If a preference relation, $\preceq$, on $\mathcal{P}^*$ satisfies A1 and A2, then there exists a utility, u, defined on $\mathcal{P}$ that satisfies* (2). *Furthermore, u is uniquely determined up to change of location and scale.*

If a utility $u(P)$ satisfies (2), then for arbitrary real numbers a and $b > 0$, the utility $\hat{u}(P) = a + bu(P)$ also satisfies (2). Thus the uniqueness of u up to change of location and scale the strongest uniqueness that can be obtained.

For a proof see Blackwell and Girshick (1954) and for extensions see Fishburn (1988). The proof is constructive. We may choose p and q in $\mathcal{P}^*$ arbitrarily, say $p \prec q$, and define $u(p) = 0$ and $u(q) = 1$. This merely fixes the location and scale. Then for any p' such that $p \prec p' \prec q$, we may define $u(p') = \text{glb}\{\lambda : p' \prec \lambda q + (1-\lambda)p\}$. For p' not preferenced between p and q, $u(p')$ may be defined by extrapolation. For example, if $p \prec q \prec p'$, we first find $\tau = \text{glb}\{\lambda : q \prec \lambda p' + (1-\lambda)p\}$, and then define $u(p') = 1/\tau$. The resulting function, u, satisfies (2).

One may conclude from Theorem 1 that if a person has a preference relation on $\mathcal{P}^*$ that satisfies Axioms 1 and 2, then that person is *acting* as if his preferences were based on a utility defined on $\mathcal{P}$ and that of two lotteries in $\mathcal{P}^*$ he prefers the one with the larger expected utility. Ordinarily, the person does not really think in terms of a utility function and is unaware of its existence. However, a utility function giving rise to his preferences may be approximated by eliciting his preferences from a series of questions.

Countable Lotteries and Bounded Utility. It is sometimes desirable to extend the notion of a lottery to countable probability distributions, i.e. distributions giving all their weight to a countable number of points. If this is done, it must generally be assumed that utility is bounded. The reason for this is that if Eq. (1) is to be satisfied for countable lotteries and if u is unbounded, then there will exist lotteries p such that $u(p) = \pm\infty$. This may be seen as follows.

Suppose we have a utility function $u(P)$ on $\mathcal{P}$, and suppose that for a lottery, p, that chooses $P_1, P_2, \ldots$ with respective probabilities $\lambda_1, \lambda_2, \ldots$ such that $\sum_{n=1}^{\infty} \lambda_n = 1$, the extension of u to countable lotteries satisfies

$$u(p) = \sum_{n=1}^{\infty} \lambda_n u(P_n). \tag{1'}$$

If $u(P)$ is unbounded, say unbounded above, then we could find a sequence, $P_1, P_2, \ldots$, such that $u(P_n) \geq 2^n$. Then if we consider the lottery, p, that chooses P_n with probability 2^{-n} for $n = 1, 2, \ldots$, we would have

$$u(p) = \sum_{n=1}^{\infty} 2^{-n} u(P_n) \geq \sum_{n=1}^{\infty} 2^{-n} 2^n = \infty.$$

Then p would be a lottery that is infinitely more desirable than P_1, say, contradicting Axiom A2.

Since the extension of utility to countable lotteries seems innocuous, it is generally considered that utility indeed should be considered to be bounded.

Exercises

1. Does every preference given by a utility as in (1) satisfy A1 and A2?
2. Take $\mathcal{P} = \{P_1, P_2\}$, and give an example of a preference on $\mathcal{P}^*$ satisfying A2 but not A1.
3. Take $\mathcal{P} = \{P_1, P_2, P_3\}$, and give an example of a preference on $\mathcal{P}^*$ satisfying A1 but not A2.

Appendix 2 Owen's Proof of the Minimax Theorem

We now set out to prove the minimax theorem for finite games, that every matrix game has a value. The following non-constructive proof is due essentially to Guillermo Owen.

Theorem. *Every finite matrix game has a value.*

Proof. We will show that if a matrix does not have a value, then there is a submatrix (of smaller size) that does not have a value. Then by repeated application of this, we can reduce the size of the matrix without a value down to one row or one column, where we know the game has a value.

Let A be an $m \times n$ matrix, and let $\underline{V}(A)$ and $\overline{V}(A)$ denote the lower and upper values of the game with matrix A. Let $\hat{\mathbf{p}}$ and $\hat{\mathbf{q}}$ denote minimax strategies for Players I and II respectively, in the game with matrix A. Then

$$\sum_{j=1}^{n} a_{ij}\hat{q}_j \le \overline{V}(A) \quad \text{for } i = 1, \ldots, m \tag{1}$$

$$\sum_{i=1}^{m} \hat{p}_i a_{ij} \ge \underline{V}(A) \quad \text{for } j = 1, \ldots, n. \tag{2}$$

If equality holds for all the inequalities in (1) and (2), then

$$\underline{V}(A) = \sum_{j=1}^{n} \underline{V}(A)\hat{q}_j = \sum_j \sum_i \hat{p}_i a_{ij} \hat{q}_j = \sum_i \hat{p}_i \overline{V}(A) = \overline{V}(A) \tag{3}$$

so that the game A has a value.

Assume A does not have a value. Then at least one of the inequalities in (1) and (2) is strict. Without loss of generality, assume this occurs in (2) when $j = n$:

$$\sum_{i=1}^{m} \hat{p}_i a_{in} > \underline{V}(A). \tag{4}$$

This implies that $n \geq 2$. Define A' to be the submatrix of A obtained by deleting the nth column. We will show that the following two conditions are satisfied.

$$\underline{V}(A') \leq \underline{V}(A), \tag{5}$$

$$\overline{V}(A') \geq \overline{V}(A). \tag{6}$$

From this we may conclude that A' does not have a value because

$$\underline{V}(A') \leq \underline{V}(A) < \overline{V}(A) \leq \overline{V}(A').$$

To see (6), let $\hat{\mathbf{p}}'$ and $\hat{\mathbf{q}}'$ be the players' minimax strategies for A', and let $\mathbf{q}^0$ denote the vector $\mathbf{q}^0 = (\hat{q}'_1, \ldots, \hat{q}'_{n-1}, 0)$. Then,

$$\overline{V}(A') = \max_i \sum_{j=1}^{n-1} a_{ij}\hat{q}'_j = \max_i \sum_{j=1}^{n-1} a_{ij}q^0_j \geq \min_q \max_i \sum_{j=1}^{n} a_{ij}q_j = \overline{V}(A). \tag{7}$$

(Intuitively, it cannot help Player II when playing A to be restricted to strategies that do not use column n.)

To prove (5), let ϵ be a small positive number and consider the strategy

$$\mathbf{p}^\epsilon = (1-\epsilon)\hat{\mathbf{p}} + \epsilon\hat{\mathbf{p}}'.$$

Suppose (5) is not true, $\underline{V}(A') > \underline{V}(A)$. We will show that $\sum_{j=1}^{n} p_i^\epsilon a_{ij} > \underline{V}(A)$ for all j provided ϵ is sufficiently small, contradicting the definition of $\underline{V}(A)$ as the best Player I can guarantee. For $j = 1, \ldots, n-1$,

$$\sum_{i=1}^{m} p_i^\epsilon a_{ij} = (1-\epsilon)\sum_{i=1}^{m} \hat{p}_i a_{ij} + \epsilon \sum_{i=1}^{m} \hat{p}'_i a_{ij} \geq (1-\epsilon)\underline{V}(A) + \epsilon\underline{V}(A') > \underline{V}(A)$$

and for $j = n$,

$$\sum_{i=1}^{m} p_i^\epsilon a_{in} = (1-\epsilon)\sum_{i=1}^{m} \hat{p}_i a_{ij} + \epsilon \sum_{i=1}^{m} \hat{p}'_i a_{in} = \sum_{i=1}^{m} \hat{p}_i a_{in} + \epsilon \sum_{i=m}^{m} (\hat{p}'_i - \hat{p}_i) a_{in}.$$

Now since $\sum_{i=1}^{m} \hat{p}_i a_{in}$ is strictly greater than $\underline{V}(A)$, we can make $\sum_{i=1}^{m} p_i^\epsilon a_{in}$ also strictly greater than $\underline{V}(A)$ by choosing ϵ sufficiently small.

Appendix 3 Contraction Maps and Fixed Points

Definition 1. A metric space, (X, d), consists of a non-empty set X of points together with a function d from $X \times X$ to the reals satisfying the four following properties for all x, y and z in X:

(1) $d(x, x) = 0$ for all x.
(2) $d(x, y) > 0$ for all $x \neq y$.
(3) $d(x, y) = d(y, x)$ for all x and y.
(4) $d(x, y) \leq d(x, z) + d(z, y)$.

Definition 2. A metric space, (X, d), is said to be *complete* if every Cauchy sequence in X converges to a point in X. In other words, for any sequence $x_n \in X$ for which $\max_{n>m} d(x_m, x_n) \to 0$ as $m \to \infty$, there exists a point $x^* \in X$ such that $d(x_n, x^*) \to 0$ as $n \to \infty$.

Examples. Euclidean space in Ndimensions, R^N, with points $\mathbf{x} = (x_1, \ldots, x_N)$, is an example of a complete metric space under any of the following metrics.
1. Euclidean distance, $\|\mathbf{x}\| = \sqrt{x_1^2 + x_2^2 + \cdots + x_N^2}$.
2. L_1 distance: $\|\mathbf{x}\| = |x_1| + |x_2| + \cdots + |x_N|$.
3. Sup norm: $\|\mathbf{x}\| = \max\{|x_1|, |x_2|, \ldots, |x_N|\}$.

These metrics are equivalent in the sense that if for one of them $\|\mathbf{x}_n\| \to 0$ for a sequence $\mathbf{x}_n \in X$, then $\|\mathbf{x}_n\| \to 0$ for the others also. In what follows, we take a fixed metric space (X, d).

Definition 3. A map $T : X \to X$ is a contraction map if there is a positive constant $c < 1$, called the contraction factor, such that

$$d(Tx, Ty) \le c\,d(x, y)$$

for all $x \in X$ and $y \in X$.

Definition 4. A point, $x \in X$, is said to be a fixed point of a map $T : X \to X$, if $Tx = x$.

Contraction Mapping Theorem. A contraction map, T, on a complete metric space (X, d) has a unique fixed point, x_0. Moreover, for any $y \in X$, $d(T^n y, x_0) \to 0$ as $n \to \infty$.

Proof. Let $c < 1$ denote the contraction factor of T. Then, $d(T^2 y, Ty) \le c\,d(Ty, y)$, and inductively, $d(T^{n+1}y, T^n y) \le c^n d(Ty, y)$ for every n. The inequality,

$$\begin{aligned} d(T^{n+1}y, y) &\le d(T^{n+1}y, T^n y) + \cdots + d(Ty, y) \\ &\le (c^n + \cdots + 1) d(Ty, y) \\ &\le d(Ty, y)/(1 - c), \end{aligned} \qquad (*)$$

for all $n > 0$, implies that for all $n > m$,

$$d(T^{n+1}y, T^m y) \le \frac{d(T^{m+1}y, T^m y)}{1 - c} \le \frac{c^m d(Ty, y)}{1 - c}.$$

This shows that $\{T^n y\}$ is a Cauchy sequence. Hence, there is a point $x_0 \in R^d$ such that $T^n y \to x_0$ as $n \to \infty$. That x_0 is a fixed point follows from

$$\begin{aligned} d(Tx_0, x_0) &\le d(Tx_0, T^n y) + d(T^n y, x_0) \\ &\le c\,d(T^{n-1}y, x_0) + d(T^n y, x_0) \to 0, \end{aligned}$$

as $n \to \infty$, so that $d(Tx_0, x_0) = 0$. The fixed point must be unique because if z is another fixed point, then $d(x_0, z) = d(Tx_0, Tz) \le c\,d(x_0, z)$ shows that $d(x_0, z) = 0$. □

Corollary 1. *For any* $y \in X$, $d(T^n y, x_0) \le c^n d(y, x_0)$, *where* x_0 *is the fixed point of* T.

Proof. $d(T^n y, x_0) = d(T^n y, T^n x_0) \le c^n d(y, x_0)$. □

Corollary 2. *For any* $y \in X$, $d(x_0, Ty) \le \frac{c}{1-c} d(Ty, y)$, *where* x_0 *is the fixed point of* T.

Proof. From (*), letting $n \to \infty$, we have $d(x_0, y) \le d(Ty, y)/(1-c)$. Therefore, $d(x_0, Ty) \le c\,d(x_0, y) \le c\,d(Ty, y)/(1-c)$. □

Corollary 1 gives a bound on the rate of convergence of $T^n y$ to x_0. Corollary 2 gives an upper bound on the distance of Ty to x_0 based on the distance from y to Ty.

Proof of Theorem 1 of Sec. 12.5. This proof is based on the Contraction Mapping Theorem and the following simple lemma.

Lemma 1. *If* **A** *and* **B** *are two matrices of the same dimensions, then*

$$|\text{Val}(\mathbf{A}) - \text{Val}(\mathbf{B})| \le \max_{i,j} |a_{ij} - b_{ij}|.$$

Proof. Let $z = \max_{i,j} |a_{ij} - b_{ij}|$. Then $\text{Val}(\mathbf{A}) + z = \text{Val}(\mathbf{A} + z) \ge \text{Val}(\mathbf{B})$, because $b_{ij} \le a_{ij} + z$ for all i and j. Similarly, $\text{Val}(\mathbf{B}) + z \ge \text{Val}(\mathbf{A})$, completing the proof.

To prove Theorem 1, let T be the map from R^N to R^N defined by $T\mathbf{x} = \mathbf{y}$ where $y_k = \text{Val}\big(\mathbf{A}^{(k)}(\mathbf{x})\big)$ for $k = 1, \dots, N$, where $\mathbf{A}^{(k)}(\mathbf{x})$ is the matrix of Eq. (10) of Sec 12.5. We show that under the sup norm metric, $\|\mathbf{x}\| = \max_k\{|x_k| : k = 1, \dots, N\}$, T is a contraction map with contraction factor $c = 1 - s$, where s is the smallest of the stopping probabilities, assumed to be positive by (6) of Sec. 12.5. Using Lemma 1,

$$\begin{aligned}
\|T\mathbf{x} - T\mathbf{y}\| &= \max_k \left| \text{Val}\left(a_{ij}^{(k)} + \sum_{\ell=1}^{N} P_{ij}^{(k)}(\ell) x_\ell \right) \right. \\
&\qquad \left. - \text{Val}\left(a_{ij}^{(k)} + \sum_{\ell=1}^{N} P_{ij}^{(k)}(\ell) y_\ell \right) \right| \\
&\le \max_{k,i,j} \left| \sum_{\ell=1}^{N} P_{ij}^{(k)}(\ell) x_\ell - \sum_{\ell=1}^{N} P_{ij}^{(k)}(\ell) y_\ell \right| \\
&\le \max_{k,i,j} \sum_{\ell=1}^{N} P_{ij}^{(k)}(\ell)\, |x_\ell - y_\ell| \\
&\le \left(\max_{k,i,j} \sum_{\ell=1}^{N} P_{ij}^{(k)}(\ell) \right) \|\mathbf{x} - \mathbf{y}\| \\
&= (1 - s)\|\mathbf{x} - \mathbf{y}\|.
\end{aligned}$$

Since $c = 1 - s$ is less than 1, the Contraction Mapping Theorem implies that there is a unique vector $\mathbf{v}$ such that $T\mathbf{v} = \mathbf{v}$. But this is exactly Eq. (9) of Theorem 1.

We must now show that the suggested stationary optimal strategies guarantee the value. Let $\mathbf{x}^* = (\mathbf{p}_1, \ldots, \mathbf{p}_N)$ denote the suggested stationary strategy for Player I where $\mathbf{p}_k$ is optimal for him for the matrix $\mathbf{A}^{(k)}(\mathbf{v})$. We must show that in the stochastic game that starts at state k, this gives an expected return of at least $v(k)$ no matter what Player II does. Let n be arbitrary and consider the game up to stage n. If at stage n, play is forced to stop and the payoff at that stage is $a_{ij}^{(h)} + \sum_{\ell=1}^{N} P_{ij}^{(h)}(\ell)v(l)$ rather than just $a_{ij}^{(h)}$, assuming the state at that time were h, then $\mathbf{x}^*$ would be optimal for this multistage game and the value would be $v(k)$. Hence in the infinite stage game, Player I's expected payoff for the first n stages is at least

$$v(k) - (1-s)^n \max_{h,i,j} \sum_{\ell=1}^{N} P_{ij}^{(h)}(\ell)v(\ell) \geq v(k) - (1-s)^n \max_{\ell} v(\ell)$$

and for the remaining stages is bounded below by $(1-s)^n M/s$, as in (7). Therefore, Player I's expected payoff is at least

$$v(k) - (1-s)^n \max_{\ell} v(\ell) - (1-s)^n M/s.$$

Since this is true for all n and n is arbitrary, Player I's expected payoff is at least $v(k)$. By symmetry, Player II's expected loss is at most $v(k)$. This is true for all k, proving the theorem.

Appendix 4 Existence of Equilibria in Finite Games

We give a proof of Nash's Theorem based on the celebrated Fixed Point Theorem of L. E. J. Brouwer. Given a set C and a mapping T of C into itself, a point $\mathbf{z} \in C$ is said to be a fixed point of T, if $T(\mathbf{z}) = \mathbf{z}$.

The Brouwer Fixed Point Theorem. Let C be a nonempty, compact, convex set in a finite dimensional Euclidean space, and let T be a continuous map of C into itself. Then there exists a point $\mathbf{z} \in C$ such that $T(\mathbf{z}) = \mathbf{z}$.

The proof is not easy. You might look at the paper of Y. Takeuchi and T. Suzuki, "An easily verifiable proof of the Brouwer fixed point theorem", arXiv:1109.4604.

Now consider a finite n-person game with the notation of Sec. 15.1. The pure strategy sets are denoted by $X_1, \ldots, X_n$, with X_k consisting of $m_k \geq 1$ elements, say $X_k = \{1, \ldots, m_k\}$. The space of mixed strategies of Player k is given by X_k^*,

$$X_k^* = \left\{ \mathbf{p}_k = (p_{k,1}, \ldots, p_{k,m_k}) : p_{k,i} \geq 0 \text{ for } i = 1, \ldots, m_k, \right.$$

$$\left. \text{and } \sum_{i=1}^{m_k} p_{k,i} = 1 \right\}. \tag{1}$$

For a given joint pure strategy selection, $\mathbf{x} = (i_1, \ldots, i_n)$ with $i_j \in X_j$ for all j, the payoff, or utility, to Player k is denoted by $u_k(i_1, \ldots, i_n)$ for $k = 1, \ldots, n$. For a given joint mixed strategy selection, $(\mathbf{p}_1, \ldots, \mathbf{p}_n)$ with $\mathbf{p}_j \in X_j^*$ for $j = 1, \ldots, n$, the corresponding expected payoff to Player k

is given by $g_k(\mathbf{p}_1, \ldots, \mathbf{p}_n)$,

$$g_k(\mathbf{p}_1, \ldots, \mathbf{p}_n) = \sum_{i_1=1}^{m_1} \cdots \sum_{i_n=1}^{m_n} p_{1,i_1} \cdots p_{n,i_n} u_k(i_1, \ldots, i_n). \tag{2}$$

Let us use the notation $g_k(\mathbf{p}_1, \ldots, \mathbf{p}_n|i)$ to denote the expected payoff to Player k if Player k changes strategy from p_k to the pure strategy $i \in X_k$,

$$g_k(\mathbf{p}_1, \ldots, \mathbf{p}_n|i) = g_k(\mathbf{p}_1, \ldots, \mathbf{p}_{k-1}, \delta_i, \mathbf{p}_{k+1}, \ldots, \mathbf{p}_n), \tag{3}$$

where δ_i represents the probability distribution giving probability 1 to the point i. Note that $g_k(\mathbf{p}_1, \ldots, \mathbf{p}_n)$ can be reconstructed from the $g_k(\mathbf{p}_1, \ldots, \mathbf{p}_n|i)$ by

$$g_k(\mathbf{p}_1, \ldots, \mathbf{p}_n) = \sum_{i=1}^{m_k} p_{k,i} g_k(\mathbf{p}_1, \ldots, \mathbf{p}_n|i). \tag{4}$$

A vector of mixed strategies, $(\mathbf{p}_1, \ldots, \mathbf{p}_n)$, is a strategic equilibrium if for all $k = 1, \ldots, n$, and all $i \in X_k$,

$$g_k(\mathbf{p}_1, \ldots, \mathbf{p}_n|i) \le g_k(\mathbf{p}_1, \ldots, \mathbf{p}_n). \tag{5}$$

Theorem. *Every finite n-person game in strategic form has at least one strategic equilibrium.*

Proof. For each k, X_k^* is a compact convex subset of m_k-dimensional Euclidean space, and so the product, $C = X_1^* \times \cdots \times X_n^*$, is a compact convex subset of a Euclidean space of dimension $\sum_{i=1}^n m_i$. For $\mathbf{z} = (\mathbf{p}_1, \ldots, \mathbf{p}_n) \in C$, define the mapping $T(\mathbf{z})$ of C into C by

$$T(\mathbf{z}) = \mathbf{z}' = (\mathbf{p}_1', \ldots, \mathbf{p}_n') \tag{6}$$

where

$$p_{k,i}' = \frac{p_{k,i} + \max(0, g_k(\mathbf{p}_1, \ldots, \mathbf{p}_n|i) - g_k(\mathbf{p}_1, \ldots, \mathbf{p}_n))}{1 + \sum_{j=1}^{m_k} \max(0, g_k(\mathbf{p}_1, \ldots, \mathbf{p}_n|j) - g_k(\mathbf{p}_1, \ldots, \mathbf{p}_n))}. \tag{7}$$

Note that $p_{k,i} \ge 0$, and the denominator is chosen so that $\sum_{i=1}^{m_k} p_{k,i}' = 1$. Thus $\mathbf{z}' \in C$. Moreover the function $f(\mathbf{z})$ is continuous since each $g_k(\mathbf{p}_1, \ldots, \mathbf{p}_n)$ is continuous. Therefore, by the Brouwer Fixed Point

Theorem, there is a point, $\mathbf{z}' = (\mathbf{q}_1, \ldots, \mathbf{q}_n) \in C$ such that $T(\mathbf{z}') = \mathbf{z}'$. Thus from (7)

$$q_{k,i} = \frac{q_{k,i} + \max(0, g_k(\mathbf{z}'|i) - g_k(\mathbf{z}'))}{1 + \sum_{j=1}^{m_k} \max(0, g_k(\mathbf{z}'|j) - g_k(\mathbf{z}'))} \tag{8}$$

for all $k = 1, \ldots, n$ and $i = 1, \ldots, m_n$. Since from (4) $g_k(\mathbf{z}')$ is an average of the numbers $g_k(\mathbf{z}'|i)$, we must have $g_k(\mathbf{z}'|i) \leq g_k(\mathbf{z}')$ for at least one i for which $q_{k,i} > 0$, so that $\max(0, g_k(\mathbf{z}'|i) - g_k(\mathbf{z}')) = 0$ for that i. But then (8) implies that $\sum_{j=1}^{m_k} \max(0, g_k(\mathbf{z}'|j) - g_k(\mathbf{z}')) = 0$, so that $g_k(\mathbf{z}'|i) \leq g_k(\mathbf{z}')$ for all k and i. From (5) this shows that $\mathbf{z}' = (\mathbf{q}_1, \ldots, \mathbf{q}_n)$ is a strategic equilibrium. □

Remark. From the definition of $T(\mathbf{z})$, we see that $\mathbf{z} = (p_1, \ldots, p_n)$ is a strategic equilibrium if and only if $\mathbf{z}$ is a fixed point of T. In other words, the set of strategic equilibria is given by $\{\mathbf{z} : T(\mathbf{z}) = \mathbf{z}\}$. If we could solve the equation $T(\mathbf{z}) = \mathbf{z}$ we could find the equilibria. Unfortunately, the equation is not easily solved. The method of iteration does not ordinarily work because T is not a contraction map.

Solutions to Exercises of Part I

Solutions to Chap. 1

1. To make your opponent take the last chip, you must leave a pile of size 1. So 1 is a P-position, and then 2, 3, and 4 are N-positions. Then 5 is a P-position, etc. The P-positions are 1, 5, 9, 13, ..., i.e. the numbers equal to 1 mod 4.

2. (a) The target positions are now 0, 7, 14, 21, etc.; i.e. anything divisible by 7.

(b) With 31 chips, you should remove 3, leaving 28.

3. (a) The target sums are 3, 10, 17, 24, and 31. If you start by choosing 3 and your opponent chooses 4, and this repeats four times, then the sum is 28, but there are no 3's left. You must choose 1 or 2 and he can then make the sum 31 and so win.

(b) Start with 5. If your opponent chooses 5 to get in the target series, you choose 2, and repeat 2 every time he chooses 5. When the sum is 26, it is his turn and there are no 5's left, so you will win. But if he ever departs from the target series, you can enter the series and win.

4. (a) The P-positions are the even numbers, $\{0, 2, 4, \ldots\}$.

(b) The P-positions are $\{0, 2, 4, 9, 11, 13, 18, 20, 22, \ldots\}$, the nonnegative integers equal to 0, 2, or 4 mod 9.

(c) The P-positions are $\{0, 3, 6, 9, \ldots\}$, the nonnegative integers divisible by 3.

(d) In (a), 100 is a P-position. In (b), 100 is an N-position since $100 = 1$ mod 9. It can be put into a P-position by subtracting 1 (or 6). In (c), 100 is an N-position. It can be put into a P-position by subtracting 1 (or 4 or 16 or 64).

5. The P-positions are those (m, n) with both m and n odd integers. If m and n are both odd, then any move will require putting an odd number of chips in two boxes; one of the two boxes would contain an even number of chips. If one of m and n is even, then we can empty the other box and put an odd number of chips in each box.

6. (a) In solving such problems, it is advisable to start the investigation with simpler positions and work up to the more difficult ones. Here are the simplest P-positions.

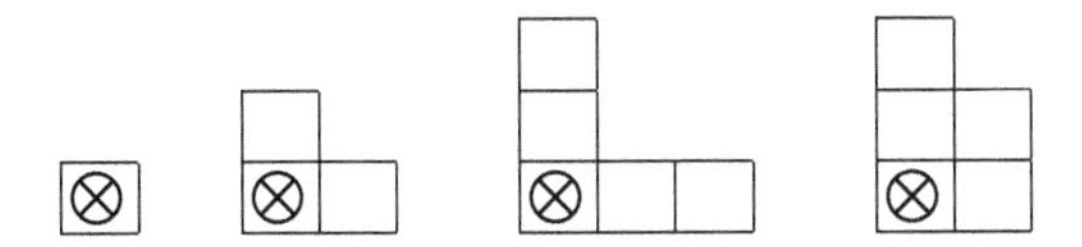

The last position shows that chomping at (3,1) is a winning move for the first player.

(b) The proof uses an argument, called "strategy stealing" that is useful in other problems as well. Consider removing the upper right corner. If this is a winning move, we are done. If not, then the second player has a winning reply. But whatever that reply is, the first player could have used it instead of the move he chose. (He could "steal" the second player's move.) Thus, in either case, the first player has a winning first move.

7. (a) Write the integer n in binary, $44 = (101100)_2$. A strategy that wins if it can be started out is to remove the smallest power of 2 in this expansion, in this case $4 = (100)_2$. Then the next player must leave a position for which this strategy can be continued. Another optimal first move for the first player is to remove $12 = (1100)_2$ chips. The only initial values of n for which the second player can win are the powers of 2: $n = 1, 2, 4, 8, 16, \ldots$.

(b) A strategy that wins is to remove the smallest Fibonacci number in the Zeckendorf expansion of n, (if possible). To see this, we note two things. First, if you do this, your opponent will be unable to take the smallest Fibonacci number of the Zeckendorf expansion of the result, because it is greater than twice what you took. Second, if your opponent takes less than the smallest Fibonacci number in the Zeckendorf expension, you can again follow this strategy.

To prove this last sentence, suppose your opponent cannot take the smallest Fibonacci number in the Zeckendorf expansion of n. Let F_{n_0} represent this number, and suppose he takes $x < F_{n_0}$. The difference has a Zeckendorf expansion, $F_{n_0} - x = F_{n_1} + \cdots + F_{n_k}$, where F_{n_k} is the smallest. We must show $F_{n_k} \leq 2x$, i.e. that you can take F_{n_k}. We do this by contradiction. Suppose $2x < F_{n_k}$. Then x is less than the next lower

Fibonacci number. This implies that when x is replaced by its Zeckendorf expansion, $x = F_{n_{k+1}} + \cdots + F_{n_\ell}$, we have

$$F_{n_0} = F_{n_1} + \cdots + F_{n_k} + x = F_{n_1} + \cdots + F_{n_k} + F_{n_{k+1}} + \cdots + F_{n_\ell}$$

which gives a second Zeckendorf expansions of F_{n_0}. This contradicts unicity.

For $n = 43 = 34 + 8 + 1$, the strategy requires that we take 1 chip. (Another optimal initial move is to remove 9 chips, leaving 34, since twice 9 is still smaller than 34.) The only initial values of n for which the second player can win are the Fibonacci numbers themselves: $n = 1, 2, 3, 5, 8, \ldots$.

8. (a) If the first player puts an S in the first square, the second player can win by putting an S in the last square. Then no matter what letter the first player puts in either empty square, the second player can complete an SOS.

(b) Player I can win by placing an S in the central square. Then if Player II plays on the left, say, without allowing I to win immediately, Player I plays an S in the last square. Now neither player can play on the right. But after Player II and then I play innocuously on the left, Player II must play on the right and lose.

(c) Call a square *x-rated* if no matter which letter a player places in the square, the other player can win immediately. It is not hard to show that the *only* way to make an x-rated square is to have it and another x-rated square between two S's as in (a). Thus, x-rated squares come in pairs. So, if n is even (like 2000) and if neither player makes an error allowing the opponent to win in one move, then after an even number of moves only x-rated squares will remain. It will then be Player I's turn and he must fill an x-rated square and so lose. However, Player II must make sure there is at least one x-rated pair. But this is easy to do if n is large say greater than 14. Just play an S in a square with at least three or four empty spots on either side. On your next move you will be able to make an x-rated pair on one side or the other. Generally, Player I wins for odd $n \geq 7$, and Player II wins for even $n \geq 16$ is even.

(d) The case $n = 14$ is special. Player I begins by playing an O at position 7. Then if Player II plays an S at position 11, Player I plays an O at position 13, say, and then Player II cannot play an S at position 8 because Player I could win immediately with an S at position 6. The position is drawn. Player I can prevent Player II from making any x-rated squares.

Solutions to Chap. 2

1. (a) $27 \oplus 17 = 10$.

(b) If $38 \oplus x = 25$, then $x = 38 \oplus 25 = 63$.

(c) Since $y = 17 \oplus x$ and $z = 13 \oplus x$, we have $y \oplus z = 17 \oplus 13 \oplus x \oplus x = 17 \oplus 13 = 28$.

2. (a) The unique winning move is to remove 4 chips from the pile of 12 leaving 8.

(b) There are three winning moves; removing 8 chips from the pile of 17 or the pile of 19, or the pile of 23.

(c) Exactly the same answer as for (a) and (b).

3. We may identify a coin on a square labelled n with a nim pile of size n and a move of that coin to the left to a square labelled k as removing $n-k$ chips from the nim pile. Since the coins do not interact, this is exactly nim. The next player wins the displayed diagram by moving the coin on square 9 to square 0 (or moving the coin on square 10 to square 3, or by moving the coin on square 14 to square 7).

4. (a) Suppose there is an H in place n.

(1) Turning this H to T without turning over another coin corresponds to completely removing a pile of n chips.

(2) Turning this H to T and some T in place k to H, where $k < n$, corresponds to removing $n-k$ chips from a pile of n.

(3) Turning this H to T and some H in place k to T, where $k < n$, corresponds to removing two piles of sizes n and k. But this is equivalent to removing $n-k$ chips from the pile of size n, thus creating two piles of size k, which effectively cancel because $k \oplus k = 0$.

(b) Since $2 \oplus 5 \oplus 9 \oplus 10 \oplus 12 = 8$, we must reduce the 9, 10 or 12 by 8. One method is to turn the H in place 9 to T and the T in place

1 to H. Another would be to turn the H in place 10 to T and the H in place 2 to T.

5. The player who moves first wins. A row with n spaces between the checkers corresponds to a nim pile with n chips. So the given position corresponds to a nim position with piles of sizes 4, 2, 3, 5, 3, 6, 2, and 1. The nim sum of these numbers is 6. You can win, for example, by moving the checker in the sixth row six squares toward the other, making the nim sum 0. Now if the opponent moves away from you in some row, you can move in the same row to keep the nim sum the same. (Such a move is called reversible.) If he moves toward you in some row, the nim sum is no longer 0, so you can find some row such that moving toward him reduces the nim sum to 0. In this way, the game will eventually end and you will be the winner.

6. Any move from $(x_1, x_2, x_3, \ldots, x_n)$ in staircase nim changes exactly one of the numbers, $x_1, x_3, \ldots, x_k$. Moreover, any nim move from $(x_1, x_3, \ldots, x_k)$ can be achieved as a staircase nim move from $(x_1, x_2, x_3, \ldots, x_n)$ by reducing one of the numbers, $x_1, x_3, \ldots, x_k$. Therefore a winning strategy is to keep the odd numbered stairs as a P-position in nim.

7. (a) When expanded in base 2 and added without carry modulo 3, we find 2212. To change the first (most significant) column to a 0, we must reduce two numbers that are 8 or greater. We may change the 10 to a 5 and the 13 to a 5.

$$\begin{array}{rrcrr}
4 = & 100_2 & & 4 = & 100_2 \\
8 = & 1000_2 & & 8 = & 1000_2 \\
8 = & 1000_2 & & 8 = & 1000_2 \\
9 = & 1001_2 & & 9 = & 1001_2 \\
10 = & 1010_2 & & 5 = & 101_2 \\
13 = & 1101_2 & & 5 = & 101_2 \\
\hline
\text{sum mod } 3 = & 2212 & \longrightarrow & & 0000
\end{array}$$

(b) Let $x_i = \sum_{j=0}^{m} x_{ij} 2^j$ be the base 2 expansion of x_i, where each x_{ij} is either 0 or 1 and m is sufficiently large. Let $\mathcal{P}$ be the set of all $(x_1, \ldots, x_n)$ such that for all j, $s_j \equiv \sum_{j=0}^{m} x_{ij} = 0 \pmod{k+1}$. (We refer to the vector s as the nim$_k$-sum of the x's. Note $0 \le s_j \le k$ for all j.) We show that $\mathcal{P}$ is the set of P-positions by following the proof of Theorem 1.

(1) *All terminal positions are in* $\mathcal{P}$. This is clear since $(0, \ldots, 0)$ is the only terminal position.

(2) *Every move from a position in $\mathcal{P}$ is to a position not in $\mathcal{P}$.* Suppose that $s_j = 0$ for all j, and that at most k of the x_i are reduced. Find the leftmost column j that is changed by one of these changes. If only one x_i had a 1 in position j, then s_j would be changed to k. If two x_i, then s_j would be changed to $k-1$, etc. But at most k changes are made, so that s_j is changed into a number between 1 and k. Thus the move cannot be in $\mathcal{P}$.

(3) *From each position not in $\mathcal{P}$, there is a move to a position in $\mathcal{P}$.* The difficulty of finding a winning move is to select which piles of chips to reduce. The problem of finding how many chips to remove from each of the selected piles is easy and there are usually many solutions. The algorithm below finds which piles to select.

Find the leftmost column j with a nonzero s_j, and select any $t = s_j$ of the x_i with $x_{ij} = 1$. If $t = k$, you are done.

Let s' denote the nim$_k$-sum of the remaining x's, and find the leftmost column $j' < j$ such that $1 \le s'_{j'} < k - t$. If there is no such j', you are done and the t selected x's may be used. Otherwise, select any $t' = s'_m$ of the remaining x's with a 1 in position j' of their binary expansion. Then set $t = t + t'$, $j = j'$, and repeat this paragraph.

(c) Move as you would in normal Nim$_k$ until you would move to a position with all piles of size 1. Then move to leave 1 mod $k+1$ piles instead of 0 mod $k+1$ piles.

Solutions to Chap. 3

1.

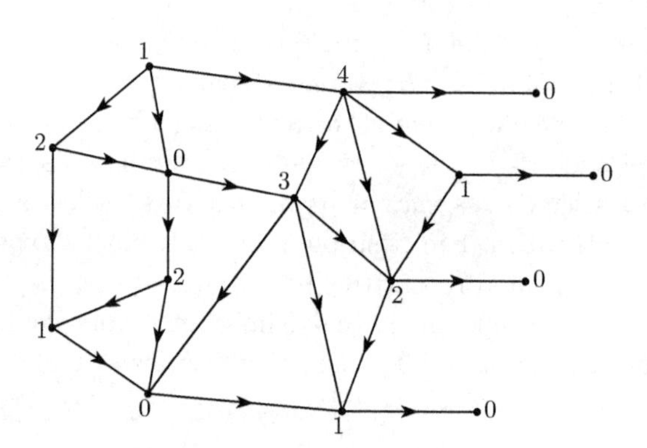

The Sprague-Grundy function.

2. The first few values of the SG function are as follows:

x	0	1	2	3	4	5	6	7	8	9	10	11	12	...
$g(x)$	0	1	0	1	2	3	2	0	1	0	1	2	3	...

Then pattern for the first 7 nonnegative integers repeats forever. We have

$$g(x) = \begin{cases} 0 & \text{if } x = 0 \text{ or } 2 \bmod 7 \\ 1 & \text{if } x = 1 \text{ or } 3 \bmod 7 \\ 2 & \text{if } x = 4 \text{ or } 6 \bmod 7 \\ 3 & \text{if } x = 5 \bmod 7. \end{cases}$$

3. The first few values of the SG function are as follows:

x	0	1	2	3	4	5	6	7	8	9	10	11	12	13	14	15	16	17	...
$g(x)$	0	0	1	0	2	1	3	0	4	2	5	1	6	3	7	0	8	4	...

One may describe this function recursively as follows.

$$g(x) = \begin{cases} x/2 & \text{if } x \text{ is even} \\ g((x-1)/2) & \text{if } x \text{ is odd.} \end{cases}$$

One may find $g(x)$ as follows. Take $x+1$ and factor out 2 as many times as possible (i.e. write $x+1 = 2^n y$ where y is an odd number). Then $g(x) = (y-1)/2$.

One may also write it as

$$g(x) = \begin{cases} 0 & \text{if } x = 2^n - 1 \\ 1 & \text{if } x = 2^n 3 - 1 \\ 2 & \text{if } x = 2^n 5 - 1 \\ \vdots & \vdots \\ k & \text{if } x = 2^n(2k+1) - 1 \\ \vdots & \vdots \end{cases} \qquad \text{for } n = 0, 1, 2, \ldots$$

4. (a) The first few values of the SG function are as follows:

x	0	1	2	3	4	5	6	7	8	9	10	11	12	13	14	15	16	...
$g(x)$	0	1	2	1	3	1	2	1	4	1	2	1	3	1	2	1	5	...

It may be represented mathematically as

$$g(x) = k + 1 \qquad \text{where } 2^k \text{ is the largest power of 2 dividing } x$$

(b) The Sprague-Grundy fnction for Aliquot is simply 1 less than the Sprague-Grundy function for Dim$^+$; so for $x \geq 1$, $g(x) = k$ where 2^k is the largest power of 2 dividing x.

5. The Sprague-Grundy function is

7	8	6	9	0	1	4	5
6	7	8	1	9	10	3	4
5	3	4	0	6	8	10	1
4	5	3	2	7	6	9	0
3	4	5	6	2	0	1	9
2	0	1	5	3	4	8	6
1	2	0	4	5	3	7	8
0	1	2	3	4	5	6	7

For larger boards, the entries seem to become chaotic, but Wythoff found that the zero entries have coordinates $(0,0),(1,2),(3,5),(4,7)$, $(6,10),(8,13),(9,15),(11,18),\ldots$ with differences $0,1,2,3,4,5,6,7,8,\ldots$, the first number in each pair being the smallest number that hasn't yet appeared. He also showed that the nth pair is $(\lfloor n\tau\rfloor, \lfloor n\tau^2\rfloor)$, for $n = 0,1,2,\ldots$, where τ is the golden ratio $(1+\sqrt{5})/2$.

6. (a) The Sprague-Grundy values are

5ω	$5\omega+1$	$5\omega+2$	$5\omega+3$	$5\omega+4$	$5\omega+5$	$5\omega+6$
4ω	$4\omega+1$	$4\omega+2$	$4\omega+3$	$4\omega+4$	$4\omega+5$	$4\omega+6$
3ω	$3\omega+1$	$3\omega+2$	$3\omega+3$	$3\omega+4$	$3\omega+5$	$3\omega+6$
2ω	$2\omega+1$	$2\omega+2$	$2\omega+3$	$2\omega+4$	$2\omega+5$	$2\omega+6$
ω	$\omega+1$	$\omega+2$	$\omega+3$	$\omega+4$	$\omega+5$	$\omega+6$
0	1	2	3	4	5	6

(b) The nim-sum of these transfinite Sprague-Grundy values follows the rule:

$$(x_1\omega + y_1) \oplus (x_2\omega + y_2) = (x_1 \oplus x_2)\omega + (y_1 \oplus y_2).$$

Therefore the Sprague-Grundy value of the given position is

$$(4\omega) \oplus (2\omega + 1) \oplus (\omega + 2) \oplus (5) = 7\omega + 6.$$

Since this is not zero, the position is an N-position. It can be moved to a P-position by moving the counter at 4ω down to $3\omega + 6$. There is no upper bound to how long the game can last, but every game ends in a finite number of moves.

(c) Yes.

7. Suppose S consists of n numbers. Then no Sprague-Grundy value can be greater than n since the set $\{g(x-y) : y \in S\}$ contains at most n numbers. Let x_n be the largest of the numbers in S. There are exactly $(n+1)^{x_n}$ sequences of length x_n consisting of the integers from 0 to n. Therefore, when by time $(n+1)^{x_n} + 1$ there will have been two identical such sequences in the Sprague-Grundy sequence. From the second time on, the Sprague-Grundy sequence will proceed exactly the same as it did the first time.

8. We have $g(x) = \text{mex}\{g(x-y) : y \in S\}$, and $g^+(x) = \text{mex}\{0, \{g^+(x-y) : y \in S\}\}$. We will show $g^+(x) = g(x-1) + 1$ for $x \geq 1$ by induction on x.

It is easily seen to be true for small values of x. Suppose it is true for all $x < z$. Then,

$$g^+(z) = \text{mex}\{0, \{g^+(z-y) : y \in S\}\} = \text{mex}\{0, \{1 + g(x-y-1) : y \in S\}\}$$
$$= 1 + \text{mex}\{g(x-y-1) : y \in S\} = 1 + g(x-1).$$

9. (a) The Sprague-Grundy function does not exist for this graph. However, the terminal vertex is a P-position and the other vertex is an N-position.

(b) The Sprague-Grundy function exists here. However, backward induction does not succeed in finding it. The terminal vertex has SG-value 0, the vertex above it has SG-value 2, and of the two vertices above, the one on the left has SG-value 1 and the one on the right has SG-value 0. Those vertices of SG-value 0 are P-positions and the others are N-positions.

(c) The node at the bottom right, call it α, obviously has Sprague-Grundy value 0. But every other node can move to some node whose Sprague-Grundy value we don't know. Here is how we make progress. Consider the node, call it β, at the middle of the southwest edge. It can move to only two positions. But neither of these positions can have Sprague-Grundy value 0 since they can both move to α. So β must have Sprague-Grundy value 0. Continuing in a similar manner, we find:

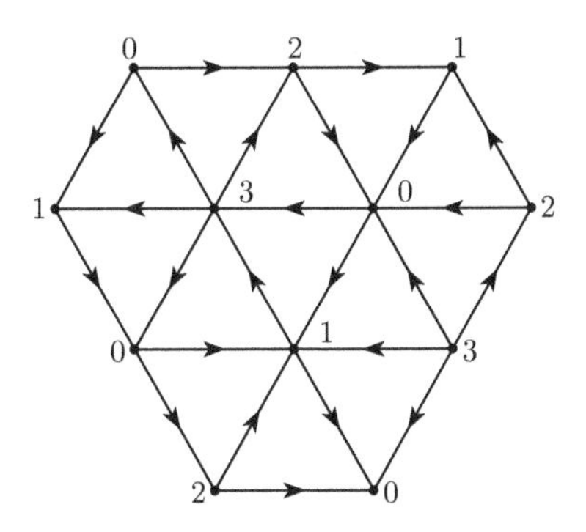

10. The P-positions for this game are the same as in Nim, because the new move changes a position of nim-sum zero to a position of nim-sum not zero.

To see this, suppose the number, say m, to be subtracted from three piles is divisible by 2^k but not 2^{k+1}. Then the *least* significant digit in the base 2 representation of m is at position $k+1$. Subtracting m from any greater number will change position $k+1$ in its base 2 representation from 1 to 0 or from 0 to 1. If the nim-sum of the n numbers was 0 before the removal of m from each pile, it will be non-zero after. This argument works if one is allowed to remove the same number of chips from any odd number of piles.

Solutions to Chap. 4

1. Remove any even number, or 1 chip if it is the whole pile. The SG-values of the first few numbers are

x	0	1	2	3	4	5	6	7	8	9	10	11	12	13	14	15	16	...
$g(x)$	0	1	1	0	2	2	3	3	4	4	5	5	6	6	7	7	8	...

The general rule is $g(0) = 0$, g(1)=1, $g(2) = 1$, g(3)=0, and $g(x) = \lfloor x/2 \rfloor$ for $x \geq 4$, where $\lfloor x \rfloor$ represents the greatest integer less than or equal to x, sometimes called the floor of x.

2. Remove any multiple of 3 if it is not the whole pile, or the whole pile if it contains 2 (mod 3) chips. The SG-values of the first few numbers are

x	0	1	2	3	4	5	6	7	8	9	10	11	12	13	14	15	16	...
$g(x)$	0	0	1	0	1	2	1	2	3	2	3	4	3	4	5	4	5	...

The general rule is $g(0) = 0$ and for $k \geq 0$,

$$g(3k+1) = k,$$
$$g(3k+2) = k+1,$$
$$g(3k+3) = k$$

3. There are three piles of sizes 18, 17, and 7 chips. The first pile uses the rules of Exercise 1, the second pile uses the rules of Exercise 2, and the third pile uses the rules of nim. The respective SG-values are 9, 6, and 7, with nim-sum $(1001)_2 \hat{+} (0110)_2 \hat{+} (0111)_2 = (1000)_2 = 8$. The can be put into a position of nim-sum 0 by moving the first pile to a position of SG-value 1. This can be done by removing 16 chips from the pile of 18, leaving 2, which has SG-value 1.

4. (a) The given position represents 2 piles of sizes 1 and 11. From Table I.4.1, the SG-values are 1 and 6, whose nim-sum is 7. Since the nim-sum is not zero, this is an N-position.

(b) We must change the SG-value 6 to SG-value 1. This may be done by knocking down pin number 6 (or pin number 10), leaving a position corresponding to 3 piles of sizes 1, 3, and 7, with SG-values 1, 3, and 2 respectively. This is a P-position since the nim-sum is 0.

5. Remove one chip and split if desired, or two chips without splitting.

(a) The SG-values of the first few numbers are

x	0	1	2	3	4	5	6	7	8	9	10	11	12	13	14	15	16	...
$g(x)$	0	1	2	3	0	1	2	3	0	1	2	3	0	1	2	3	0	...

We have $g(x) = x \pmod 4$, $0 \le g(x) \le 3$. This is periodic of period 4.

(b) Since 15 has SG-value 3, the moves to a P-position are those that remove 1 chip and split into two piles the nim-sum of whose SG-values is 0. For example, the move to two piles of sizes 1 and 13 is a winning move.

6. Remove two or more chips and split if desired, or one chip if it the whole pile. The SG-values of the first few numbers are

x	0	1	2	3	4	5	6	7	8	9	10	11	12	13	14	15	16	...
$g(x)$	0	1	1	2	2	3	4	4	5	6	6	7	8	8	9	10	10	...

The general rule is

$$\begin{aligned} g(3k) &= 2k, \\ g(3k+1) &= 2k, \\ g(3k+2) &= 2k+1 \end{aligned}$$

for $k \ge 0$, except for $g(1) = 1$.

7. Remove any number of chips equal to 1 (mod 3) and split if desired. The SG-values of the first few numbers are

x	0	1	2	3	4	5	6	7	8	9	10	11	12	13	14	15	16	...
$g(x)$	0	1	0	1	2	3	2	3	4	5	4	5	6	7	6	7	8	...

The general rule is, for $k \geq 0$,

$$\begin{aligned} g(4k) &= 2k, \\ g(4k+1) &= 2k+1, \\ g(4k+2) &= 2k, \\ g(4k+3) &= 2k+1. \end{aligned}$$

8. (a) The loops divide the plane into regions. A move in a region with n dots divides that region into two regions with a and b dots, where $a+b$ is less than n but where a and b are otherwise arbitrary. We claim that a region with n dots has SG-value n, i.e. $g(n) = n$. (This may be seen by induction: Clearly $g(0) = 0$ since 0 is terminal. If $g(k) = k$ for all $k < n$, then $g(n) \geq n$ since all SG-values less than n can be reached in one move without splitting the region into two. But if a region of n dots is split into regions of size a and b with $a+b < k$, then since $a \oplus b \leq a+b$, n cannot be obtained as the SG-value of a follower of n.) Thus a region of n dots corresponds to a nim pile of n chips.

(b) The given position corresponds to nim with three piles of sizes 3, 4 and 5. Since the nim-sum is 2, this is an N-position. An optimal move must reduce the 3 to a 1. This is achieved by drawing a loop through exactly two of the three free dots at the bottom of the figure.

9. (a) A loop in this game takes away 1 or 2 dots from a region and splits the region into two parts one of which may be empty of dots. This is exactly the same as the rules for Kayles.

(b) Using Table 4.1, $g(5) \oplus g(4) \oplus g(3) = 4 \oplus 1 \oplus 3 = 6$ so this is an N-position. An optimal move is to draw a closed loop through a dot from the innermost 5 dots such that exactly three dots stay inside the loop.

10. (a)

1	2	3	4	5	6	7	8	9	10	11	12	13
0	0	1	0	2	1	0	2	1	0	2	1	3

(b) The SG-values of 5, 8, and 13 are 2, 2, and 3 respectively. The winning first moves are (1) splitting 5 into 2 and 3, (2) splitting 8 into 2 and 6, and (3) splitting 13 into 5 and 8.

11. (a) $g(S_n) = \begin{cases} 0 & \text{if } n = 0 \\ 1 & \text{if } n \text{ is odd} \\ 2 & \text{if } n \text{ is even}, n \geq 2. \end{cases}$

(b) When played on a line with n edges, the rules of the game are: (1) You may remove one chip if it is the whole pile, or (2) you may remove two chips from any pile and if desired split that pile into two parts. In the notation of *Winning Ways*, Chap. 4, this game is called 0.37 (or 0.6 if one counts vertices rather than edges). The Sprague-Grundy values up to $n = 10342$ have been computed without finding any periodic pattern. It is generally believed that none exists. Here are the first few values.

n	0	1	2	3	4	5	6	7	8	9	10	11	...
$g(L_n)$	0	1	2	0	1	2	3	1	2	3	4	0	...

(c) $g(C_n) = \begin{cases} 0 \text{ if } g(L_{n-2}) > 0 \\ 1 \text{ if } g(L_{n-2}) = 0 \end{cases}$. Because of (b), there seems to be no periodicity in the appearance of the 1's. But we can say that $g(C_n) = 0$ if n is even.

(d) Let $DS_{m,n}$ denote the stars S_m and S_n joined by an (additional) edge (so that $DS_{0,n} = S_{n+1}$, and $DS_{1,1} = L_3$). For $n \geq 0$,

$$g(DS_{0,n}) = g(DS_{n,0}) = \begin{cases} 1 & \text{if } n \text{ is even} \\ 2 & \text{if } n \text{ is odd.} \end{cases}$$

For $m \geq 1$ and $n \geq 1$,

$$g(m,n) = \begin{cases} 0 & \text{if } m+n \text{ is even} \\ 3 & \text{if } m+n \text{ is odd.} \end{cases}$$

(e) The first player wins the square lattice (i) by taking the central vertex and reducing the position to C_8 with Sprague-Grundy value 0 from (c). The second player wins the tic-tac-toe board by playing symmetrically about the center of the graph. To generalize to larger centrally symmetric graphs, we need to define the symmetry for an arbitrary graph, (V, E). Here is one way.

Suppose there exists a one-to-one map, g, of V onto V such that
(1) (graph preserving) $\{v_1, v_2\} \in E$ implies $\{g(v_1), g(v_2)\} \in E$
(2) (pairing) $u = g(v)$ implies $v = g(u)$
(3) (no fixed vertex) $v \neq g(v)$ for all $v \in V$
(4) (no fixed edge) $\{v_1, v_2\} \in E$ implies $\{v_1, v_2\} \neq \{g(v_2), g(v_1)\}$.

The second player wins such symmetrically paired graphs without fixed vertices or fixed edges, by playing symmetrically. Can the second player always win if we allow exactly one fixed edge in the mapping?

Solutions to Chap. 5

1. (a) In Turning Turtles, the positions are labelled starting at 1, so the heads are in positions 3, 5, 6 and 9. The position has SG-value $3 \oplus 5 \oplus 6 \oplus 9 = 9$, so a winning move is to turn over the coin at position 9.

(b) In Twins, the labelling starts at 0, so the heads at in positions 2, 4, 5 and 8. The position has SG-value $2 \oplus 4 \oplus 5 \oplus 8 = 11$. A winning move is to turn over the coins at positions 3 and 8.

(c) For the subtraction set $S = \{1, 3, 4\}$, the Sprague-Grundy sequence is

position x :	0	1	2	3	4	5	6	7	8	9	10	11	12	13	14...
$g(x)$:	0	1	0	1	2	3	2	0	1	0	1	2	3	2	0...

The labelling starts at 0 so the heads are in positions 2, 4. 5 and 8, with a combined SG-value $0 \oplus 2 \oplus 3 \oplus 1 = 0$. This is a P-position.

(d) The labelling starts at 0 so the heads are in positions 2, 4, 5 and 8. In nim, this has SG-value $2 \oplus 4 \oplus 5 \oplus 8 = 11$. It can be moved to a position of SG-value 0 by turning over the coins at 3 and 8. Since this leaves an even number of heads, it is a P-position in Mock Turtles. The Mock Turtle did not need to be turned over.

2. (a) The maximum number of moves the game can last is n.

(b) Let T_n denote the maximum number of moves the game can last. This satisfies the recursion, $T_n = T_{n-1} + T_{n-2} + 1$ for $n > 2$ with initial values $T_1 = 1$ and $T_2 = 2$. We see that $T_n + 1$ is just the Fibonacci sequence, $2, 3, 5, 8, 13, 21 \ldots$. So T_n is the sequence $1, 2, 4, 7, 12, 20, \ldots$.

(c) This time T_n satisfies the recursion, $T_n = T_{n-1} + \cdots + T_1 + 1$ with initial condition $T_1 = 1$. So T_n is the sequence, $1, 2, 4, 8, 16, 32, \ldots$.

3. (a) Suppose we start the labelling from 0. Then a single heads in positions 0 or 1 is a terminal position and so receives SG-value 0. Continuing as in

Mock Turtles, we find

position x :	0	1	2	3	4	5	6	7	8	9	10	11	12	13	14...
$g(x)$:	0	0	1	2	4	7	8	11	13	14	16	19	21	22	25...

This is just the SG-sequence for Mock Turtles moved over two positions.

(b) To get nim out of this, we should have started labelling the positions of the coins from -2. The first two coins on the left are dummies. It doesn't matter whether they are heads or tails. The third coin on the left is the Mock Turtle. The P-positions in Triplets are exactly the P-positions in Mock Turtles when the first two coins on the left are ignored.

4. The SG-sequence for Rulerette is easily found to be

position x :	1	2	3	4	5	6	7	8	9	10	11	12	13	14	15	16...
$g(x)$:	0	1	0	2	0	1	0	4	0	1	0	2	0	1	0	8...

g(x) is half of the SG-value of x for Ruler except for x odd when $g(x) = 0$.

5. This becomes an impatient subtraction game mentioned in Exercise 3.8. The Sprague-Grundy function, $g^+(x)$ of this game is just $g(x-1)+1$, where $g(x)$ is the Sprague-Grundy function of the subtraction game.

6. (a) We have $6 \otimes 21 = 6 \otimes (16 \oplus 5) = (6 \otimes 16) + (6 \otimes 5) = 96 \oplus 8 = 104$.

(b) We have $25 \otimes 40 = (16 \oplus 9) \otimes (32 \oplus 8) = (16 \otimes 32) \oplus (16 \otimes 18) \oplus (9 \otimes 32) \oplus (9 \otimes 8)$.

Then using $16 \otimes 32 = 16 \otimes 16 \otimes 2 = 24 \otimes 2 = (16 \oplus 8) \otimes 2 = 32 \oplus 12 = 44$, and $9 \otimes 32 = 9 \otimes 16 \otimes 2 = 13 \otimes 16 = 224$,

we have $25 \otimes 40 = 44 \oplus 128 \oplus 224 \oplus 5 = 79$.

(c) $1 \oslash 14 = 13$, so $15 \oslash 14 = 15 \otimes 13 = 12$.

(d) Since $14 \otimes 14 = 8$, we have $\sqrt{8} = 14$.

(e) $x^2 \oplus x \oplus 6 = 0$ is the same as $x \otimes (x \oplus 1) = 6$. Looking at Table 5.2, we see this occurs for $x = 14$ or $x = 15$.

7. (a) Suppose there exists a move in Turning Corners from (v_1, v_2) into a position of SG-value u. Then there is a $u_1 < v_1$ and a $u_2 < v_2$ such that $(u_1 \otimes u_2) \oplus (v_1 \otimes u_2) \oplus (u_1 \otimes v_2) = u$. Since $u_1 < g_1(x)$, there exists a move in G_1 to an SG-value u_1, turning over the coins, say, at positions $x_1, x_2, \ldots, x_m, x$, where all $x_i < x$. Similarly there exists a move in G_2 to an SG-value u_2 turning over coins, say, at positions $y_1, y_2, \ldots, y_n, y$, where all $y_j < y$. This implies

$$\begin{aligned} g_1(x_1) \oplus g_1(x_2) \oplus \cdots \oplus g_1(x_m) &= u_1 \quad \text{and} \\ g_2(y_1) \oplus g_2(y_2) \oplus \cdots \oplus g_2(y_m) &= u_2. \end{aligned} \tag{1}$$

Then the move, $\{x_1, \ldots, x_m, x\} \times \{y_1, \ldots, y_n, y\}$ in $G_1 \times G_2$ results in SG-value

$$\begin{aligned} &\left(\sum^* \sum^* g_1(x_i) \otimes g_2(y_j)\right) \oplus \left(\sum^* g_1(x_i) \otimes g_2(y)\right) \oplus \left(\sum^* g_1(x) \otimes g_2(y_j)\right) \\ &= ((g_1(x_1) \oplus \cdots \oplus g_1(x_m)) \otimes (g_2(y_1) \oplus \cdots \oplus g_2(y_n))) \\ &\quad \oplus (g_1(x) \otimes (g_2(y_1) \oplus \cdots \oplus g_2(y_n))) \\ &\quad \oplus (g_1(x_1) \oplus \cdots \oplus g_1(x_m)) \otimes g_2(y)) \\ &= (u_1 \otimes u_2) \oplus (v_1 \otimes u_2) \oplus (u_1 \otimes v_2) = u \end{aligned} \tag{2}$$

where $\sum^*$ represents nim-sum. Conversely, for any move, $\{x_1, \ldots, x_m, x\} \times \{y_1, \ldots, y_n, y\}$, in $G_1 \times G_2$, we find u_1 and u_2 from (1). Then the same equation (2) shows that the corresponding move in Turning Corners has the same SG-value.

(b) We may conclude that the mex of the SG-values of the followers of (x, y) in $G_1 \times G_2$ is the same as the mex of the SG-values of the followers of (v_1, v_2) in Turning corners, implying $g_1(x) \otimes g_2(y) = v_1 \otimes v_2$.

8. (a) The table is

	1	2	1	4	1	2	1	8
1	1	2	1	4	1	2	1	8
2	2	3	2	8	2	3	2	12
4	4	8	4	6	4	8	4	11
7	7	9	7	4	10	9	7	15
8	8	12	8	11	8	12	8	13

(b) The given position has SG-value $2 \oplus 13 = 15$. A winning move must change the SG-value 13 to 2. In Turning corners, the move from (8,8) that that changes the SG-value 13 to 2 is the move with north west corner at (3,3). A move in Mock Turtles that changes $x = 5$ with $g_1(x) = 8$ into a position of SG-value 2, is the move that turns over 5, 2 and 1. A move in Ruler that change $y = 8$ with $g_2(8) = 9$ into a position with SG-value 2 is the move that turns over 8, 7, and 6. Therefore a winning move in the given position is $\{1, 2, 5\} \times \{6, 7, 8\}$. This gives

$$\begin{array}{cccccccc} T & H & T & T & T & H & H & H \\ T & T & T & T & T & H & H & H \\ T & T & T & T & T & T & T & T \\ T & T & T & T & T & T & T & T \\ T & T & T & T & T & H & H & T \end{array}$$

which has SG-value 0.

9. (a) Since the game is symmetric, the SG-value of heads at (i, j) is the same as the SG-value of heads at (j, i). This implies that the SG-value of the initial position is 0. It is a P-position for all n. A simple winning strategy is to play symmetrically. If your opponent makes a move with (i, j) as the south east coin, you make the symmetric move at (j, i). Such a play keeps the game symmetric without heads along the diagonal. A similar strategy works if the position is symmetric with no heads on the diagonal. This also holds true in any tartan game that is the square of some coin turning game.

(b) The SG-values of off-diagonal elements cancel, so the SG-value of the game is the sum of the SG-values on the diagonal of Table 5.3. For $n = 1, 2, \ldots$, these are $1, 2, 3, 5, 4, 7, 6, 11, 10, 9, 8, 14, 15, 12, 13, \ldots$. One can show that this hits all positive integers without repeating, and is never 0. So this is a first player win. There is a simple strategy that shows this. Namely, turn over all the coins and then apply the strategy found in part (a). A similar strategy works if the position is symmetric and all heads on the diagonal (at least one) are contiguous.

10. The SG-sequence for G_1 is

$$\begin{array}{rcccccccccccccc} \text{position } x: & 1 & 2 & 3 & 4 & 5 & 6 & 7 & 8 & 9 & 10 & \ldots & 98 & 99 & 100 \\ g(x): & 0 & 1 & 2 & 3 & 4 & 0 & 1 & 2 & 3 & 4 & \ldots & 2 & 3 & 4. \end{array}$$

For G_2, it is

$$\begin{array}{rcccccccccccc} \text{position } x: & 1 & 2 & 3 & 4 & 5 & 6 & 7 & 8 & \ldots & 97 & 98 & 99 & 100 \\ g(x): & 1 & 2 & 1 & 4 & 1 & 2 & 1 & 8 & \ldots & 1 & 2 & 1 & 4. \end{array}$$

The coin at (100,100) has SG-value $4 \otimes 4 = 6$ and the coin at (4,1) has SG-value $3 \otimes 1 = 3$. You can win by turning over the 8 coins at positions (x, y) with $x = 98, 100$ and $y = 97, 98, 99, 100$.

Solution to Chap. 6

1. The SG-value of the three-leaf clover is 2. The SG-value of the girl is 3. The SG-value of the dog is 2. And the SG-value of the tree is 5. So there exists a winning move on the tree that reduces the SG-value to 3. The unique winning move is to hack the left branch of the rightmost branch completely away.

Solutions to Exercises of Part II

Solutions to Chap. 7

1. The new payoff matrix is

$$\begin{array}{|cc|} -1 & +2 \\ +2 & -4 \end{array}$$

If Player I uses the mixed strategy $(p, 1-p)$, the expected payoff is $-1p+2(1-p)$ if II uses column 1, and $-2p-4(1-p)$ if II uses column 2. Equating these, we get $-p+2(1-p) = 2p-4(1-p)$ and solving for p gives $p = 2/3$. Use of this strategy guarantees that Player I wins 0 on the average no matter what II does. Similarly, if II uses the same mixed strategy $(2/3, 1/3)$, II is guaranteed to win 0 on the average no matter what I does. Thus, 0 is the value of the game. Since the value of a game is zero, the game is fair by definition.

2. The payoff matrix is

$$\begin{array}{c} \\ \text{black Ace} \\ \text{red 8} \end{array} \begin{array}{c} \begin{array}{cc} \text{red 2} & \text{black 7} \end{array} \\ \begin{pmatrix} -2 & 1 \\ 8 & -7 \end{pmatrix} \end{array}$$

Solving $-2p+8(1-p) = p-7(1-p)$ gives $p = 5/6$ as the probability that Player I should use the black Ace. Similarly, $q = 5/9$ is the probability that Player II should use the red 2. The value is $-1/3$.

3. If Professor Moriarty stops at Canterbury with probability p and continues to Dover with probability $1-p$, then his average payoff is $100p$ if Holmes stops at Canterbury, and is $-50p+100(1-p)$ if Holmes continues to Dover. Equating these payoffs gives $250p = 100$, or $p = 2/5$. Use of this mixed strategy guarantees Moriarty an average payoff of $100p = 40$.

On the other hand, if Holmes stops at Canterbury with probability q and continues to Dover with probability $1-q$, then his average payoff is $100q-50(1-q)$ if Moriarty stops at Canterbury, and is $100(1-q)$ if Holmes continues to Dover. Equating these gives $q=3/5$. Use of this strategy holds Moriarty to an average payoff of 40.

The value of the game is 40, and so the game favors Moriarty. But, as related by Dr. Watson in *The Final Problem*, Holmes outwitted Moriarty once again and held the diabolical professor to a draw.

4. Without the side payment, the game matrix in cents is

$$\begin{array}{cc} & \begin{array}{cc} 1 & 2 \end{array} \\ \begin{array}{c} 1 \\ 2 \end{array} & \begin{pmatrix} 55 & 10 \\ 10 & 110 \end{pmatrix} \end{array}$$

Let p be the probability that Player I (Alex) uses row 1. Equating his payoffs if Player II uses cols 1 or 2 gives $55p+10(1-p)=10p+110(1-p)$, or $145p=100$ or $p=20/29$. If Player I uses $(20/29, 9/29)$, his average payoff is $55(20/29)+10(1-9/29)=1190/29$. Since this is $41\frac{1}{29}$, a side payment of 42 cents overcompensates slightly. With the side payment, the game is in Olaf's favor by 28/29 of one cent.

Solutions to Chap. 8

1. The value is $-4/3$. The mixed strategy, (2/3,1/3), is optimal for I, and the mixed strategy (5/6,1/6) is optimal for II.

2. If $t \leq 0$, the strategy pair $\langle 1,1 \rangle$ is a saddle-point, and the value is $v(t) = 0$. If $0 \leq t \leq 1$, the strategy pair $\langle 2,1 \rangle$ is a saddle-point, and the value is $v(t) = t$. If $t > 1$, there is no saddle-point; I's optimal strategy is $((t-1)/(t+1), 2/(t+1))$, II's optimal strategy is $(1/(t+1), t/(t+1))$, and the value is $v(t) = 2t/(t+1)$.

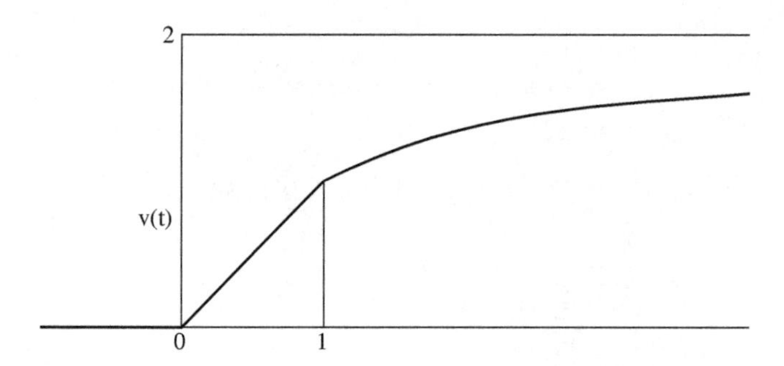

3. Suppose that $\langle x, y \rangle$ and $\langle u, v \rangle$ are saddle-points. Look at the four numbers $a_{x,y}$, $a_{x,v}$, $a_{u,v}$, and $a_{u,y}$. We must have $a_{x,y} \leq a_{x,v}$ since $a_{x,y}$ is the minimum in its row. Also, $a_{x,v} \leq a_{u,v}$ since $a_{u,v}$ is the maximum of its column. Keep going: $a_{u,v} \leq a_{u,y}$ since $a_{u,v}$ is the minimum of its row and $a_{u,y} \leq a_{x,y}$ since $a_{x,y}$ is the maximum of its column. We have

$$a_{x,y} \leq a_{x,v} \leq a_{u,v} \leq a_{u,y} \leq a_{x,y}.$$

Since this begins and ends with the same number, we must have equality throughout: $a_{x,y} = a_{x,v} = a_{u,v} = a_{u,y} = a_{x,y}$. (This argument also works if $x = u$ or $y = v$.)

4. (a) Column 2 dominates column 1; then row 3 dominates row 4; then column 4 dominates column 3; then row 1 dominates row 2. The resulting submatrix consists of rows 1 and 3 vs. columns 2 and 4. Solving this 2 by 2 game and moving back to the original game we find that the value is 3/2, I's optimal strategy is $p = (1/2, 0, 1/2, 0)$ and II's optimal strategy is $q = (0, 3/8, 0, 5/8)$.

(b) Column 2 dominates column 4; then (1/2)row 1 + (1/2)row 2 dominates row 3; then (1/2) col 1 + (1/2)col 2 dominates col 3. The resulting 2 by 2 game is easily solved. Moving back to the original game we find that the value is 30/7, I's optimal strategy is (2/7,5/7,0) and II's optimal strategy is (3/7,4/7,0,0).

5. (a) From the graph on the left, we guess that Player II uses columns 1 and 4. Solving this 2 by 2 subgame gives

$$\begin{array}{c} \\ 7/10 \\ 3/10 \end{array} \begin{array}{c} \begin{array}{cc} 1/2 & 1/2 \end{array} \\ \begin{pmatrix} 3 & 0 \\ -2 & 5 \end{pmatrix} \end{array} \qquad \text{Value} = 1.5$$

We conjecture I's optimal strategy is (0.7,0.3), II's optimal strategy is (0.5,0,0,0.5), and the value is 1.5. Let us check how well I's strategy works on columns 2 and 3. For column 2, $2(0.7) + 1(0.3) = 1.7$ and for column 3, $4(0.7) - 4(0.3) = 1.6$, both greater than 1.5. This strategy guarantees I at least 1.5 so our conjecture is verified.

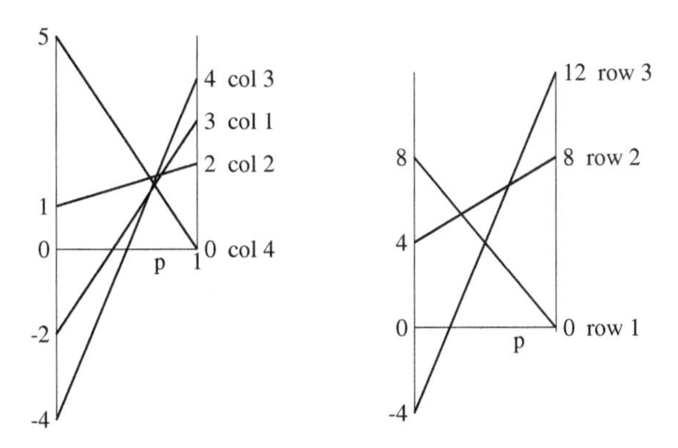

(b) (3/8)col 1 + (5/8)col 2 dominates col 3. Removing column 3 leaves a 3×2 game whose payoffs for a given q are displayed in the graph on the right. The upper envelope takes on its minimum value at the intersection of row 1 and row 2. Solving the 2×2 game in the upper left corner of the original matrix gives the solution. Player I's optimal strategy is (1/3,2/3,0), Player II's optimal strategy is (1/3,2/3,0), and the value is 16/3.

6. (a) The first row is dominated by the third; the seventh is dominated by the fifth. Then the third column is dominated by the first; the fourth is dominated by the second; the fifth column is dominated by the seventh. Then the middle row is dominated. When these three rows and columns are removed, the resulting matrix is the 4×4 identity matrix with value $v = 1/4$ and optimal strategies giving equal weight $1/4$ to each choice. This results in the optimal strategies $\mathbf{p} = (0, 0.25, 0.25, 0, 0.25, 0.25, 0)$ for I, and $\mathbf{q} = (0.25, 0.25, 0, 0, 0, 0.25, 0.25)$ for II.

(b) For all n, domination reduces the game matrix to the identity matrix. We find the value for arbitrary $n \geq 2$ to be $v_n = 1/(2k)$ for $n = 4k-2, 4k-1, 4k$, and $v_n = 1/(2k+1)$ for $n = 4k+1$. For n equal to 2 or 3, the optimal strategies are simple special cases. For $n \geq 4$, an optimal strategy for Player I is $\mathbf{p} = (p_1, p_2, \ldots, p_n)$, symmetric about its midpoint and such that for $i \leq (n+1)/2$,

$$p_i = \begin{cases} v_n & \text{if } i = 2 \text{ or } 3 \text{ mod } 4 \\ 0 & \text{if } i = 0 \text{ or } 1 \text{ mod } 4. \end{cases}$$

Similarly, an optimal strategy for Player II is $\mathbf{q} = (q_1, \ldots, q_n)$, symmetric about its midpoint and such that for $j \leq (n+1)/2$,

$$q_j = \begin{cases} v_n & \text{if } j = 1 \text{ or } 2 \text{ mod } 4 \\ 0 & \text{if } j = 0 \text{ or } 3 \text{ mod } 4. \end{cases}$$

7. If Player I uses $\mathbf{p}$ and Player II uses column 1, the average payoff is $(6/37)5 + (20/37)4 + (11/37) = 121/37$. Similarly for columns 2, 3, 4 and 5, the average payoffs are $121/37$, $160/37$, $121/37$, and $161/37$. So Player I can guarantee an average payoff of at least $121/37$ by using $\mathbf{p}$. Similarly, if Player II uses $\mathbf{q}$ and Player I uses rows 1, 2, 3, or 4, the average payoffs are $121/37$, $121/37$, $120/37$, and $121/27$ respectively. By using $\mathbf{q}$, Player II can keep the average payoff to at most $121/37$. Thus, $121/37$ is the value of the game and $\mathbf{p}$ and $\mathbf{q}$ are optimal strategies.

8. If $(52/143, 50/143.41/143)$ is optimal, then the value is the minimum of the inner product of this vector with the three columns of the matrix. This inner product with each of the three columns gives the same number, namely $96/143$, which is then the value.

9. The matrix is

$$\begin{array}{c} \\ 1 \\ 2 \\ 3 \end{array} \begin{array}{c} \begin{array}{ccc} 1 & 2 & 3 \end{array} \\ \begin{pmatrix} 0 & 1 & 2 \\ 1 & 0 & 1 \\ 2 & 1 & 0 \end{pmatrix} \end{array}$$

(1/2)row 1 + (1/2)row 3 dominates row 2; then (1/2)col 1 + (1/2)col 3 dominates row 2. Solving the resulting 2 by 2 game and moving back to the original game, we find the value is 1 and an optimal stratgey for I is (1/2,0,1/2), and an optimal strategy for II is (1/2,0,1/2). However, the pure strategy of choosing col 2 is also optimal. In fact it is better than the mixed strategy (1/2,0,1/2) whenever Player I makes the mistake of playing row 2.

10. In an $n \times n$ magic square, $\mathbf{A} = (a_{ij})$, there is a number s such that $\sum_i a_{ij} = s$ for all j, and $\sum_j a_{ij} = s$ for all i. If Player I uses the mixed strategy $\mathbf{p} = (1/n, 1/n, \ldots, 1/n)$ his average payoff is $V = s/n$ no matter what Player II does. The same goes for player II, so the value is s/n and $\mathbf{p}$ is optimal for both players. In the example, $n = 4$ and $s = 34$, so the value of the game is $17/2$ and the optimal strategy is $(1/4, 1/4, 1/4, 1/4)$.

11. (a) First, 6 dominates 4 and 5. With 4 and 5 removed, C dominates D and F; A dominates E. Also, the mixture $(3/4)A + (1/4)C$ dominates B. Then with B, D, E and F removed, 3 dominates 2 and 1. The resulting 2 by 2 game

$$\begin{array}{c} \\ 3 \\ 6 \end{array} \begin{array}{c} \begin{array}{cc} A & C \end{array} \\ \begin{pmatrix} 18 & 31 \\ 23 & 19 \end{pmatrix} \end{array}$$

is easily solved. The value is $21 + \frac{14}{17}$. Optimal for Player I is $(0, 0, 4/17, 0, 0, 13/17)$ and optimal for Player II is $(12/17, 0, 5/17, 0, 0, 0)$.

(b) Neither player was using an optimal strategy. The German choice was very poor, and the Allies were lucky. (Or did they have inside information?)

Solutions to Chap. 9

1. (a) There is a saddle at row 2, column 3. The value is 1.

(b) The inverse is $\mathbf{A}^{-1} = \frac{1}{5}\begin{pmatrix} 1 & -2 & 3 \\ 2 & 1 & 1 \\ -3 & 6 & -4 \end{pmatrix}$.

(c) The mixed strategy $(1/4, 1/2, 1/4)$, for example, is optimal for II.

(d) Equation (16) gives $\mathbf{q} = (2/5, 4/5, -1/5)$. Equations (16) are valid when $\mathbf{A}$ is nonsingular and Player I has an optimal strategy giving positive weight to each strategy. That is not the case here.

2. (a) If $d_i = 0$ for some i, then (row i, col i) is a saddle point of value zero. And row i and col i are optimal pure strategies for the players.

(b) If $d_i > 0$ and $d_j < 0$, then (row i, col j) is a saddle point of value zero. And row i and col j are optimal pure strategies for the players.

(c) If all $d_i < 0$, then the same analysis as in Sec. 9.3 holds. The value is $V = \sum_1^m 1/d_i$, and the players have the same optimal strategy, $(V/d_1, \ldots, V/d_m)$.

3. The matrix is

$$\begin{pmatrix} 2 & 0 & 0 & 0 \\ 0 & 4 & 0 & 0 \\ 0 & 0 & 8 & 0 \\ 0 & 0 & 0 & 16 \end{pmatrix}.$$

This is a diagonal game of value $V = (1/2 + 1/4 + 1/8 + 1/16)^{-1} = 16/15$. The optimal strategy for both players is $(8/15, 4/15, 2/15, 1/15)$.

4. The matrix is

$$\begin{pmatrix} 1 & 0 & 0 & 0 \\ 1/2 & 1 & 0 & 0 \\ 1/2 & 1/2 & 1 & 0 \\ 1/2 & 1/2 & 1/2 & 1 \end{pmatrix}.$$

This is a triangular game. If V is the value and if (p_1, p_2, p_3, p_4) is Player I's optimal strategy, then Eq. (12) become $V = p_4 = p_3 + (1/2)p_4 = p_2 + (1/2)(p_3 + p_4) = p_1 + (1/2)(p_2 + p_3 + p_4)$. We may solve the equations one at a time to find $p_4 = V$, $p_3 = (1/2)V$, $p_2 = (1/4)V$ and $p_1 = (1/8)V$. Since the sum of the p's is one, we find $(\frac{1}{8} + \frac{1}{4} + \frac{1}{2} + 1)V = 1$, so that $V = 8/15$. This is the value and $\mathbf{p} = (1/15, 2/15, 4/15, 8/15)$ is optimal for Player I and $\mathbf{q} = (8/15, 4/15, 2/15, 1/15)$ is optimal for II.

5. This is similar to Exercise 4. Equations (12) become:

$$\begin{aligned} p_n &= V \\ p_{n-1} - p_n &= V \\ p_{n-2} - p_{n-1} - p_n &= V \\ &\vdots \\ p_1 - p_2 - \cdots - p_{n-2} - p_{n-1} - p_n &= V \end{aligned}$$

The solution is $p_n = V$, $p_{n-1} = 2V, \ldots, p_1 = 2^{n-1}V$. Since $1 = p_1 + p_2 + \cdots + p_n = [2^{n-1} + 2^{n-2} + \cdots + 1]V = [2^n - 1]V$, we find that the value is $V = 1/(2^n - 1)$. The optimal strategy for I (and for II also) is $(2^{n-1}, 2^{n-2}, \ldots, 2, 1)/(2^n - 1)$.

6. The matrix $\mathbf{A}$ has components $a_{ij} = 0$ for $i < j$ and $a_{ij} = b^{i-j}$ for $i \geq j$. It is easy to show that

$$\begin{pmatrix} 1 & 0 & 0 & \cdots & 0 \\ b & 1 & 0 & & 0 \\ b^2 & b & 1 & & 0 \\ \vdots & & \ddots & \ddots & \\ b^{n-1} & b^{n-2} & \cdots & b & 1 \end{pmatrix} \begin{pmatrix} 1 & 0 & 0 & \cdots & 0 \\ -b & 1 & 0 & & 0 \\ 0 & -b & 1 & & 0 \\ \vdots & & \ddots & \ddots & \\ 0 & 0 & \cdots & -b & 1 \end{pmatrix} = I.$$

Therefore, we may use Theorem 9.2 to find the value V as the reciprocal of the sum of all the elements of $\mathbf{A}^{-1}$, $V = 1/(n - (n-1)b)$, and I's optimal strategy is proportional to the sums of the columns of $\mathbf{A}^{-1}$, $\mathbf{p} = (1-b, 1-b, \ldots, 1-b, 1)/(n-(n-1)b)$, and II's optimal strategy is $\mathbf{q} = (1, 1-b, 1-b, \ldots, 1-b)/(n-(n-1)b)$, proportional to the sums of the rows of $\mathbf{A}^{-1}$.

7. We may use Theorem 9.2 with $\mathbf{A}_n^{-1}$ replaced by $\mathbf{B}_n$. Since the sum of the ith row of $\mathbf{B}_n$ is 2^{i-1} (the binomial theorem), we have $\mathbf{B}_n\mathbf{1} = (1, 2, 4, \ldots, 2^{n-1})^T$, and so $\mathbf{1}^T\mathbf{B}_n\mathbf{1} = 2^n - 1$. Similarly, the sum of column j is $\sum_{k=j}^{n} \binom{k-1}{j-1} = \binom{n}{k}$ (easily proved by induction). So that $\mathbf{1}^T\mathbf{B} = (\binom{n}{1}, \binom{n}{2}, \ldots, \binom{n}{n})$. From this we may conclude that the value is $V = 1/(2^n - 1)$, the optimal strategy of Player I is $\mathbf{p} = (1, 2, \ldots, 2^{n-1})/(2^n - 1)$, and the optimal strategy of Player II is $\mathbf{q} = (\binom{n}{1}, \binom{n}{2}, \ldots, \binom{n}{n})/(2^n - 1)$.

8. (a) Assuming all strategies active, the optimal strategy for I satisfies, $p_1 = V$, $-p_1 + 2p_2 = V$ and $-p_1 + p_2 + 3p_3 = V$, from which we find $p_1 = V$, $p_2 = V$ and $p_3 = V/3$. Since $1 = p_1 + p_2 + p_3 = V + V + (1/3)V = (7/3)V$, we have $V = 3/7$ and $\mathbf{p} = (p_1, p_2, p_3) = (3/7, 3/7, 1/7)$. A similar analysis for Player II gives $\mathbf{q} = (q_1, q_2, q_3) = (5/7, 1/7, 1/7)$. Since $\mathbf{p}$ and $\mathbf{q}$ are nonnegative, these are the optimal strategies, and V is the value.

(b) If we subtract 1 from all entries of the matrix, we end up with a diagonal game with 1, 1/2, 1/3 and 1/4 along the diagonal. The value of that game is $1/10$, and the optimal strategies for both players is $(1/10, 2/10, 3/10, 4/10)$. The original game has value $11/10$ and the same optimal strategies.

(c) The last column is dominated by the first, and the bottom row is dominated by the mixture of row 1 and row 2 with probability 1/2 each. The resulting 3×3 matrix is a diagonal game with value $1/[(1/2) + (1/3) + (1/4)] = 12/13$. The optimal strategy for both players is $(6/13, 4/13, 3/13, 0)$.

9. (a) The matrix is

$$\begin{pmatrix} 0 & -2 & 1 & 1 & 1 & \cdots \\ 2 & 0 & -2 & 1 & 1 & \cdots \\ -1 & 2 & 0 & -2 & 1 & \\ -1 & -1 & 2 & 0 & -1 & \\ \vdots & \vdots & & & & \ddots \end{pmatrix}$$

(b) The game is symmetric and has value zero (if it exists). If the first five rows and columns are the active ones, the equations become

$$\begin{aligned}
2p_2 - \ p_3 - \ p_4 - \ p_5 &= 0 \\
-2p_1 \qquad + 2p_3 - \ p_4 - \ p_5 &= 0 \\
p_1 - 2p_2 \qquad + 2p_4 - \ p_5 &= 0 \\
p_1 + \ p_2 - 2p_3 \qquad + 2p_5 &= 0 \\
p_1 + \ p_2 + \ p_3 - 2p_4 \qquad &= 0
\end{aligned}$$

If we interchange (p_1, p_2) with (p_5, p_4) in these equations, we get the same set of equations. So in the solution, we must have $p_1 = p_5$ and $p_2 = p_4$. Using this, the top two equations become $p_2 = p_1 + p_3$ and $2p_3 = 3p_2 + p_1$, which together with $2p_1 + 2p_2 + p_3 = 1$ gives $p_1 = p_5 = 1/16$, $p_2 = p_4 = 5/16$ and $p_3 = 4/16$. If Player I uses $\mathbf{p} = (1/16, 5/16, 4/16, 5/16, 1/16, 0, 0, \ldots)$ on the game with general n, then Player II will never use columns 6 or greater because the average payoff to Player I would be positive. Thus, the value is zero and $\mathbf{p}$ is optimal for both players.

10. (a) The matrix is

$$\begin{array}{c} \\ 1 \\ 2 \\ 3 \\ 4 \\ 5 \\ 6 \\ \vdots \end{array}
\begin{array}{c}
\begin{array}{ccccccc} 1 & 2 & 3 & 4 & 5 & 6 & \ldots \end{array} \\
\begin{pmatrix}
0 & -1 & 2 & 2 & 2 & 2 & \ldots \\
1 & 0 & -1 & -1 & -1 & 2 & \ldots \\
-2 & 1 & 0 & -1 & -1 & -1 & \ldots \\
-2 & 1 & 1 & 0 & -1 & -1 & \ldots \\
-2 & 1 & 1 & 1 & 0 & -1 & \ldots \\
-2 & -2 & 1 & 1 & 1 & 0 & \ldots \\
\vdots & \vdots & \vdots & \vdots & \vdots & \vdots &
\end{pmatrix}
\end{array}$$

One can see that columns $6, 7, \ldots$ are all dominated by column 1. Similarly for rows. This reduces the game to a 5×5 matrix. Columns 3 and 4 are dominated by column 5. This reduces the game to 3×3.

(b) The game restricted to rows and columns 1, 2 and 5 has matrix $\begin{pmatrix} 0 & -1 & 2 \\ 1 & 0 & -1 \\ -2 & 1 & 0 \end{pmatrix}$, whose solution (see (25)) is $(1/4, 1/2, 1/4)$. It is easy to check that the mixed strategy $(1/4, 1/2, 0, 0, 1/4, 0, \ldots)$ gives Player I an average payoff of at least 0 for every pure strategy of Player II. So this strategy is optimal for Player I, and by symmetry Player II as well.

11. (a) The matrix is skew-symmetric so this is a symmetric game. So the value is 0. To find an optimal strategy for I, we try (p_1, p_2, p_3) against the columns. The first column gives $-p_2 + 2p_3 = 0$ (since 0 is the value), and the second gives $p_1 - 3p_3 - 0$. We have $p_1 = 3p_3$ and $p_2 = 2p_3$. Then since the probabilities sum to 1, we have $3p_3 + 2p_3 + p_3 = 1$ or $p_3 = 1/6$. Then, $p_1 = 1/2$ and $p_2 = 1/3$. The optimal strategy for both players is $(1/2, 1/3, 1/6)$.

(b) This is a Latin square game, so $(1/3, 1/3, 1/3)$ is optimal for both players and the value is $v = (0 + 1 - 2)/3 = -1/3$.

(c) (1/4)row 1 + (1/4)row 2 + (1/4)row 3 + (1/4)row 4 dominates row 5. After removing row 5, the matrix is a Latin square. So $(1/4, 1/4, 1/4, 1/4)$ is optimal for II, and $(1/4, 1/4, 1/4, 1/4, 0)$ is optimal for I. The value is $v = (1 + 4 - 1 + 5)/4 = 9/4$.

12. The answer given by the Matrix Game Solver gives the same value and optimal strategy for Player I as in the text, but gives the optimal strategy for Player II as (3/90,32/90,48/90,7/90). This shows that although there may be a unique invariant optimal strategy, there may be other noninvariant optimal strategies as well. The simplex method only finds basic feasible solutions, and so will not find the invariant optimal solution (1/18,4/9,4/9,1/18), because it is not basic.

In (40), the middle row is strictly dominated by (3/4) the top row plus (1/4) the bottom row. Our solution and the one found by the Matrix Game Solver both give zero weight to (3,1) and (1,3) for Player I.

13. (a) The reduced matrix has a saddle point.

$$\begin{array}{c} \\ (2,0)^* \\ (1,1) \end{array} \begin{array}{c} (1,0)^* \\ \begin{pmatrix} 1 \\ 1 \end{pmatrix} \end{array}$$

So the value is 1, $(1,1)$ (or $(2,0)^*$) is optimal for Player I and $(1,0)^*$ is optimal for Player II.

(b) The reduced matrix is

$$\begin{array}{c} \\ (3,0)^* \\ (2,1)^* \end{array} \begin{array}{c} \quad (2,0)^* \quad (1,1) \\ \begin{pmatrix} 3/2 & 1/2 \\ 0 & 2 \end{pmatrix} \end{array}$$

The value is 1. An optimal strategy for Player I is to use $(3,0)^*$ with probability 2/3, and $(2,1)^*$ with probability 1/3. This corresponds to playing $(3,0)$ and $(0,3)$ with probability 1/3 each, and $(2,1)$ and $(1,2)$ with probability 1/6 each. An optimal strategy for Player II is to use (1.1) with probability 1/2, and $(2,0)$ and $(0,2)$ with probability 1/4 each.

14. (a) In the matrix below, row 4 is dominated by (1/2) row 2 + (1/2) row 3. Then col 2 is dominated by (1/2) col 1 + (1/2) col 3. Then row 2 is dominated by (2/3) row 1 + (1/3)row 3.

$$\begin{array}{c} \\ (4,0,0)^* \\ (3,1,0)^* \\ (2,2,0)^* \\ (2,1,1)^* \end{array} \begin{array}{c} \begin{array}{ccc} (3,0,0)^* & (2,1,0)^* & (1,1,1) \end{array} \\ \begin{pmatrix} 4/3 & 2/3 & 0 \\ 1/3 & 1 & 1 \\ -1 & 4/3 & 3 \\ -1/3 & 1/3 & 2 \end{pmatrix} \end{array}$$

We find that (3/4,0,1/4,0) is optimal for Player I, (9/16,0,7/16) is optimal for Player II, and the value is 3/4.

(b) In the matrix below, row 4 is dominated by (1/2)row 3 + (1/2)row 5. But we might as well use the Matrix Game Solver directly.

$$\begin{array}{c} \\ (4,0,0,0)^* \\ (3,1,0,0)^* \\ (2,2,0,0)^* \\ (2,2,1,1)^* \\ (1,1,1,1) \end{array} \begin{array}{c} \begin{array}{ccc} (3,0,0,0)^* & (2,1,0,0)^* & (1,1,1,0) \end{array} \\ \begin{pmatrix} 1 & 1/4 & -1/2 \\ 1/2 & 3/4 & 1/2 \\ -1/2 & 1 & 2 \\ 1/4 & 1/2 & 3/2 \\ 1 & 0 & 1 \end{pmatrix} \end{array}$$

The value is 3/5, (0, 4/5, 0, 0, 1/5) is optimal for Player I, and (8/15, 2/5, 1/15) is optimal for Player II.

15. Consider the following strategies for Player II.

A: Start at the center square; if this is a hit continue with a 2, 4, 6, or 8 in random order each order equally likely; if this is a miss, shoot at the corners 1, 3, 7, 9 in a random, equally likely order, and when a hit occurs, choose one of the two possible middle edge squares at random, then the other.

B: Start at the four middle edge squares, 2, 4, 6, 8 in some random order; when a hit occurs, try the center next, the possible corner squares.

C: Start at the four middle edge squares, 2, 4, 6, 8 in some random order; when a hit occurs, try the possible corners next, then the center.

There are many other strategies for Player II, but they should be dominated by some mixture of these. In particular, starting at a corner square should be dominated by starting at a middle edge.

Using invariance, Player I has the two strategies, $[1,2]^*$ and $[2,5]^*$. Suppose Player I uses $[1,2]^*$ and Player II uses C. Then the first hit will occur on shot 1, 2, 3, or 4 with probability 1/4 each. After the first hit it takes on the average 1.5 more shots to get the other hit. The average number of shots then is

$$(1/4)(2.5) + (1/4)(3.5) + (1/4)(4.5) + (1/4)(5.5) = 4.$$

But if Player II starts off by shooting in the center before trying the corners, it will take one more shot on the average, namely 5. This gives the top row of the matrix below. The whole matrix turns out to be

$$\begin{array}{c} \\ [1,2]^* \\ [2,5]^* \end{array} \begin{array}{c} \begin{array}{ccc} A & B & C \end{array} \\ \begin{pmatrix} 5 & 5 & 4 \\ 3.5 & 3.5 & 5.5 \end{pmatrix} \end{array}$$

The first two columns are equivalent. Player I's optimal strategy is $(2/3, 1/3)$. This translates into choosing one of the 12 positions at random with probability 1/12 each. One optimal strategy for Player II is to randomize with equal probability between B and C. The value is 4.5.

16. Invariance reduces Player I to two strategies; choose 1 and 3 with probability 1/2 each, denoted by 1^*, and choose 3. Similarly, invariance and dominance reduces Player II to two strategies, we call A and B. For A, start with 2. For B, with probability 1/2, start with 1 and if it's not successful follow it with 3, and with probability 1/2 start with 3 if it's not successful and follow it with 1. This leads to a 2×2 game with matrix

$$\begin{array}{c} \\ 1^* \\ 2 \end{array} \begin{array}{c} \begin{array}{cc} A & B \end{array} \\ \begin{pmatrix} 2 & 3/2 \\ 1 & 3 \end{pmatrix} \end{array}$$

The value is $9/5$. An optimal strategy for I is to choose 2 with probability 1/5, and 1 or 3 equally likely with probability 2/5 each. An optimal strategy for II is guess 2 first with probability 3/5, and otherwise to guess 1 then 3, or 3 then 1 with probability 1/5 each; that is, II never guesses 1 then 2 then 3 and never guesses 3 then 2 then 1.

17. To make Player I indifferent in choosing among rows 1 through k, Player II will choose $\mathbf{q} = (q_1, \ldots, q_k, 0, \ldots, 0)$ so that $u_i \sum_{i \neq j} q_j = V_k$

for $i = 1, \ldots, k$ for some constant V_k. Using $\sum_1^k q_j = 1$, this reduces to $(1 - q_i) = V_k/u_i$. Since $\sum_1^k (1 - q_i) = k - 1$, we have

$$V_k = \frac{k-1}{\sum_1^k 1/u_i} \quad \text{and} \quad q_i = \begin{cases} 1 - V_k/u_i & \text{for } i = 1, \ldots, k \\ 0 & \text{for } i = k+1, \ldots, m \end{cases} \tag{1}$$

The q_i are nondecreasing but we must have $q_i \geq 0$, which reduces to $V_k \leq u_k$. If $k < m$, we also require that Player I will not choose rows $k+1$ to m. This reduces to $u_{k+1} \leq V_k$. Therefore, if $u_{k+1} \leq V_k \leq u_k$, Player II can achieve V_k by using $\mathbf{q}$. (It is easy to show that $V_2 < u_2$ and that $V_k \leq u_k$ implies that $V_{k-1} \leq u_{k-1}$. This shows that such a k exists and is unique.)

To make Player II indifferent in choosing among columns 1 through k, Player I will choose $\mathbf{p} = (p_1, \ldots, p_k, 0, \ldots, 0)$ so that $\sum_1^k p_i u_i - p_j u_j = V_k$ for some constant V_k and $j = 1, \ldots, k$. This shows that p_j is equal to some constant over u_j for $j = 1, \ldots, k$. Using $\sum_1^k p_j = 1$, we find

$$p_j = \begin{cases} \dfrac{1/u_j}{\sum_1^k 1/u_i} & \text{for } j = 1, \ldots, k \\ 0 & \text{for } j = k+1, \ldots, m. \end{cases} \tag{2}$$

Solving for V_k shows it indeed has the same value as above. All the p_j are nonnegative, so we only have to show that Player II will not want to choose columns $k+1, \ldots, m$. The expected payoff is the same for each of these columns, namely, $\sum_1^k p_i u_i$ which is clearly greater than V_k, so Player I can achieve at least V_k.

In summary, find the largest k in $\{2, \ldots, m\}$ such that $(k-1)/u_k \leq \sum_1^k 1/u_i$. Then the value and the optimal strategies are given by (1) and (2).

18. The payoff matrix is $\mathbf{A}_n = (a_{ij})$, where

$$a_{ij} = \begin{cases} 2 & \text{if } i = j \\ -1 & \text{if } |i - j| = 1 \\ 0 & \text{otherwise.} \end{cases}$$

It is straightforward to check that $\mathbf{A}_n \mathbf{B}_n = \mathbf{I}_n$, so that $\mathbf{A}_n^{-1} = \mathbf{B}_n$. The sum of the ith row of $\mathbf{A}_n^{-1}$ is $i(n+1-i)/2$. By symmetry these are also the column sums. Since they are all positive the game is completely mixed, and the optimal strategy, the same for both players, is proportional to $(n, 2(n-1), 3(n-2), \ldots, n)$, namely, $p(i) = 6i(n+1-i)/(n(n+1)(n+2))$.

The sum of all numbers in $\mathbf{A}_n^{-1}$ is $n(n+1)(n+2)/12$, so the value is its reciprocal, $v_n = 12/(n(n+1)(n+2))$.

19. (a) n odd: Let $\mathbf{x} = (1, 0, 1, 0, \ldots, 1)^T$. Then $\mathbf{x}^T \mathbf{A} = (2, 2, \ldots, 2)$. There are $(n+1)/2$ 1's in $\mathbf{x}$, so $\mathbf{p} = 2\mathbf{x}/(n+1)$ is a mixed strategy for I that guarantees I will win $4/(n+1)$ no matter what column II chooses. The matrix is symmetric, so the same strategy guarantees II will lose $4/(n+1)$ no matter what I does. Thus, $\mathbf{p}$ is an optimal strategy for both I and II, and $4/(n+2)$ is the value.

(b) n even: Let $k = n/2$ and $\mathbf{x} = (k, 1, k-1, 2, \ldots, 1, k)^T$. Then $\mathbf{x}^t \mathbf{A} = (2k+1, 2k+1, \ldots, 2k+1)$. The sum of the elements of $\mathbf{x}$ is $k(k+1)$ so $\mathbf{p} = \mathbf{x}/(k(k+1))$ is a mixed strategy for I that guarantees I will win $(2k+1)/(k(k+1))$ no matter what column II chooses. The same strategy guarantees II will lose $(2k+1)/(k(k+1))$ no matter what I does. Thus, $\mathbf{p}$ is an optimal strategy for both I and II, and the value is $v = (2k+1)/(k(k+1)) = 4(n+1)/(n(n+2))$.

Solutions to Chap. 10

1. (a) If Player II uses the mixed strategy, $(1/5, 1/5, 1/5, 2/5)$, I's expected payoff from rows 1, 2, and 3 are 17/5, 17/5, and 23/5 respectively. So I's Bayes strategy is row 3, giving expected payoff 23/5.

(b) If II guesses correctly that I will use the Bayes strategy against $(1/5, 1/5, 1/5, 2/5)$, she should choose column 3, giving Player I a payoff of -1.

2. (a) We have $b_{ij} = 5 + 2a_{ij}$ for all i and j. Hence, A and B have the same optimal strategies for the players, and the value of B is $\text{Val}(B) = 5 + 2\text{Val}(A) = 5$. The optimal strategy for I is $(6/11, 3/11, 2/11)$.

(b) Since we are given that $\text{Val}(A) = 0$, we may solve for the optimal $\mathbf{q}$ for II using the equations, $-q_2 + q_3 = 0$, and $2q_1 - 2q_3 = 0$. So $q_1 = q_3$ and $q_2 = q_3$. Since the probabilities sum to 1, all three must be equal to 1/3. So $(1/3, 1/3, 1/3)$ is optimal for Player II for both matrices A and B.

3. (a) Let $\epsilon > 0$ and let $\mathbf{q}$ be any element of Y^*. Then since $\sum_{j=1}^{n} q_j \to 1$ as $n \to \infty$, we have $\sum_{j=n}^{\infty} q_j \to 0$, so that there is an integer N such that $\sum_{j=N}^{\infty} q_j < \epsilon$. If Player I uses $i = N$, the expected payoff is $\sum_{j=1}^{\infty} L(N, j)q_j = \sum_{j=1}^{N-1} q_j - \sum_{j=N+1}^{\infty} q_j < 1 - 2\epsilon$. Thus for every $\mathbf{q} \in Y^*$, we have $\sup_{1 \le i < \infty} \sum_{j=1}^{\infty} L(i, j)q_j \ge 1 - 2\epsilon$. Since this is true for all $\epsilon > 0$, it is also true for $\epsilon = 0$.

(b) Since (a) is true for all $\mathbf{q} \in Y^*$, we have $\overline{V} = \inf_{\mathbf{q} \in Y^*} \sup_{1 \le i < \infty} \sum_{j=1}^{\infty} L(i, j)q_j \ge 1$. Since no payoff is greater than 1, we have $\overline{V} = 1$.

(c) The game is symmetric, so $\underline{V} = -\overline{V}$. Hence, $\underline{V} = -1$.

(d) Any strategy is minimax for Player I since any strategy guarantees an expected payoff of at least $\underline{V} = -1$.

4. The value of $\mathbf{A}$ is positive since the simple strategy $(2/3)$row 1 + $(1/3)$row 2 guarantees a positive return for Player I. But let's add 1 to $\mathbf{A}$ anyway to get $\mathbf{B}$:

$$\mathbf{B} = \begin{pmatrix} 1 & 2 & 3 \\ 3 & 0 & -1 \\ 4 & -2 & 1 \end{pmatrix}.$$

The simplex tableau is displayed below on the left. We are asked to pivot in the second column. But there is only one positive number there, so we must pivot on the first row second column. We arrive at:

	y_1	y_2	y_3	
x_1	1	②	3	1
x_2	3	0	-1	1
x_3	4	-2	1	1
	-1	-1	-1	0

$\longrightarrow$

	y_1	x_1	y_3	
y_2	$1/2$	$1/2$	$3/2$	$1/2$
x_2	3	0	-1	1
x_3	5	1	4	2
	$-1/2$	$1/2$	$1/2$	$1/2$

There is still a negative element on the bottom edge so we continue. It is unique and in the first column, so we pivot in the first column. the smallest of the ratios is $1/3$ occuring in the second row. So we pivot on the second row first column to find:

	y_1	x_1	y_3	
y_2	$1/2$	$1/2$	$3/2$	$1/2$
x_2	③	0	-1	1
x_3	5	1	4	2
	$-1/2$	$1/2$	$1/2$	$1/2$

$\longrightarrow$

	x_2	x_1	y_3	
y_2	$-1/6$	$1/2$	$5/3$	$1/3$
y_1	$1/3$	0	$-1/3$	$1/3$
x_3	$-5/3$	1	$17/3$	$1/3$
	$1/6$	$1/2$	$1/3$	$2/3$

From this we see that $\text{Val}(\mathbf{B}) = 3/2$, so that $\text{Val}(\mathbf{A}) = 1/2$. For either game $(p_1, p_2, p_3) = (3/4, 1/4, 0)$ is optimal for Player I and $(q_1, q_2, q_3) = (1/2, 1/2, 0)$ is optimal for Player II. You may use the sure-fire test to see that this is correct.

5. (a) For all $j = 0, 1, 2, \ldots$,

$$\begin{aligned}
\sum_{i=0}^{\infty} p_i A(i,j) &= \sum_{i=0}^{j-1} \frac{1}{2^{(i+1)}}(-4^i) + \sum_{i=j+1}^{\infty} \frac{1}{2^{(i+1)}}(4^j) \\
&= -\frac{1}{2}(1 + 2 + \cdots + 2^{j-1}) + 4^j \frac{1}{2^{(j+2)}}\left(1 + \frac{1}{2} + \frac{1}{4} + \cdots\right) \\
&= -\frac{1}{2}(2^j - 1) + \frac{1}{2}2^j = \frac{1}{2}.
\end{aligned}$$

(b) If both players use the mixed strategy, $\mathbf{p}$, the payoff is $\sum\sum p_i A(i,j) p_j$. The trouble is that the answer depends on the order of summation. If we sum over i first, we get $+1/2$, and if we sum over j first we get $-1/2$. In other words, Fubini's Theorem does not hold here. For Fubini's Theorem, we need $\sum\sum p_i |A(i,j)| p_j < \infty$, which is not the case here. The whole theory of using mixed strategies in games depends heavily on Utility Theory. In Utility Theory, at least as presented in Appendix 1, the utility functions are bounded. So it would seem most logical to restrict attention to games in which the payoff function, A, is bounded. That is certainly one way to avoid such examples. However, in many important problems the payoff function is unbounded, at least on one side, so one usually assumes that the payoff function is bounded below, say.

There is another way of dealing with the problem that is more germane to the example above, and that is by restricting the notion of a mixed strategy to be a probability distribution that gives weight to only a finite number of pure strategies. (Then Fubini's theorem holds because the summations are finite.) If this is done in the example, then one can easily see that the value of the game does not exist. This seems to be the "proper" solution of the game because it is just a blown-up version of the game, "the-player-that-chooses-the-larger-integer-wins".

6. We take the initial $i_1 = 1$, and find

k	i_k	$s_k(1)$	$s_k(2)$	$\underline{V}_k$	j_k	$t_k(1)$	$t_k(2)$	$\overline{V}_k$
1	1	1	**−1**	−1	2	−1	**2**	2
2	2	**1**	1	0.5	1	0	**2**	1
3	2	**1**	3	0.3333	1	1	**2**	0.6667
4	2	**1**	5	0.25	1	**2**	2	0.5

The largest of the $\underline{V}_k$, namely 0.5, is equal to the smallest of the $\overline{V}_k$. So the value of the game is $1/2$. An optimal strategy for Player I is found at $k = 2$ to be $\mathbf{p} = (0.5, 0.5)$. An optimal strategy for Player II is $\mathbf{q} = (0.75, 0.25)$ found at $k = 4$. Don't expect to find the value of a game by this method again!

7. (a) The upper left payoff of $\sqrt{2}$ was chosen so that there would be no ties in the fictitious play. So Player II knows exactly what Player I will do and will be able to guarantee a zero payoff at each future stage. If Player II's relative frequency, q_k, of column 1 by stage k goes above $1/(\sqrt{2}+1)$, Player I will play row 1, causing Player II to play column 2, thus causing q_k to decrease. Thus q_k converges to $1/(\sqrt{2}+1)$, which in fact is Player

II's optimal strategy for the game. Similarly, Player I's relative frequency, p_k, of row 1 converges to $1/(\sqrt{2}+1)$, which is his optimal strategy.

(b) Player I should play the same pure strategy as his opponent at each stage, gaining either $\sqrt{2}$ or 1 at each stage. The same argument as in (a) shows that Player II's average relative frequency of column 2 converges to $1/(\sqrt{2}+1)$, so Player I's limiting average payoff is

$$\frac{1}{\sqrt{2}+1}\cdot\sqrt{2}+\frac{\sqrt{2}}{\sqrt{2}+1}=2(2-\sqrt{2}),$$

twice the value of the game.

Solutions to Chap. 11

1. The Silver Dollar. I hides the dollar, II searches for it with random success.

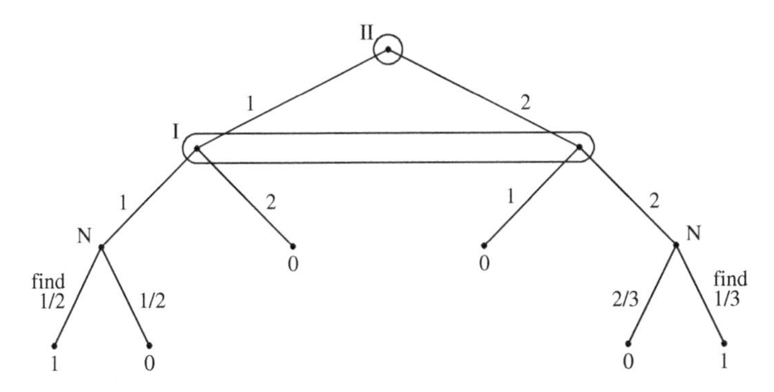

2. Two Guesses for the Silver Dollar. I hides the dollar, II searches for it twice with random success.

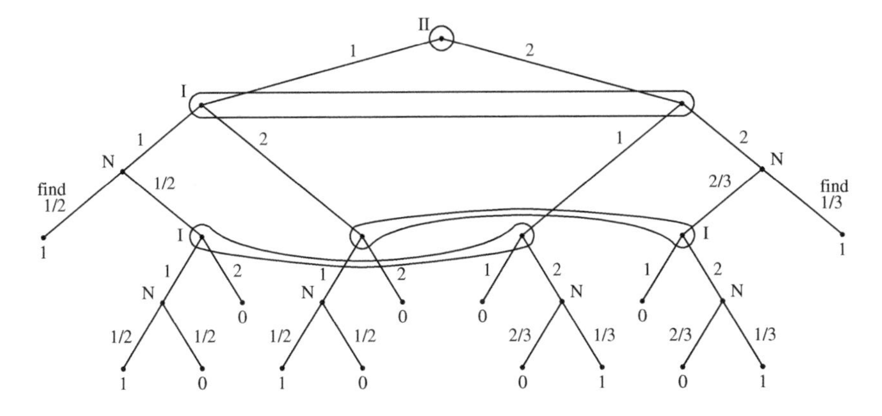

3. Guessing the Probability of a Coin. I chooses the fair (F) or biased (B) coin. II observes H or T on one toss of the coin and must guess which coin.

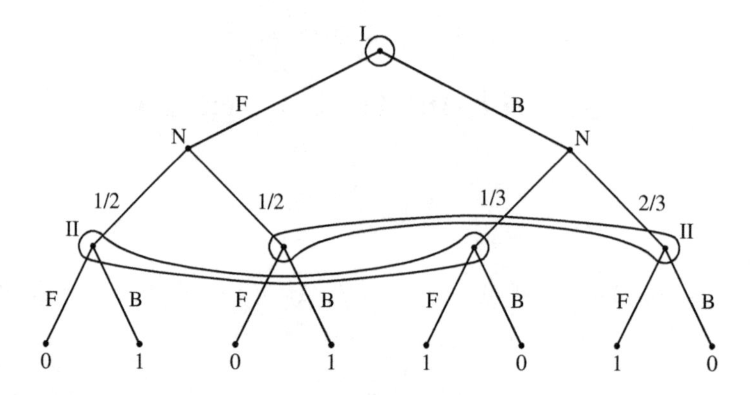

4. A Forgetful Player. A fair coin is tossed. I hears the outcome and bets 1 or 2. II guesses H or T. I forgets the toss and doubles or passes.

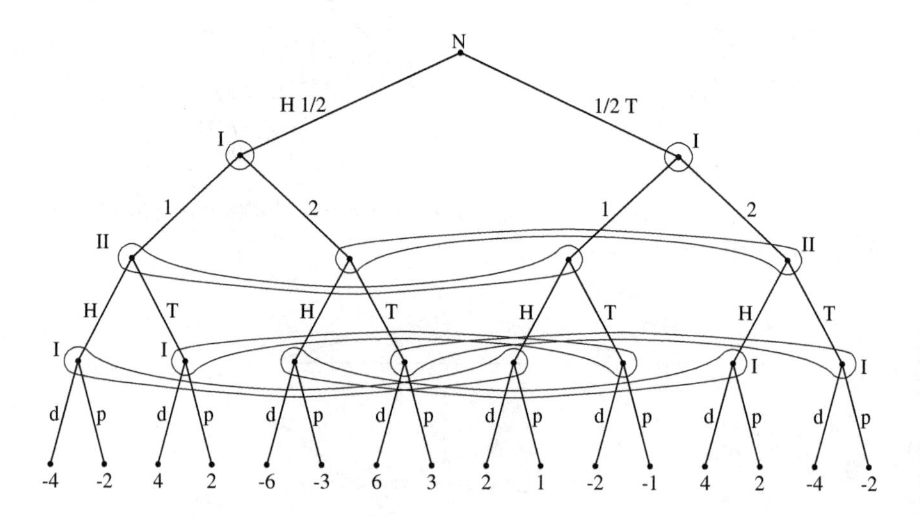

5. A One-Shot Game of Incomplete Information.

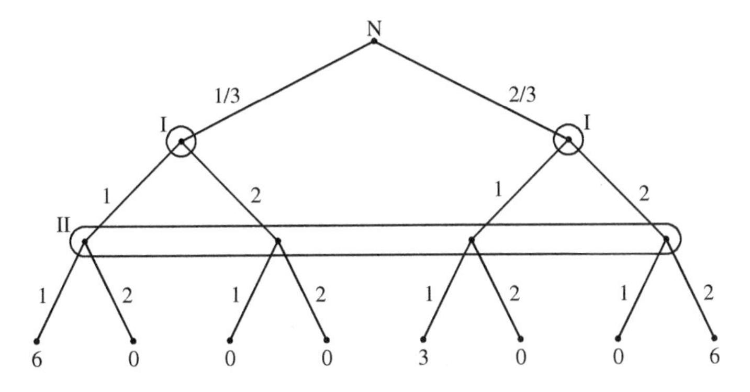

6. In Fig. 11.2, replace $1/4$ by p, $3/4$ by $1-p$, and ± 3 by $\pm(1+b)$. Again one can argue that Player I should bet with a winning card; he wins 1 if he checks, and wins at least 1 if he bets. In other words, as in the analysis in the text of the resulting 4×2 matrix, the first row dominates the third row and the second row dominates the fourth row. The top two rows of the matrix are

$$\begin{array}{c} \\ (b,b) \\ (b,c) \end{array} \begin{array}{c} \begin{array}{cc} \quad c \quad & \quad f \end{array} \\ \begin{pmatrix} (1+b)(2p-1) & 1 \\ p(2+b)-1 & 2p-1 \end{pmatrix} \end{array}$$

If $(2p-1)(1+b) \geq 1$, (that is, if $p \geq (2+b)/(2+2b)$), there is a saddle-point in the upper right corner. The value of the game is 1, Player I should always bet, and Player II should always fold.

Otherwise, (if $p < (2+b)/(2+2b)$), the game does not have a saddle-point and we can use the straightforward method for solving two by two games. It is optimal for Player I to choose row 1 with probability $pb/((2+b)(1-p))$, and row 2 otherwise. It is optimal for Player II to choose column 1 with probability $2/(2+b)$, and column 2 otherwise. The value is $(4p(1+b)-(2+b))/(2+b)$.

7. (a) The strategic (normal) form is

$$\begin{array}{c} \\ a \\ b \\ c \end{array} \begin{array}{c} \begin{array}{cccc} (d,f) & (d,g) & (e,f) & (e,g) \end{array} \\ \begin{pmatrix} -1 & -1 & 1 & 3 \\ 1 & 0 & 1 & 2 \\ 1 & 1 & -1 & 1 \end{pmatrix} \end{array}$$

(b) Column 2 dominates columns 1 and 4. Then row 2 dominates row 1. The resulting 2×2 matrix is $\begin{pmatrix} 0 & 1 \\ 1 & -1 \end{pmatrix}$, with value 1/3. The optimal mixed strategy for Player I is $(0, 2/3, 1/3)$. The optimal mixed strategy for Player 2 is $(0, 2/3, 1/3, 0)$.

8. (a) The strategic (normal) form is

$$\begin{array}{c} \\ (A,C) \\ (A,D) \\ (B,C) \\ (B,D) \end{array} \begin{array}{c} \begin{array}{cccc} (a,c) & (a,d) & (b,c) & (b,d) \end{array} \\ \begin{pmatrix} 1 & 1 & 0 & 0 \\ 0 & 2 & 1 & 3 \\ 3/2 & 1/2 & -1/2 & -3/2 \\ 1/2 & 3/2 & 1/2 & 3/2 \end{pmatrix} \end{array}$$

(b) Column 3 dominates column 2. The mixture, $(2/3)$row 2 + $(1/3)$ row 3, dominates row 4. The mixture, $(1/3)$row 2 + $(2/3)$row 3, dominates row 1. The resulting 2×3 matrix is $\begin{pmatrix} 0 & 1 & 3 \\ 3/2 & -1/2 & -3/2 \end{pmatrix}$. The first two columns of this matrix are active. The value is $1/2$. An optimal mixed strategy for Player I in the original game is $(0, 2/3, 1/3, 0)$. An optimal mixed strategy for Player II is $(1/2, 0, 1/2, 0)$. It is interesting to note that Player I also has an optimal pure strategy, namely row 4.

9. (a)

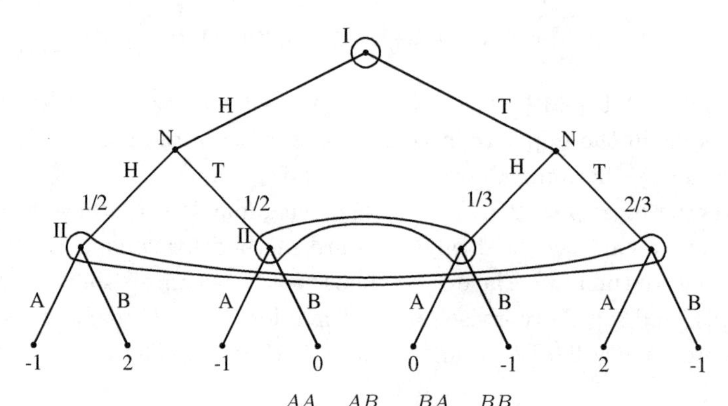

(b) The matrix is $\begin{array}{c} \\ H \\ T \end{array} \begin{array}{c} \begin{array}{cccc} AA & AB & BA & BB \end{array} \\ \begin{pmatrix} -1 & -1/2 & 1/2 & 1 \\ 4/3 & 1 & -2/3 & -1 \end{pmatrix} \end{array}$.

(c) Optimal for I $=$ $(4/7, 3/7)$. Optimal for II $=$ $(1/3, 0, 2/3, 0)$. Value $= 0$.

10. (a) The matrix is $\begin{array}{c} \\ 1 \\ 2 \end{array} \begin{array}{c} \begin{array}{cc} 1 & 2 \end{array} \\ \begin{pmatrix} 1/2 & 0 \\ 0 & 1/3 \end{pmatrix} \end{array}$. The mixed strategy $(2/5, 3/5)$ is optimal for both I and II. The value is $1/5$.

(b) The matrix is $\begin{array}{c} \\ 11 \\ 12 \\ 21 \\ 22 \end{array}\begin{array}{c} \begin{array}{cc} 1 & 2 \end{array} \\ \begin{pmatrix} 3/4 & 0 \\ 1/2 & 1/3 \\ 1/2 & 1/3 \\ 0 & 5/9 \end{pmatrix} \end{array}$. An optimal strategy for I is $(0,0,10/13,3/13)$. The optimal strategy for II is $(4/13,9/13)$. The value is $5/13$.

(c) The matrix is $\begin{array}{c} \\ F \\ B \end{array}\begin{array}{c} \begin{array}{cccc} FF & FB & BF & BB \end{array} \\ \begin{pmatrix} 0 & 1/2 & 1/2 & 1 \\ 1 & 1/3 & 2/3 & 0 \end{pmatrix} \end{array}$. Optimal for I is $(4/7,3/7)$. Optimal for II is $(1/7,6/7,0,0)$. The value is $3/7$.

(d) The matrix is 64 by 4, much too large write out by hand. However, simple arguments show that most of Player I's pure strategies are dominated. First some notation. We denote Player I's pure strategies by a six-tuple, $ab;wxyz$, where a and b are 1 or 2 (the amount bet) for information sets I_1 and I_2 respectively, and each of w, x, y and z are p or d (pass or double) for information sets I_3, I_4, I_5 and I_6 respectively. Thus, for example, $12;pdpp$ represents the strategy: Bet 1 with heads and 2 with tails; his partner passes unless 1 is bet and Player II guesses heads, in which case he doubles.

If Player I uses a strategy starting 12, then his partner upon hearing a bet of 1 and a guess of heads should pass rather than double since that means a loss of 2 rather than 4. Similarly, on hearing a bet of 1 and a guess of tails, his partner should double. Continuing in this way, we see that the strategy $12;pddp$ dominates all strategies beginning 12.

Similarly, we may see that the strategy $21;dppd$ dominates all strategies beginning 21, the strategy $11;pdxx$ (where x stands for "any") dominates all strategies beginning 11, and $22;xxpd$ dominates all strategies beginning 22. Thus, dominance reduces the game to the following 4×4 matrix.

$$\begin{array}{c} \\ 11;pdxx \\ 12;pddp \\ 21;dppd \\ 22;xxpd \end{array}\begin{array}{c} \begin{array}{cccc} HH & HT & TH & TT \end{array} \\ \begin{pmatrix} -1/2 & -1/2 & 1 & 1 \\ 1 & -2 & 4 & 1 \\ -1/2 & 4 & -2 & 5/2 \\ -1/2 & 1 & -1/2 & 1 \end{pmatrix} \end{array}.$$

Now, we can see that col 1 dominates col 4. Moreover an equiprobable mixture of rows 2 and 3 dominate rows 1 and 4. This reduces the game to a 2×3 matrix which is easily solvable. For the above matrix, (0,3/5,2/5,0) is optimal for I, (4/5,1/5,0,0) is optimal for II and the value is $2/5$.

We can describe Player I's strategy as follows. 60% of the time, Player I bets low on heads and high on tails, and his partner doubles when Player

II is wrong. The other 40% of the time, Player I bets high on heads and low on tails, and his partner doubles when Player II is wrong.

(e) The matrix is
$$\begin{array}{c} \\ 11 \\ 12 \\ 21 \\ 22 \end{array}\begin{array}{c} \begin{array}{cc} 1 & 2 \end{array} \\ \begin{pmatrix} 4 & 0 \\ 2 & 4 \\ 2 & 0 \\ 0 & 4 \end{pmatrix} \end{array}.$$
An optimal strategy for I is $(1/3, 2/3, 0, 0)$. An optimal strategy for II is $(2/3, 1/3)$. The value is $8/3$.

11. Suppose Player I uses f with probability p_1 and c with probability p_2 (and so g with probability $1 - p_1$ and d with probability $1 - p_2$). Suppose Player II uses a with probability q (and b with probability $1 - q$). The average payoff is then

$$\begin{aligned} v &= q(p_1 - (1 - p_1)(1 - p_2)) + (1 - q)(-p_1 p_2 + 2(1 - p_1)) \\ &= q(4p_1 + p_2 - 3) + 2 - 2p_1 - p_1 p_2. \end{aligned}$$

Therefore, Player II will choose $q = 0$ if $4p_1 + p_2 \geq 3$, and $q = 1$ if $4p_1 + p_2 < 3$. Against this, the best Player I can do is to use $p_1 = 1/2$ and $p_2 = 1$ (or $p_1 = 3/4$ and $p_2 = 0$) giving an average payoff of $1/2$. If q is announced, then Player I will use $p_1 = 1$ and $p_2 = 0$ if $q \geq 2/3$, and $p_1 = 0$ and $p_2 = 1$ if $q < 2/3$. Against this, the best Player II can do is $q = 2/3$, which gives Player I an average payoff of $2/3$. Therefore, the value of the game does not exist if behavioral strategies must be used.

12. (a)

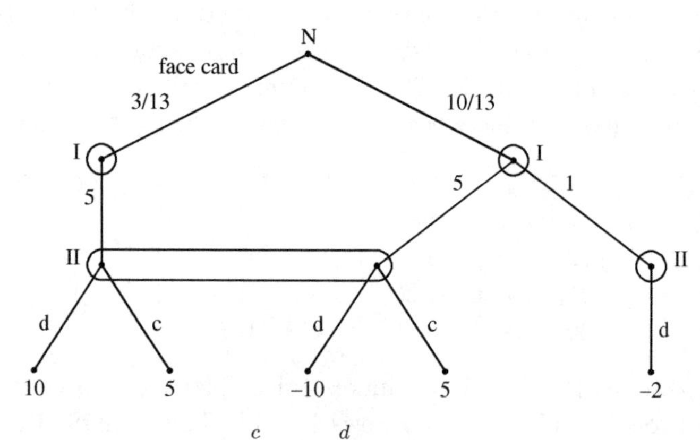

(b) The matrix is
$$\begin{array}{c} \\ 5 \\ 1 \end{array}\begin{array}{c} \begin{array}{cc} c & d \end{array} \\ \begin{pmatrix} 5 & -70/13 \\ -5/13 & 10/13 \end{pmatrix} \end{array}.$$

(c) The strategy $(1/10, 9/10)$ is optimal for I, $(8/15, 7/15)$ is optimal for II and the value is $2/13$.

Solutions to Chap. 12

1. (a) G_1 has a saddle point. The value is 3, the pure strategy $(1,0)$ is optimal for I, and $(0,1)$ is optimal for II.

The value of G_2 is 3, the strategy $(0.4, 0.6)$ is optimal for I, and $(0.5, 0.5)$ is optimal for II.

The value of G_3 is -1, the strategy $(0.5, 0.5)$ is optimal for I and for II.

The game G is thus equivalent to a game with matrix $\begin{pmatrix} 0 & 3 \\ 3 & -1 \end{pmatrix}$.

Hence, the value of G is $9/7$, and the strategy $(4/7, 3/7)$ is optimal for both players.

(b1) Interchanging the second and third columns, we have a matrix of the form $G = \begin{pmatrix} 0 & G_1 \\ G_2 & 0 \end{pmatrix}$, where $G_1 = \begin{pmatrix} 6 & 2 \\ 3 & 5 \end{pmatrix}$ and $G_2 = \begin{pmatrix} 5 & 2 \\ 1 & 4 \end{pmatrix}$. Since $\text{Val}(G_1) = 4$ and $\text{Val}(G_2) = 3$, we find $\text{Val}(G) = 12/7$.

(b2) The matrix has the form, $G = \begin{pmatrix} G_1 & 2 \\ 1 & G_2 \end{pmatrix}$, where $G_2 = \begin{pmatrix} 6 & 3 \\ 4 & 7 \end{pmatrix}$ and $G_1 = \begin{pmatrix} 3 & 1 & 5 \\ 1 & 3 & 5 \\ 4 & 4 & 1 \end{pmatrix} = \begin{pmatrix} G_3 & 5 \\ 4 & 1 \end{pmatrix}$, where $G_3 = \begin{pmatrix} 3 & 1 \\ 1 & 3 \end{pmatrix}$. Since $\text{Val}(G_3) = 2$, we have $\text{Val}(G_1) = 3$, and then since $\text{Val}(G_2) = 5$, we have $\text{Val}(G) = 13/6$.

2. The games $G_{m,n}$ are defined by the induction

$$G_{m,n} = \begin{array}{c} \\ \text{inspect} \\ \text{wait} \end{array} \begin{array}{c} \begin{array}{cc} \text{act} & \text{wait} \end{array} \\ \begin{pmatrix} 1 & G_{m-1,n-1} \\ 0 & G_{m,n-1} \end{pmatrix} \end{array}$$

$$\text{for} \quad n = m+1, m+2, \ldots \quad \text{and} \quad m = 1, 2, \ldots$$

with boundary conditions $G_{0,n} = (0)$. Let $V_{m,n} = \text{Value}(G_{m,n})$. We have $V_{m,n} = 1$ for $m \geq n$ and $V_{0,n} = 0$ for $n \geq 1$. Also from the Example given in the text, we have $V_{1,n} = 1/n$. We compute the next few values,

$$V_{2,3} = \text{Value}\begin{pmatrix}1 & 1/2\\ 0 & 1\end{pmatrix} = \frac{2}{3}, \quad V_{2,4} = \text{Value}\begin{pmatrix}1 & 1/3\\ 0 & 2/3\end{pmatrix} = \frac{2}{4}$$

and perhaps a few more, and then we conjecture $V_{m,n} = m/n$ for $0 \le m \le n$. Let us check this conjecture by induction. It is true for $m = 0$ and for $m = n$. Suppose that $0 < m < n$ and suppose the conjecture is true for all smaller values. Then,

$$\begin{aligned} V_{m,n} &= \text{Value}\begin{pmatrix}1 & V_{m-1,n-1}\\ 0 & V_{m,n-1}\end{pmatrix} \\ &= \text{Value}\begin{pmatrix}1 & (m-1)/(n-1)\\ 0 & m/(n-1)\end{pmatrix} \\ &= \frac{m}{n}. \end{aligned}$$

and the conjecture is verified. The optimal strategy for I in $G_{m,n}$ is the mixed strategy $(m/n, (n-m)/n)$, and the optimal strategy for II is $(1/n, (n-1/n)$. It is interesting to note that to play optimally Player II does not need to keep track of how many times I has searched for him.

3. The induction is

$$G_n = \frac{1}{2}\begin{pmatrix} \overset{1}{G_{n-1}} & \overset{2}{0} \\ 0 & G_{n-2} \end{pmatrix} \quad \text{for } n = 2, 3, \ldots$$

and boundary conditions $G_0 = (1)$ and $G_1 = (1)$. Let $V_n = \text{Value}(G_n)$. The recursion for V_n becomes

$$V_n = \text{Value}\begin{pmatrix}V_{n-1} & 0\\ 0 & V_{n-2}\end{pmatrix} = \frac{V_{n-1}V_{n-2}}{V_{n-1}+V_{n-2}} \quad \text{for } n = 2, 3, \ldots$$

with boundary conditions $V_0 = 1$ and $V_1 = 1$. Taking reciprocals of this equation, we find

$$\frac{1}{V_n} = \frac{1}{V_{n-1}} + \frac{1}{V_{n-2}} \quad \text{for } n = 2, 3, \ldots.$$

This is the recursion for the Fibonacci sequence. Since $1/V_0 = F_0$ and $1/V_1 = F_1$, we must have $1/V_n = F_n$. Hence we have $V_n = 1/F_n$

for all n and we may compute the optimal strategy for I and II to be $(F_{n-1}/F_n, F_{n-2}/F_n)$ for $n = 2, 3, \ldots$.

4. Use of the first row implies $v_n \geq n + 2$. Use of the first column implies $v_n \leq n+3$. Since $n+2 \leq v_n \leq n+3$, none of the games have saddle points. So for $n = 0, 1, \ldots$,

$$v_n = \text{Val}\begin{pmatrix} n+3 & n+2 \\ n+1 & v_{n+1} \end{pmatrix} = n + 3 - \frac{2}{v_{n+1} - n}.$$

Let $w_n = v_n - n + 1$ for $n = 0, 1, \ldots$. Then the w_n satisfy

$$w_n = 4 - \frac{2}{w_{n+1}}.$$

In fact, the w_n are the values of the games G'_n where

$$G'_n = 1 + \begin{pmatrix} 3 & 2 \\ 1 & G'_{n+1} \end{pmatrix} \quad \text{for } n = 0, 1, \ldots.$$

In game G'_n, I receives 1 from II and then the players choose row and column; if the players choose the second row second column, then I receives 1 from II and they next play G'_{n+1}. It may be seen that each of the games G'_n has the same structure. It is as if the players were playing the recursive game G' where

$$G' = 1 + \begin{pmatrix} 3 & 2 \\ 1 & G' \end{pmatrix}.$$

So all the games G'_n should have the same values. If so, denoting the common value by w, we would have $w = 4 - (2/w)$, or $w^2 - 4w + 2 = 0$. This has a unique solution in the interval $3 \leq w \leq 4$, namely $w = 2 + \sqrt{2}$. From this we have $v_n = n + 1 + \sqrt{2}$. The optimal strategies are the same for all games, namely,

$$\left(\frac{1+\sqrt{2}}{2+\sqrt{2}}, \frac{1}{2+\sqrt{2}}\right) = (0.707\cdots, 0.293\cdots)$$

is optimal for I for all games G_n

$$\left(\frac{\sqrt{2}}{2+\sqrt{2}}, \frac{2}{2+\sqrt{2}}\right) = (0.414\cdots, 0.586\cdots)$$

is optimal for II for all games G_n.

5. The game matrix of $G_{1,n}$ reduces to

$$\begin{pmatrix} 1-\dfrac{n}{n+1}V_{n-1,1} & \dfrac{n}{n+1}-\dfrac{n}{n+1}V_{n-1,1} \\ 0 & 1 \end{pmatrix}.$$

(a) So Player I's optimal strategy uses odds $1 : 1/(n+1) = n+1 : 1$; i.e. he should bluff with probability $1/(n+2)$.
(b) Player II's optimal odds are $\frac{1}{n+1}+\frac{n}{n+1}V_{n-1,1} : 1-\frac{n}{n+1}V_{n-1,1} = 1+nV_{n-1,1} : n+1-nV_{n-1,1}$; i.e. she should call with probability $(n+1-nV_{n-1,1})/(n+2) = V_{1,n}$.

6. (a) If $Q \geq 2$, the top row forever is optimal for I, the second column is optimal for II, and the value is $v = 2$. If $0 \leq Q \leq 2$, the top row forever is optimal for I, the first column forever is optimal for II, and the value is $v = Q$. If $Q \leq 0$, the bottom row is optimal for I, the first column forever is optimal for II, and the value is $v = 0$.

(b) If $Q \geq 1$, the value is $v = 1$, $(1,0,0)^\infty$ (i.e. the top row forever) is optimal for I, and $(0,1/2,1/2)^\infty$ is optimal for II. If $Q \leq 1$, the value is still $v = 1$, $(1,0,0)^\infty$ is optimal for II, and $(1-\epsilon,\epsilon/2,\epsilon/2)^\infty$ is ϵ-optimal for I (actually $\epsilon/(2-\epsilon)$-optimal)

7. Since in game G_3, I can choose the second row and II can choose the second column, we have $0 \leq v_3 \leq 1$. But since $v_3 \leq 1$, the most that I can hope to achieve in G_2 is $\mathrm{Val}\begin{pmatrix}1 & 0\\ 0 & 2\end{pmatrix} = 2/3$, so we have $0 \leq v_2 \leq 2/3$. Similarly, $1/2 \leq v_1 \leq 1$. So none of the game matrices have saddle points and we can write

$$v_1 = \frac{1}{2-v_2} \qquad v_2 = \frac{2v_3}{2+v_3} \qquad v_3 = \frac{1}{2-v_1}.$$

Substitution of the first equation into the third and then the result into the second yields the quadratic equation, $v_2 = (4-2v_2)/(8-5v_2)$, or $5v_2^2 - 10v_2 + 4 = 0$. Solving this gives $v_2 = (5-\sqrt{5})/5$ as the root less than 1. From this we can find $v_1 = (5-\sqrt{5})/4$ and $v_3 = 3-\sqrt{5}$. The complete solution is

G_1: $\mathrm{Val}(G_1) = v_1 = (5-\sqrt{5})/5 = 0.691\cdots$
Optimal for I = Optimal for II = $(v_1, 1-v_1) = (0.691\cdots, 0.309\cdots)$

G_2: $\mathrm{Val}(G_2) = v_2 = (5-\sqrt{5})/5 = 0.553\cdots$
Optimal for I = Optimal for II = $(\frac{5+\sqrt{5}}{10}, \frac{5-\sqrt{5}}{10})$
$= (0.724\cdots, 0.276\cdots)$

G_3: $\text{Val}(G_3) = v_3 = 3 - \sqrt{5} = 0.764\cdots$
Optimal for I = Optimal for II = $(v_3, 1-v_3) = (0.764\cdots, 0.236\cdots)$

independent of Q.

8. Whatever the values of G_1, G_2 and G_3, the game G_1 is a latin square game and so has optimal strategies $(1/3, 1/3, 1/3)$ for both players. So we must have $v_1 = (v_1 + v_2 + v_3)/3$, or equivalently $2v_1 = v_2 + v_3$. From the form of G_2 and G_3, we see that $0 \le v_2 \le 2$ and $0 \le v_3 \le 1$. Hence, $0 \le v_1 \le 3/2$. We now see that G_2 does not have a saddle point, and that $0 \le v_2 < 1$ so that G_3 does not have a saddle point either. We arrive at the three equations,

$$2v_1 = v_2 + v_3, \quad v_2 = \frac{2v_1}{2 + v_1}, \quad v_3 = \frac{1}{2 - v_2}.$$

Eliminating v_1 and v_3 leads to a quadratic equation for v_2, namely $v_2^2 + 2v_2 - 1 = 0$. This has one positive root, namely $v_2 = \sqrt{2} - 1$. From this we can find $v_1 = (4\sqrt{2} - 2)/7$ and $v_3 = (3 + \sqrt{2})/7$. The complete solution is

G_1: $\text{Val}(G_1) = v_1 = (4\sqrt{2} - 2)/7 = 0.522\cdots$
Optimal for I = Optimal for II = $(1/3, 1/3, 1/3)$

G_2: $\text{Val}(G_2) = v_2 = \sqrt{2} - 1 = 0.414\cdots$
Optimal for I = Optimal for II = $(0.793\cdots, 0.207\cdots)$

G_3: $\text{Val}(G_3) = v_3 = (3 + \sqrt{2})/7 = 0.631\cdots$
Optimal for I = Optimal for II = $(v_3, 1 - v_3) = (0.631\cdots, 0.369\cdots)$

independent of Q.

9. This is a recursive game of the form

$$G = \begin{pmatrix} 0.8 + 0.2(-G^T) & 0.5 + 0.5(-G^T) \\ 0.6 + 0.4(-G^T) & 0.7 + 0.3(-G^T) \end{pmatrix}$$

and the value, v, of the game satisfies

$$v = \text{Val}\begin{pmatrix} 0.8 - 0.2v & 0.5 - 0.5v \\ 0.6 - 0.4v & 0.7 - 0.3v \end{pmatrix}$$

The game is in favor of the server, so the value is between zero and one and the game does not have a saddle point. The optimal strategy for the server is to serve (high, low) with probabilities proportional to

$(0.1+0.1v, 0.3+0.3v)$, namely $(1/4, 3/4)$. The optimal strategy for the receiver is to receive (near,far) with probabilitites proportional to $(0.2+0.2v, 0.2+0.2v)$, namely, $(1/2, 1/2)$. Using the second of these equalizing strategies, the value may be found to be $v=(1/2)(0.8-0.2v)+(1/2)(0.5-0.5v)=0.65-0.35v$. Solving for v gives $v=13/27=0.481\cdots$.

10. (a) We may think of the basic game, G, as the one in which player I chooses a number k to be the number of times he tosses the coin before challenging II. In this, player II has no choice and the matrix G is an $\infty \times 1$ matrix, which is to say an infinite dimensional column vector. The probability of tossing k heads in a row is p^k. Counting 1 for a win and -1 for a loss, the expected payoff given that I tosses k heads in a row is $p^k(-1)+(1-p^k)=1-2p^k$. Thus the kth component of G is $p^k(1-2p^k)+(1-p^k)(-G^T)$. Whatever the value $v=\text{Val}(G)$, Player I will choose k to maximize this. We have the equation

$$v=\max_k(p^k(1-2p^k)+(1-p^k)(-v)).$$

Clearly $v>0$, so there is a finite integer k at which the maximum is taken on, call it k_0. Then $v=p^{k_0}(1-2p^{k_0})/(2-p^{k_0})$ and since v takes on its maximum value at k_0, we have

$$v=\max_k\left(\frac{p^k(1-2p^k)}{2-p^k}\right).$$

When $p=0.5$, evaluating $p^k(1-2p^k)/(2-p^k)$ at $k=1,2,3,4,\ldots$ gives 0, $1/14=0.0714\cdots$, $1/20=0.05$, $7/248=0.0282\cdots$, and so on, with a clear maximum at $k=2$.

(b) For arbitrary p, there is a maximum value attainable by v. Replace p^k by y in the formula for v and write it as $f(y)=y(1-2y)/(2-y)$. Calculus gives $f'(y)=(2y^2-8y+2)/(2-y)^2$, so the function $f(y)$ has a unique maximum on the interval (0,1) attained when $y^2-4y+1=0$. The root of this equation in the interval (0,1) is $y=2-\sqrt{3}$, and the value attained there is $V^*=f(2-\sqrt{3})=7-4\sqrt{3}=0.0718\cdots$. This is quite close to $0.0714\cdots$ attainable when $p=0.5$. If $p=2-\sqrt{3}$, then V^* is attainable with $k=1$. As $p\to\infty$, it becomes easier and easier to choose k so that p^k is close to $2-\sqrt{3}$ so the value converges to V^*.

11. The first row shows the value is at least 1, and the first column shows the value is at most 4. So $1\le v\le 4$. Then we see by "down-up-down-up" that the game does not have a saddle-point, so

$$v = \mathrm{Val}\begin{pmatrix} 4 & 1+(v/3) \\ 0 & 1+(2v/3) \end{pmatrix} = \frac{4+(8v/3)}{4+(v/3)}.$$

This leads to the quadratic equation, $v^2 + 4v - 12 = 0$, which has solutions $v = -2 \pm 4$. Since v is positive, we have $v = 2$ as the value. The matrix becomes $\begin{pmatrix} 4 & 1+(2/3) \\ 0 & 1+(4/3) \end{pmatrix}$. Player I's stationary optimal strategy is $(1/2, 1/2)$, and Player II's stationary optimal strategy is $(1/7, 6/7)$.

12. (a) We have

$$v(1) = \mathrm{Val}\begin{pmatrix} 2 & 2+(v(2)/2) \\ 0 & 4+(v(2)/2) \end{pmatrix}$$

$$v(2) = \mathrm{Val}\begin{pmatrix} -4 & 0 \\ -2+(v(1)/2) & -4+(v(1)/2) \end{pmatrix}.$$

It may be difficult to guess that the matrices do not have saddle-points, so let us assume they do not and check later to see if this assumption is correct. If neither matrix has a saddle-point, then the equations become,

$$v(1) = \frac{8+v(2)}{4} \qquad v(2) = \frac{16-2v(1)}{-6}.$$

Solving these equations simultaneously, we find $v(1) = 16/11$ and $v(2) = -24/11$. With these values the matrices above become

$$\begin{pmatrix} 2 & 2-(12/11) \\ 0 & 4-(12/11) \end{pmatrix} \qquad \begin{pmatrix} -4 & 0 \\ -2+(8/11) & -4+(8/11) \end{pmatrix}.$$

Since these do not have saddle-points, our assumption is valid and $v(1) = 16/11$ and $v(2) = -24/11$ are the values. In $G^{(1)}$, the optimal stationary strategies are $(8/11, 3/11)$ for I and $(1/2, 1/2)$ for II. In $G^{(2)}$, the optimal stationary strategies are $(1/3, 2/3)$ for I and $(6/11, 5/11)$ for II.

(b) Starting with $v_0(1) = 0$ and $v_0(2) = 0$, we find

$$v_1(1) = \mathrm{Val}\begin{pmatrix} 2 & 2 \\ 0 & 4 \end{pmatrix} = 2, \qquad v_1(2) = \mathrm{Val}\begin{pmatrix} -4 & 0 \\ -2 & -4 \end{pmatrix} = -8/3,$$

and then

$$v_2(1) = \mathrm{Val}\begin{pmatrix} 2 & 2/3 \\ 0 & 8/3 \end{pmatrix} = 4/3, \qquad v_2(2) = \mathrm{Val}\begin{pmatrix} -4 & 0 \\ -1 & -3 \end{pmatrix} = -2,$$

compared to the true values, $v(1) = 1.4545\ldots$ and $v(2) = -2.1818\ldots$.

Solutions to Chap. 13

1. (a) The value is 0. Player II has a pure optimal strategy, namely $y = 0$, since $A(x, 0) = 0$ for all $x \in X$. Player I's optimal strategy, $\mathbf{p}^*$, is 1 and -1 with equal probability 1/2, since if $\mathbf{q} \in Y_F^*$,

$$A(\mathbf{p}^*, \mathbf{q}) = (1/2)\sum_j jq_j - (1/2)\sum_j jq_j = 0$$

because the sums are finite.

(b) Let $q_j = (1/4)(1/|j|)$ for $|j|$ some power of 2 (i.e. $j = \pm1, \pm2, \pm4, \pm8, \ldots$), and let $p_j = 0$ for all other j. Then $\sum_j q_j = 1$ and

$$\begin{aligned} A(-1, \mathbf{q}) = A(1, \mathbf{q}) &= \left(\cdots - \frac{1}{4} - \frac{1}{4} - \frac{1}{4}\right) + \left(\frac{1}{4} + \frac{1}{4} + \frac{1}{4} + \cdots\right) \\ &= -\infty + \infty. \end{aligned}$$

2. It is easier to use Method 2. The set S is the convex hull of the curve (y, e^{-y}) as y goes from 0 to 1. The value occurs at the intersection of the curve $y_2 = e^{-y_1}$ and the line $y_1 = y_2$, namely, it is the solution of the equation $v = e^{-v}$. This is about $v = 0.5671$. The optimal strategy for II is $y = v$. The slope of the tangent line to the curve $y_2 = e^{-y_1}$ at the point $y_1 = v$ is $-e^{-v} = -v$. The normal to this is the negative of the reciprocal, namely $1/v$. The optimal strategy of Player I takes x_1 and x_2 in proportions $v : 1$. This is the mixed strategy $(v/(1+v), 1/(1+v)) = (0.3619, 0.6381)$.

3. The value occurs where the wedge W_v first hits the curve $(y_1 - 3)^2 + 4(y_2 - 2)^2 = 4$. This occurs when $y_1 = y_2 = v$ and v is the smaller root of the equation $(v-3)^2 + 4(v-2)^2 = 4$. This leads to the equation,

$5v^2-22v+21=0$, or $(5v-7)(v-3)=0$, so $v=7/5$ is the value. Player II's optimal pure strategy is (v,v). The slope of the curve $(y_1-3)^2+4(y_2-2)^2=4$ at the point (v,v) is $y_2'=\frac{-(y_1-3)}{4(y_2-2)}=\frac{-(v-3)}{4(v-2)}$. The normal is $\frac{4(2-v)}{(3-v)}=\frac{3}{2}$. So Player I's optimal strategy is $(2/5,3/5)$.

4. (a) For the upper semi-continuous payoff, the value is $1/2$. An optimal strategy for Player I is to choose 0 and 1 with probability $1/2$ each. Player II has no optimal strategies. For any $0<\epsilon<1/2$, an ϵ-optimal strategy for Player II is choose $1/2-\epsilon$ and $1/2+\epsilon$ with probability $1/2$ each.

(b) For the lower semi-continuous payoff, the value is 0. An optimal pure strategy for Player II is to choose $y=1/2$. Player I also has optimal strategies. In fact, all his strategies are optimal!

5. This is essentially the game, He-who-chooses-the-smaller-positive-number-wins.

6. The game is symmetric, so if the value, if it exists, is zero and the players have identical optimal strategies. The pure strategy $x=1$ for Player I dominates the pure strategies x such that $0\le x\le 1-2b$. Similarly for Player II. The game, played on the remaining square, $[1-2b,1]^2$, is a latin square game of the form of Example 1 of Sec. 13.2. So the optimal strategy for both players is the uniform distribution on the interval $[1-2b,1]$, and the value is zero. If $b=0$, the optimal strategies are the pure strategies, $x=1$ and $y=1$.

7. (a) This is similar to the estimation problem of Example 1 of Sec. 13.3. The upper envelope is $\max\{A(0,y),A(1,y)\}=\max\{y^2,2(1-y)^2\}$. This has a minimum when $y^2=2(1-y)^2$. This reduces to $y^2-4y+2=0$ whose solution in $[0,1]$ is $y^*=2-\sqrt{2}=0.586\cdots$. This is Player II's pure optimal strategy. The slope of the curve $pA(0,y)+(1-p)A(1,y)$ at $y=y^*$ is zero when $p\,2y^*+(1-p)\,4(1-y^*)=0$. This occurs when $p=2-\sqrt{2}$. So Player I's optimal strategy is: mix $x=0$ and $x=1$ with probabilities $2-\sqrt{2}$ and $\sqrt{2}-1$, respectively. Numerically this is $(0.586\cdots,0.414\cdots)$.

(b) This is a convex-concave game so both players have optimal pure strategies. If y_0 is an optimal pure strategy for Player II, then x_0 must maximize $A(x,y_0)$. As a function of x this is a line of slope $e^{-y_0}-y_0$. So

$$x_0=\begin{cases}0 & \text{if } e^{-y_0}<y_0\\ \text{any} & \text{if } e^{-y_0}=y_0\\ 1 & \text{if } e^{-y_0}>y_0.\end{cases}$$

We are bound to have a solution to this equation if $e^{-y_0} = y_0$. So $y_0 = 0.5671\cdots$. But y must minimize $A(x_0, y)$, whose derivative, $-x_0e^{-y}+1-x_0$ must be zero at y_0. This gives $x_0(e^{-y_0}+1) = 1$. Since $e^{-y_0} = y_0$, we have $x_0 = 1/(1+y_0) = 0.6381\cdots$.

8. (a) The payoff is convex in $\mathbf{y}$ for every $\mathbf{x}$, so Player II has an optimal pure strategy. Player I will hide in one of the three corners to make it as hard as possible for Player II to come close. Therefore Player II will choose the point equidistant from $(-1,0)$, $(1,0)$, and $(0,2)$. This will be the point $(0,a)$ such that the distance from $(0,a)$ to $(1,0)$ is equal to the distance from $(0,a)$ to $(0,2)$. This gives $1+a^2 = (2-a)^2$, or $a = 3/4$. So II's optimal strategy is $\mathbf{y} = (0,3/4)$, and the value of the game is $(2-a)^2 = 25/16$. By symmetry, I's optimal strategy gives equal probability, say p, to $(-1,0)$ and $(1,0)$ and probability $1-2p$ to $(0,2)$, such that $p(-1,0)+p(1,0)+(1-2p)(0,2) = (0,3/4)$. This gives $p = 5/16$. Player I gives probability $5/16$ to each of $(-1,0)$ and $(1,0)$ and probability $6/16$ to $(0,2)$.

(b) Find the smallest sphere, $\{\mathbf{y} : \|\mathbf{y}-\mathbf{y}_0\|^2 \le r^2\}$ containing S. Then, r^2 is the value, and Player II has the optimal pure strategy $\mathbf{y}_0$. Let $X_0 = \{\mathbf{x} \in S : \|\mathbf{x}-\mathbf{y}_0\|^2 = r^2\}$. Then, there exist $\mathbf{x}_1, \ldots, \mathbf{x}_k$ in X_0 (for some $k \le n+1$) and probabilities $p_1, \ldots, p_k$ adding to one, such that $\sum_1^k p_i\mathbf{x}_i = \mathbf{y}_0$. An optimal mixed strategy for Player I is to choose $\mathbf{x}_i$ with probability p_i.

9. Let us assume that the "good" strategies are those that give all mass to some interval, $[a,b]$, with $0 < a < 1 < b$, and let us search for a strategy F with a density $f(x)$ for which $A(F,G)$ is constant over good strategies, G. Let $\Phi(y) = \int_a^y xf(x)\,dx$. Then

$$\begin{aligned} A(F,G) &= \int_a^b \int_a^y (y+x)f(x)\,dx\,dG(y) - 1 \\ &= \int_a^b [yF(y)+\Phi(y)]\,dG(y) - 1. \end{aligned} \tag{2}$$

Since $\int_a^b dG(y) = 1$ and $\int_a^b y\,dG(y) = 1$, the payoff (2) will be constant for all good G, provided $yF(y)+\Phi(y)$ is linear in $y \in (a,b)$, for some α. This means the second derivative of $yF(y)+\Phi(y)$ must be zero:

$$3yf(y) + 2f'(y) = 0 \tag{3}$$

whose solution is

$$f(y) = cy^{-3/2} \qquad \text{for } y \in (a, b), \tag{4}$$

for some constant $c > 0$. We have

$$F(y) = 2c[a^{-1/2} - y^{-1/2}] \quad \text{and} \quad \Phi(y) = 2c[y^{1/2} - a^{1/2}] \qquad \text{for } a < y < b. \tag{5}$$

For f to be a density, the constant c must satisfy

$$2c\left(\frac{1}{\sqrt{a}} - \frac{1}{\sqrt{b}}\right) = 1. \tag{6}$$

For $\mathrm{E}(X) = 1$, we must have

$$2c(\sqrt{b} - \sqrt{a}) = 1. \tag{7}$$

We see from (6) and (7) that $b = 1/a$, and $c = \sqrt{b}/(2(b-1))$. Therefore from (5),

$$yF(y) + \Phi(y) = \frac{2c}{\sqrt{a}}(y - a) = \frac{b}{b-1}\left(y - \frac{1}{b}\right). \tag{8}$$

The value of this game is zero. The distribution with density

$$f(y) = \sqrt{b}/(2(b-1))y^{-3/2} \qquad \text{for } 1/b < y < b \tag{9}$$

is an optimal pure strategy for both players.
Proof. Suppose Player I uses $f(x)$ of (9). Since there can be no ties, we must show that if Player II uses any distribution function $G(y)$ on $[0, b]$ having mean 1, then $\mathrm{E}((Y + X)\mathrm{I}(X < Y)) \geq 1$. But,

$$\begin{aligned}
E((Y + X)\mathrm{I}(X < Y)) &= \int_{1/b}^{b}\int_{1/b}^{y}(y + x)f(x)\,dx\,dG(y) \\
&= \int_{1/b}^{b}[yF(y) + \Phi(y)]\,dG(y) \\
&= \int_{1/b}^{b}\frac{b}{b-1}\left(y - \frac{1}{b}\right)dG(y) \\
&\geq \int_{0}^{b}\frac{b}{b-1}\left(y - \frac{1}{b}\right)dG(y) \\
&= \frac{b}{b-1}\left(1 - \frac{1}{b}\right) = 1.
\end{aligned}$$

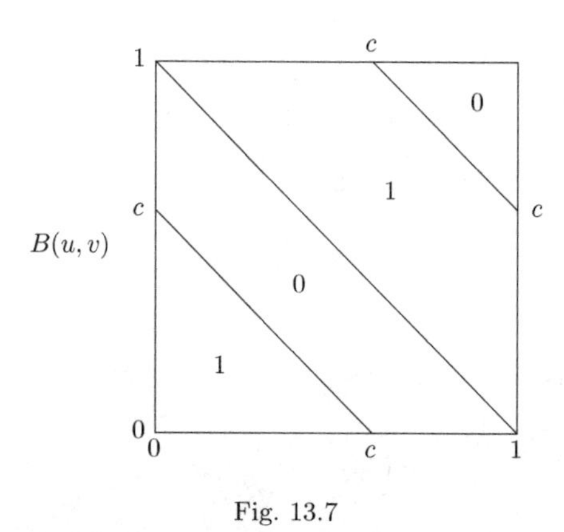

Fig. 13.7

This also shows that if Player II puts any mass below $1/b$, then Player I's expected payoff is positive.

10. The payoff function, $B(u, v)$, is displayed in Fig. 13.7.

This is a Latin Square type game with $\int_0^1 B(u, v)du$ equal to the constant c for all v, and similarly $\int_0^1 B(u, v)\, dv = c$ for all u. Thus, the optimal mixed strategy for both players is the uniform distribution on (0,1), and the value of the game is $c = 0.60206\ldots$.

This implies that in the original game the optimal strategy for I is to choose U from a uniform distribution on (0,1) and let $X = 10^U$, and similarly for II. The value is c.

It is interesting to note that if the payoff is changed so that Player I wins 1 if and only if the first significant digit is in some set, such as $\{2, 4, 7, 8\}$, the optimal strategies of the players remain the same. Only the value changes.

11. (a)

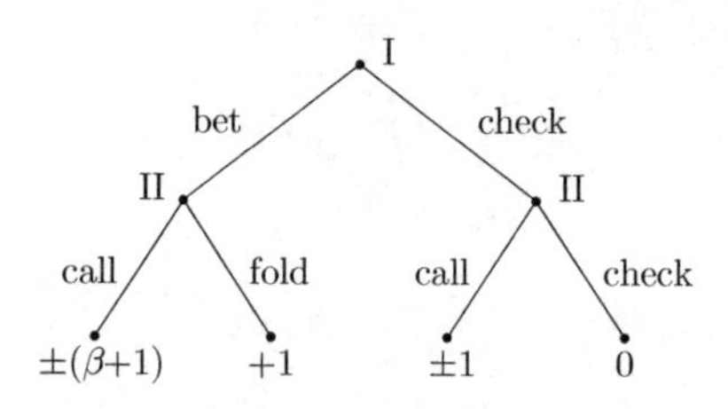

(b) Indifference at a: $c - (\beta + 1)(1 - c) = 0 \cdot d - (1 - d)$.
Indifference at b: $c + (\beta + 1)(b - c - (1 - b)) = 0 \cdot d + (b - c - (1 - b))$.

Indifference at c: $(\beta+1)(-a+1-b) = a+1-b$.
Indifference at d: $d-a-(b-d) = 0$.

Simplifying gives

$$\begin{aligned} d &= (\beta+2)c-\beta \\ d &= \beta(1+c-2b) \\ \beta(1-b) &= (\beta+2)a \\ d &= (a+b)/2 \end{aligned}$$

(c) At $\beta = 2$, I get $a = 5/33$, $b = 23/33$, $c = 20/33$, and $d = 14/33$.

The value is: $v = (\beta+1)[(1-b)(b-c) - a(1-c)] + ac + (1-b)c - (1-b)(b-a) - (d-a)^2$. This turns out to be negative for all $\beta > 0$. In particular for $\beta = 2$, $v = -2/33$; the game favors Player II.

12. It is optimal for Player I to play the points $x_0 = 0$ and $x_1 = 0.7$, and for Player II to play the points $y_0 = 0$ and $y_1 = 0.5$ (but any $0.4 < y_1 < 0.6$ works as well). This leads to the following 2×2 game with matrix

$$\begin{array}{c} \\ x_0 \\ x_1 \end{array} \begin{array}{c} \begin{array}{cc} y_0 & \;\; y_1 \end{array} \\ \begin{pmatrix} 0.00 & 0.40 \\ 0.60 & 0.24 \end{pmatrix} \end{array}$$

Methods of Sec. 8.2 may be used to solve this game. The value is $v = 6/19$. Player I bets nothing with probability $9/19$, and bets \$70 with probability $10/19$. Player II bets nothing with probability $4/19$, and \$50 with probability $15/19$. Player I wins with probability $v = 0.316\ldots$.

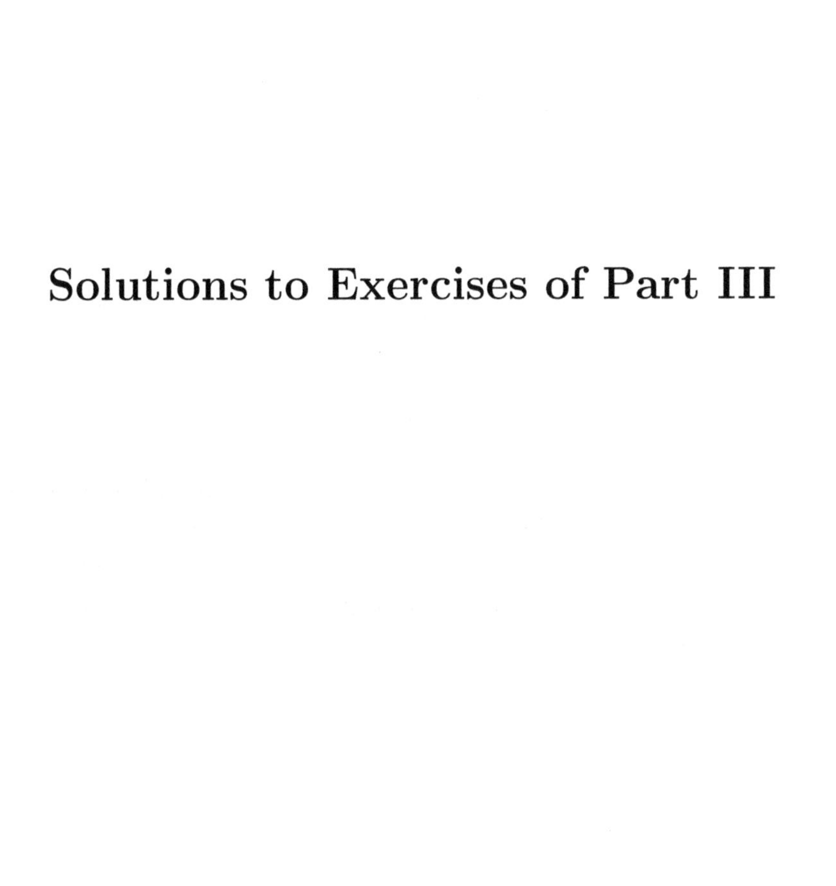

Solutions to Exercises of Part III

Solutions to Chap. 14

1. The bimatrix is $\begin{matrix} & c & d \\ a & (13/4, 3) & (22/4, 3/4) \\ b & (4, 10/4) & (21/4, 2) \end{matrix}$.

2. (a) Player I's maxmin strategy is $(1, 0)$ (i.e. row 1) guaranteeing him the safety level $v_{\text{I}} = 1$. Player II's maxmin strategy is $(1, 0)$ (i.e. column 1) guaranteeing her the safety level $v_{\text{II}} = 1$.

(b) Player I's maxmin strategy is $(1/2, 1/2)$ guaranteeing him the safety level $v_{\text{I}} = 5/2$. Player II's maxmin strategy is $(3/5, 2/5)$ guaranteeing her the safety level $v_{\text{II}} = 8$.

3. (a) There are many ways to draw the Kuhn tree. Here is one. The payoffs are in units of 100 dollars.

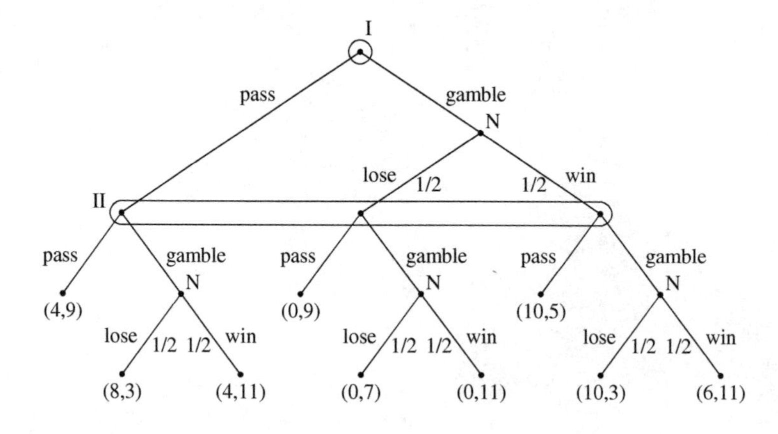

(b) The bimatrix is:

$$\begin{array}{c} \\ \text{gamble} \\ \text{pass} \end{array} \begin{array}{c} \begin{array}{cc} \text{gamble} & \text{pass} \end{array} \\ \begin{pmatrix} (4,8) & (5,7) \\ (6,7) & (4,9) \end{pmatrix} \end{array}$$

(c) Player I's safety level is $v_{\mathrm{I}} = 14/3$. Player II's safety level is $v_{\mathrm{II}} = 23/3$. Both maxmin strategies are $(2/3, 1/3)$.

4. Let Q denote the proportion of students in the class (excluding yourself) who choose row 2. If you choose row 2 you win $Q \cdot 6$ on the average. If you choose row 1, you win 4. So you should choose row 2 only if you predict that at least 2/3 of the rest of the class will choose row 2.

In my classes, only between 15% and 35% of the students chose row 2. If your classes are like mine, you should choose row 1.

Solutions to Chap. 15

1. Let $\mathbf{p}_0$ denote the maxmin strategy of Player I, and let $(\mathbf{p},\mathbf{q})$ be any strategic equilibrium. Then, $v_I \le \mathbf{p}_0'\mathbf{A}\mathbf{q}$ since use of $\mathbf{p}_0$ guarantees Player I at least $v_{\rm I}$ no matter what Player II does. But also $\mathbf{p}_0'\mathbf{A}\mathbf{q} \le \mathbf{p}'\mathbf{A}\mathbf{q}$ since $\mathbf{p}$ is a best response to $\mathbf{q}$. This shows $v_{\rm I} \le \mathbf{p}'\mathbf{A}\mathbf{q}$. Then $v_{\rm II} \le \mathbf{p}'\mathbf{A}\mathbf{q}$ follows from symmetry.

2. (a) The safety levels are $v_{\rm I} = 2$ and $v_{\rm II} = 16/5$. The corresponding MM strategies are $(0,1)$ (the second row) for Player I, and $(1/5,4/5)$ (the equalizing strategy on $\mathbf{B}$) for Player II. The unique SE is the PSE in the lower left corner with payoff $(2,4)$. It may be found by removing strictly dominated rows and columns.

(b) The safety levels are $v_{\rm I} = 2$ and $v_{\rm II} = 5/2$. The corresponding MM strategies are $(0,1)$ for Player I, and $(1/2,1/2)$ for Player II. There are no pure SE's, and the unique SE is the one using the equalizing strategies $(1/4,3/4)$ for Player I on Player II's payoff matrix, and $(1/2,1/2)$ for Player II on Player I's payoff matrix. The vector payoff is $(5/2,5/2)$. Note that Player II's equalizing strategy is not an optimal strategy on Player I's matrix.

(c) The safety levels are $v_{\rm I} = 0$ and $v_{\rm II} = 0$. The corresponding MM strategies are $(1,0)$ for Player I, and $(1,0)$ for Player II. There is no PSE. The unique SE is the one using equalizing strategies, $(3/4,1/4)$ for Player I and $(1/2,1/2)$ for Player II, with payoff vector $(0,0)$.

3. (a) The bimatrix is

$$\begin{array}{r} \\ \text{chicken} \\ \text{iron nerves} \end{array} \begin{array}{c} \begin{array}{cc} \text{chicken} & \text{iron nerves} \end{array} \\ \begin{pmatrix} (1,1) & (-1,2) \\ (2,-1) & (-2,-2) \end{pmatrix} \end{array}$$

(b) I's matrix has a saddle point with value $v_{\mathrm{I}} = -1$, achievable if I uses the top row. Similarly, II's MM strategy is the first column, with value $v_{\mathrm{II}} = -1$. Thus the safety levels are -1 for both players. Yet if both players play their MM strategies, the happy result is that they both receive $+1$.

(c) There are two PSE's: the lower left corner and the upper right corner. The third SE involves mixed strategies and may be found using equalization. The mixed strategy (1/2,1/2) for II is an equalizing strategy for I's matrix (even though it is not optimal there). Against this strategy, the average payoff to I is zero. Similarly, the strategy (1/2,1/2) for I is equalizing for II's matrix, giving an average payoff of zero. Thus, $((1/2, 1/2), (1/2, 1/2))$ is a mixed SE with payoff vector, (0,0).

4. (a)

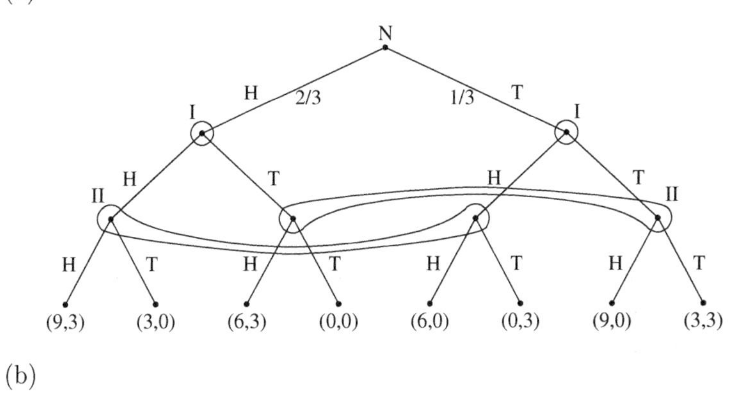

(b)

	HH	HT	TH	TT
HH	$(8, 2^*)$	$(8^*, 2^*)$	$(2, 1)$	$(2, 1)$
HT	$(9^*, 2)$	$(7, 3^*)$	$(5, 0)$	$(3^*, 1)$
TH	$(6, 2)$	$(2, 0)$	$(4, 3^*)$	$(0, 1)$
TT	$(7, 2^*)$	$(1, 1)$	$(7^*, 2^*)$	$(1, 1)$

(c) There are two PSE's, those with double asterisks.

5. (a) We star the entries of the A matrix that are maxima of their column and entries of the B matrix that are maxima of their row.

$-3, -4$	$2^*, -1$	$0, 6^*$	$1^*, 1$
$2^*, 0$	$2^*, 2^*$	$-3, 0$	$1^*, -2$
$2^*, -3$	$-5, 1^*$	$-1, -1$	$1^*, -3$
$-4, 3^*$	$2^*, -5$	$1^*, 2$	$-3, 1$

The only doubly starred entry occurs in the second row, second column, and hence the unique PSE is $\langle 2, 2\rangle$.

(b) Starring the entries in a similar manner leads to the matrix

$$\begin{array}{cccc} 0,\ 0 & 1^*, -1 & 1^*,\ 1^* & -1,\ 0 \\ -1,\ 1^* & 0,\ 1^* & 1^*,\ 0 & 0^*,\ 0 \\ 1^*,\ 0 & -1, -1 & 0,\ 1^* & -1, 1^* \\ 1^*, -1 & -1,\ 0^* & 1^*, -1 & 0^*, 0^* \\ 1^*,\ 1^* & 0,\ 0 & -1, -1 & 0^*,\ 0 \end{array}$$

We find there are three doubly starred squares, and hence three PSE's, namely, $\langle 5, 1\rangle$ and $\langle 1, 3\rangle$ and $\langle 4, 4\rangle$.

6. (a) $v_{\rm I} = 0$ and $v_{\rm II} = 2/3$.

(b) There is a unique PSE at row 2, column 1, with payoff vector (0,1).

(c) The mixed strategy $(1/3, 2/3)$ is the unique equalizing strategy for Player I. Column 1 is an equalizing strategy for II, but so is the mixture, $(0, 2/3, 1/3)$. More generally, any mixture of the form $(1-p, 2p/3, p/3)$ for $0 \le p \le 1$ is an equalizing strategy for II. Therefore, any of the strategy pairs, $(1/3, 2/3)$ for I and $(1-p, 2p/3, p/3)$ for $0 \le p \le 1$ for II, gives a strategic equilibrium. There are also some non-equalizing strategy pairs forming a strategic equilibrium, namely $(p, 1-p)$ for $0 \le p \le 1/3$ for I, and column 1 for II.

7. We are given $a_{1j} < \sum_{i=2}^m x_i a_{ij}$ for all j, where $x_i \ge 0$ and $\sum_{i=2}^m x_i = 1$. Suppose $(\mathbf{p}^*, \mathbf{q}^*)$ is a strategic equilibrium. Then

$$\sum_j \sum_i p_i^* a_{ij} q_j^* \ge \sum_j \sum_i p_i a_{ij} q_j^* \quad \text{for all } \mathbf{p} = (p_1, \ldots, p_m). \qquad (*)$$

We are to show $p_1^* = 0$. Suppose to the contrary that $p_1^* > 0$. Then

$$\begin{aligned} \sum_j \sum_i p_i^* a_{ij} q_j^* &= \sum_j \left[p_1^* a_{1j} q_j^* + \sum_{i=2}^m p_i^* a_{ij} q_j^* \right] \\ &< \sum_j \left[p_1^* (\sum_{i=2}^m x_i a_{ij}) q_j^* + \sum_{i=2}^m p_i^* a_{ij} q_j^* \right] \quad \text{(strict inequality)} \\ &= \sum_j \sum_{i=2}^m (p_1^* x_i + p_i^*) a_{ij} q_j^* = \sum_j \sum_i p_i a_{ij} q_j^*, \end{aligned}$$

where $p_1 = 0$ and $p_i = p_1^* x_i + p_i^*$ for $i = 2, \ldots, m$. But The p's are nonnegative and add to one, so this contradicts (*).

8. (a) We have

$$A = \begin{pmatrix} 3 & 2 & 3 \\ 6 & 0 & 3 \\ 4 & 3 & 4 \end{pmatrix}, \quad B = \begin{pmatrix} 4 & 3 & 2 \\ 1 & 2 & 3 \\ 6 & 4 & 5 \end{pmatrix}.$$

In A, the third row second col is a saddle point. So $v_I = 3$ and the third row is a maxmin strategy for Player I. In B, the row 3 is dominated by row 1, and col 2 is an equal probability mixture of col 1 and col 3. With these removed, the resulting 2 by 2 matrix has value $v_{II} = 2.5$. The maxmin strategy for Player II is $(1/4, 0, 3/4)$. (Another maxmin strategy for II is $(0, 1/2, 1/2)$.)

(b) There are no PSE's. In A, row 1 is strictly dominated by row 3 and may be removed from consideration. Then in B, col 2 is strictly dominated by col 3 and may be removed. In the resulting 2×2 bimatrix game, there is a unique SE. It is given by equalizing strategies, $(1/3, 2/3)$ for I and $(1/3, 2/3)$ for II. In the original game, the unique SE is $(0, 1/3, 2/3)$ for Player I and $(1/3, 0, 2/3)$ for Player II. The equilibrium payoff is $(4, 4\frac{1}{3})$.

9. (a) At II's information set, a dominates b, so that vertex is worth $(1, 0)$. Then at I's information set, B dominates A, so the PSE found by backward induction is (B, a), having payoff $(1, 0)$. This is a subgame perfect PSE.

(b)

$$\begin{array}{c} \\ A \\ B \end{array} \begin{array}{c} \begin{array}{cc} \quad a \quad & \quad b \quad \end{array} \\ \begin{pmatrix} (0,1) & (0,1) \\ (1,0) & (-10,-1) \end{pmatrix} \end{array}$$

(c) There are two PSE's, the lower left and the upper right. The lower left, (B, a), is the subgame perfect PSE. The upper right, (A, b), corresponds to the PSE where Player I plays A because he believes Player II will play b. This is not subgame perfect because at Player II's vertex, it is not an equilibrium for Player II to play b.

10. There were 18 answers for Player I and 17 for Player II. The data is as follows.

I:	Stop at	No.	Score	II:	Stop at	No.	Score
	(1, 1)	10	17		(0, 3)	8	34
	(98, 98)	2	882		(97, 100)	2	806
	(99, 99)	4	887		(98, 101)	7	804
	never	2	880				

Scores for Player I ranged from 17 for those who selected $(1, 1)$, to 887 for those who selected (99, 99). Scores for Player II ranged from 34 for those who chose (0, 3) to 806 for those who chose (97, 100). Total scores ranged from 51 to 1693. Those who scored above 1000 received 5 points. Those who scored between 500 and 1000 received 3 points. Those who scored less than 100 received 1 point.

Solutions to Chap. 16

1. (a) I's strategy space is $X = [0, \infty)$ and II's strategy space is $Y = [0, \infty)$. If I chooses $q_1 \in X$ and II chooses $q_2 \in Y$, the payoffs to I and II are

$$u_1(q_1, q_2) = q_1(a - q_1 - q_2)^+ - c_1 q_1, \quad u_2(q_1, q_2) = q_2(a - q_1 - q_2)^+ - c_2 q_2$$

respectively. To find a PSE, we set derivatives to zero:

$$\frac{\partial}{\partial q_1} u_1(q_1, q_2) = a - 2q_1 - q_2 - c_1 = 0, \quad \frac{\partial}{\partial q_2} u_2(q_1, q_2) = a - q_1 - 2q_2 - c_2 = 0.$$

The unique solution is (q_1^*, q_2^*), where

$$q_1^* = (a + c_2 - 2c_1)/3, \quad q_2^* = (a + c_1 - 2c_2)/3.$$

Since we have assumed $c_1 < a/2$ and $c_2 < a/2$, both these production points are positive. Thus (q_1^*, q_2^*) is a PSE. Its payoff vector is $((a + c_2 - 2c_1)^2/9, (a + c_1 - 2c_2)^2/9)$.

(b) I's profit is $v_1(x, y) = x(17 - x - y) - x - 2 = x(16 - x - y) - 2$. II's profit is $v_2(x, y) = y(17 - x - y) - 3y - 1 = y(14 - x - y) - 1$. For fixed y, I should choose x so that $\partial v_1/\partial x = 16 - 2x - y = 0$. For fixed x, II should choose y so that $\partial v_2/\partial y = 14 - x - 2y = 0$. The equilibrium point is achieved if these two equations are satisfied simultaneously. This gives $x = 6$ and $y = 4$. The equilibrium payoff is $(36 - 2, 16 - 1) = (34, 15)$.

2. We assume $c < a$ — otherwise no company will produce anything. The payoff functions are

$$u_i(q_1, q_2, q_3) = q_i P(q_1 + q_2 + q_3) - c q_i = q_i[(a - q_1 - q_2 - q_3)^+ - c]$$

for $i = 1, 2, 3$. Assuming $q_1 + q_2 + q_3 < a$, there will be equilibrium production if the following three equations are satisfied:

$$\frac{\partial}{\partial q_i} u_i(q_1, q_2, q_3) = a - q_i - q_1 - q_2 - q_3 - c = 0$$

for $i = 1, 2, 3$. This solution is easily found to be $q_i = (a - c)/4$ for $i = 1, 2, 3$. This is the equilibrium production. The total production is $(3/4)(a - c)$, compared to $(2/3)(a - c)$ for the duopoly production, and $(1/2)(a - c)$ for the monopoly production.

3. The profit functions are

$$u_1(p_1, p_2) = (a - p_1 + bp_2)^+ (p_1 - c) \quad \text{and}$$
$$u_2(p_1, p_2) = (a - p_2 + bp_1)^+ (p_2 - c).$$

Knowing Player I's choice of p_1, Player II would choose p_2 to maximize $u_2(p_1, p_2)$. As in the Bertrand model with differentiated products, we find

$$\frac{\partial}{\partial p_2} u_2(p_1, p_2) = a - 2p_2 + bp_1 + c = 0 \quad \text{and hence}$$
$$p_2(p_1) = (a + bp_1 + c)/2.$$

Knowing Player II will use $p_2(p_1)$, Player I would choose p_1 to maximize $u_1(p_1, p_2(p_1))$. We have

$$\frac{\partial}{\partial p_1} u_1(p_1, p_2(p_1)) = a - p_1 + (b/2)(2 + bp_1 + c)$$
$$- (p_1 - c)(2 - b^2)/2 = 0.$$

Hence, solving for p_1 and substituting into p_2 gives

$$p_1^* = \frac{a(2 + b) + c(2 + b - b^2)}{2(2 - b^2)} \quad \text{and}$$

$$p_2^* = \frac{a + c}{2} + \frac{b}{2} \cdot \frac{a(2 + b) + c(2 + b - b^2)}{2(2 - b^2)}$$

as the PSE.

Both p_1^* and p_2^* are greater than $(a+c)/(2-b)$, so both players charge more than in the Bertrand model. Surprisingly, both players receive more from the sequential PSE than they do from the PSE of the Bertrand model. However, Player I receives more than Player II. (This model is suspect. Do not assume these results hold in general.)

4. (a) $u(Q) = QP(Q) - Q$, so $u'(Q) = P(Q) + QP'(Q) - 1 = (3/4)Q^2 - 10Q + 25$. This quadratic function has roots $Q = 10/3$ and $Q = 10$. The maximum of $u(Q)$ on the interval $[0, 10]$ is at $Q = 10/3$, so this is the monopoly production. The monopoly price is $P(10/3) = 109/9$ and the return of this production is $u(10/3) = 1000/27 = 37+$.

(b) $u_1(q_1, q_2) = q_1 P(q_1 + q_2) - q_1$, so

$$\frac{\partial}{\partial q_1} u_1\left(q_1, \frac{5}{2}\right) = P\left(q_1 + \frac{5}{2}\right) + q_1 p'\left(q_1 + \frac{5}{2}\right) - 1$$
$$= \frac{3}{4}\left(q_1^2 - 10q_1 + \frac{75}{4}\right).$$

This has roots $q_1 = 5/2$ and $q_1 = 15/2$. The maximum occurs at $q_1 = 5/2$, and for $q_1 > 15/2$, $u_1(q_1, 5/2) = 0$. This shows that the optimal reply to $q_2 = 5/2$ is $q_1 = 5/2$. But the situation is symmetric, so the optimal reply of firm 2 to $q_1 = 5/2$ of firm 1, is $q_2 = 5/2$ also. This shows that $q_1 = q_2 = 5/2$ is a PSE.

5. (a) If Firm 2 knows Firm 1 is producing q_1, then Firm 2 will produce $q_2 \in [0, a]$ to maximize $q_2(a - q_1 - q_2)^+ - c_2 q_2$. This gives

$$q_2(q_1) = \begin{cases} (a - q_1 - c_2)/2 & \text{if } q_1 < a - c_2 \\ 0 & \text{if } q_1 \geq a - c_2 \end{cases}$$

as in Eq. (13). Therefore Firm 1 wull choose to produce $q_1 \in [0, a]$ to maximize the payoff

$$u_1(q_1) = \begin{cases} q_1(a - 2c_1 + c_2)/2 - q_1^2/2 & \text{if } q_1 < a - c_2 \\ q_1(a - c_1) - q_1^2 & \text{if } q_1 \geq a - c_2. \end{cases}$$

The two functions, $f_1(q) = q(a - 2c_1 + c_2)/2 - q^2/2$ and $f_2(q) = q(a - c_1) - q^2$, are quadratic and agree at $q = 0$ and at $q = a - c_2$. Since the difference, $f_1(q) - f_2(q)$ is also quadratic, and the slope of $f_1(q)$ at 0 is less than the slope of $f_2(q)$ at 0, we have $f_1(q) < f_2(q)$ for all $q \in (0, a - c_2)$ and $f_1(q) > f_2(q)$ for $q \in (a - c_2, a)$. Therefore, if $f_1'(a - c_2) \leq 0$, that is if $c_2 \leq (a + 2c_1)/3$, then the maximum of $u_1(q_1)$ occurs at $f_1'(q_1) = 0$ or at $q_1 = 0$, namely at $q_1 = (a - 2c_1 + c_2)^+$. If $f'(a - c_2) > 0$ and $f_2'(a - c_2) \leq 0$, that is, if $c_2 > (1a + 2c_1)/3$ and $c_2 < (a + c_1)/2$, the maximum occurs at $q_1 = a - c_2$. If $f'(a - c_2) > 0$ and $f_2'(a - c_2) > 0$, that is, if $c_2 > (a + c_1)/2$, the maximum occurs at $f_2'(q_1) = 0$, namely at $q_1 = (a - c_1)/2$.

In summary we have four cases:

(1) If $c_1 > (a + c_2)/2$, then $q_1 = 0$ and $q_2 = (a - c_2)/2$.

(2) If $c_1 < (a + c_2)/2$ and $c_2 < (a + 2c_1)/3$, then $q_1 = (a - 2c_1 + c_2)/2$ and $q_2 = (a - q_1 - c_2)/2$.

(3) If $c_2 > (a + 2c_1)/3$ and $c_2 < (a + c_1)/2$, then $q_1 = a - c_2$ and $q_2 = 0$.

(4) If $c_2 > (a + c_1)/2$, then $q_1 = (a - c_1)/2$ and $q_2 = 0$.

(b) The payoff functions are

$$u_1(q_1, q_2) = q_1(17 - q_1 - q_2) - q_1 - 2,$$

$$u_2(q_1, q_2) = q_2(17 - q_1 - q_2) - 3q_2 - 1.$$

The optimal production for Firm 2 satisfies $(\partial/\partial q_2)u_2(q_1, q_2) = 17 - q_1 - 3 - 2q_2 = 0$. So $q_2 = (14 - q_1)/2$. Then,

$$u_1(q_1, q_2(q_1)) = q_1(10 - (q_1/2)) - q_1 - 2$$

from which we find the equilibrium productions to be

$$q_1 = 9 \quad \text{and} \quad q_2 = 2.5,$$

the equilibrium price is $P = 17 - 9 - (5/2) = 11/2$, and the equilibrium payoffs are

$$u_1 = 9 \cdot 11/2 - 9 - 2 = 37.5 \quad \text{and}$$

$$u_2 = 2.5 \cdot 11/2 - 3 \cdot 2.5 = 4.25.$$

6. The three payoffs are

$$u_1(q_1, q_2, q_3) = q_1(a - c - q_1 - q_2 - q_3),$$

$$u_2(q_1, q_2, q_3) = q_2(a - c - q_1 - q_2 - q_3),$$

$$u_3(q_1, q_2, q_3) = q_3(a - c - q_1 - q_2 - q_3).$$

Setting $\partial u_3(q_1, q_2, q_3)/\partial q_3$ to zero and solving gives

$$q_3(q_1, q_2) = (a - c - q_1 - q_2)/2$$

as the optimal production for Firm 3. Then, evaluating

$$u_2(q_1, q_2, q_3(q_1, q_2)) = q_2\left(a - c - q_1 - q_2 - \frac{a - c - q_1 - q_2}{2}\right)$$

$$= q_2\left(\frac{a - c - q_1 - q_2}{2}\right)$$

and setting $\partial q_2(q_1, q_2, q_3(q_1, q)_2))/\partial q_3$ to zero and solving gives

$$q_2(q_1) = (a - c - q_1)/4$$

as the optimal production for Firm 2. We find $q_3(q_1, q_2(q_1)) = (a - c - q_1)/4$. Then, evaluating

$$\begin{aligned} u_1(q_1, q_2(q_1), q_3(q_1, q_2(q_1))) &= q_1 \left(a - c - q_1 - \frac{a - c - q_1}{2} - \frac{a - c - q_1}{4} \right) \\ &= q_1 \left(\frac{a - c - q_1}{4} \right) \end{aligned}$$

as the optimal production for Firm 1. Setting the derivative of this to zero, solving and substituting into the productions of Firm 2 and Firm 3, gives

$$q_1 = (a - c)/2, \quad q_2 = (a - c)/4, \quad q_3 = (a - c)/8,$$

as the strategic equilibrium.

7. (a) Setting the partial derivatives to zero,

$$\begin{aligned} \frac{\partial M_1}{\partial x} &= V \frac{y}{(x + y)^2} - C_1 = 0, \\ \frac{\partial M_2}{\partial y} &= V \frac{x}{(x + y)^2} - C_2 = 0, \end{aligned}$$

we see that $C_1 x = C_2 y$, from which it is easy to solve for x and y:

$$\begin{aligned} x &= V C_2/(C_1 + C_2)^2, \\ y &= V C_1/(C_1 + C_2)^2. \end{aligned}$$

The profits are

$$\begin{aligned} M_1 &= C_2^2/(C_1 + C_2)^2, \\ M_2 &= C_1^2/(C_1 + C_2)^2. \end{aligned}$$

(b) If $V = 1$, $C_1 = 1$ and $C_2 = 2$, we find

$$\begin{aligned} x &= 2/9, \quad M_1 = 4/9, \\ y &= 1/9, \quad M_2 = 1/9. \end{aligned}$$

Solutions to Chap. 17

1.

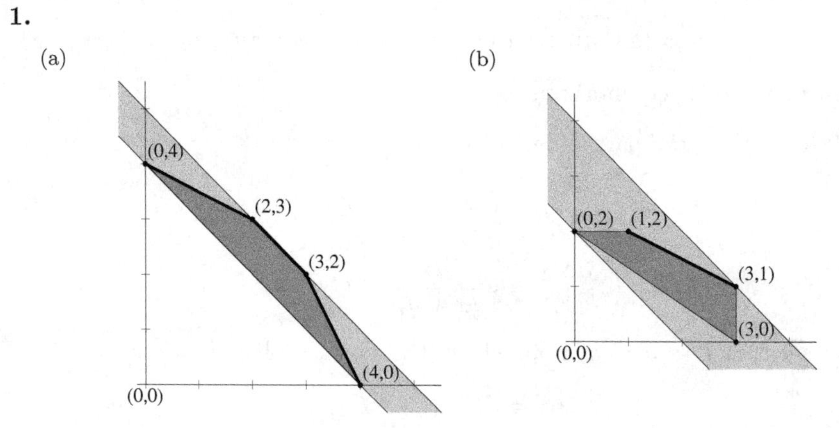

The light shaded region is the TU-feasible set. The dark shaded region is the NTU-feasible region. The NTU-Pareto optimal outcomes are the vectors along the heavy line. The TU-Pareto outcomes are the upper right lines of slope -1.

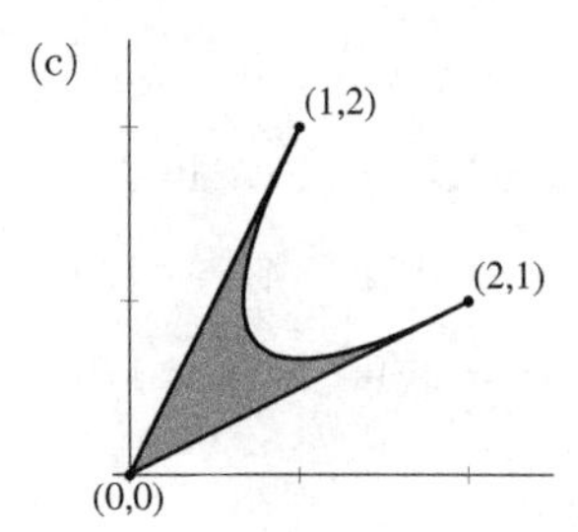

The curve joining (1, 2) and (2, 1) has parametric form $(x, y) = (1 - 2a + 3a^2, 2 - 4a + 3a^2)$ for $0 \le a \le 1$.

2. (a) The cooperative strategy is ((1,0),(1,0)) with sum $\sigma = 7$. The difference matrix $\mathbf{A} - \mathbf{B} = \begin{pmatrix} 1 & 0 \\ 0 & -3 \end{pmatrix}$ has a saddle point at the upper right with value $\delta = 0$. So $\mathbf{p}^* = (1,0)$ and $\mathbf{q}^* = (0,1)$ are the threat strategies and the disagreement point is $(0,0)$. The TU-solution is $\varphi = ((\sigma+\delta)/2, (\sigma-\delta)/2) = (7/2, 7/2)$. Since the cooperative strategy gives payoff $(4,3)$, this requires a side payment of $1/2$ from I to II.

(b) The cooperative strategy is ((0,1),(0,1)) with sum $\sigma = 9$. The difference matrix $\mathbf{A}-\mathbf{B} = \begin{pmatrix} 2 & 4 \\ 0 & -3 \end{pmatrix}$ has a saddle point at the upper left with value $\delta = 2$. So $\mathbf{p}^* = (1,0)$ and $\mathbf{q}^* = (1,0)$ are the threat strategies and the disagreement point is $(5,3)$. The TU-solution is $\varphi = ((\sigma+\delta)/2, (\sigma-\delta)/2) = (11/2, 7/2)$. Since the cooperative strategy gives payoff $(3,6)$, this requires a side payment of $5/2$ from II to I.

3. (a) The cooperative strategy is ((1,0,0,0), (0,0,1,0)) with sum $\sigma = 6$. The difference matrix is

$$\mathbf{A} - \mathbf{B} = \begin{pmatrix} 1 & 3 & -6 & 0 \\ 2 & 0 & -3 & 3 \\ 5 & -6 & 0 & 4 \\ -7 & 7 & -1 & -4 \end{pmatrix}$$

By the matrix game solver, the value is $\delta = -3/7$ and the threat strategies are $\mathbf{p}^* = (0,0,4/7,3/7)$ and $\mathbf{q}^* = (0,1/14,13/14,0)$. So the TU-solution is $\varphi = ((\sigma+\delta)/2, (\sigma-\delta)/2) = (3 - \frac{3}{14}, 3 + \frac{3}{14})$. The disagreement point is $(-27/98, 15/98)$. The cooperative strategy gives payoff $(0,6)$ so this requires a side payment of $2 + \frac{11}{14}$ from II to I.

(b) We have $\sigma = 2$. One cooperative strategy is ((0,0,0,0,1), (1,0,0,0)). The difference matrix is

$$\begin{pmatrix} 0 & 2 & 0 & -1 \\ -2 & -1 & 1 & 0 \\ 1 & 0 & -1 & -2 \\ 2 & -1 & 2 & 0 \\ 0 & 0 & 0 & 0 \end{pmatrix}$$

There is a saddle point at the lower right corner. The value is $\delta = 0$. The TU-solution is $(1,1)$. The threat strategies are $(0,0,0,0,1)$ and $(0,0,0,1)$. The disagreement point is $(0,0)$. There is no side payment.

4. (a) The set of Pareto optimal points is the parabolic arc, $y = 4 - x^2$, from $x = 0$ to $x = 2$. We seek the point on this arc that maximizes the product $xy = x(4 - x^2) = 4x - x^3$. Setting the derivative with respect to x to zero gives $4 - 3x^2 = 0$, or $x = 2/\sqrt{3}$. The corresponding value of y is $y = 4 - (4/3) = 8/3$. Hence the NTU solution is $(\bar{u}, \bar{v}) = (2/\sqrt{3}, 8/3)$.

(b) This time we seek the Pareto optimal point that maximizes the product $x(y-1) = x(3-x^2) = 3x - x^3$. Setting the derivative with respect to x to zero gives $3 - 3x^2 = 0$, or $x = 1$. The corresponding value of y is $y = 4 - 1 = 3$. Hence the NTU solution is $(\bar{u}, \bar{v}) = (1, 3)$.

5. (a) The fixed threat point is $(u^*, v^*) = (0, 0)$. The set of Pareto optimal points consists of the two line segments, from (1, 8) to (4, 6) and from (4, 6) to (6, 3). The first has slope, $-2/3$, and the second has slope, $-3/2$. The slope of the line from (0, 0) to (4, 6) is 3/2, exactly the negative of the slope of the second line. Thus, $(\bar{u}, \bar{v}) = (4, 6)$ is the NTU-solution. The equilibrium exchange rate is $\lambda^* = 3/2$.

(b) Both matrices, A and B, have saddle-points at the first row, second column. Therefore, the fixed threat point is $(u^*, v^*) = (-1, 1)$. The set of Pareto optimal points consists of the two line segments, from $(-2, 9)$ to (2, 7) and from (2, 7) to (3, 3). The first has slope, $-1/2$, and the second has slope, -4. The slope of the line from $(-1, 1)$ to (2, 7) is 2. Since this is between the negatives of the two neighboring slopes, the NTU-solution is $(\bar{u}, \bar{v}) = (2, 7)$. The equilibrium exchange rate is $\lambda^* = 2$.

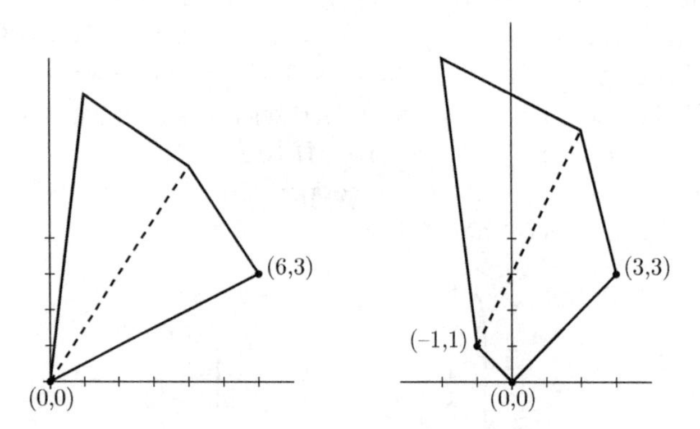

6. (a) Clearly, the NTU-solution must be on the line joining $(1, 4)$ and $(5, 2)$. For the TU-solution to be equal to the NTU-solution, we suspect that the slope of the λ-transformed line, from $(\lambda, 4)$ to $(5\lambda, 2)$, would be equal to -1. Since the slope of this line is $-2/(4\lambda)$, we have $\lambda^* = 1/2$, in which case the game matrix becomes $\begin{pmatrix} (5/2, 2) & (0, 0) \\ (0, 0) & (1/2, 4) \end{pmatrix}$. This gives $\sigma = 9/2$ and $\delta = 0$, so that the TU-solution is $(9/4, 9/4)$. The NTU-solution of the original matrix is obtained from this by dividing the first coordinate by λ^*, so that $\varphi = (9/2, 9/4)$.

(b) The lambda-transfer matrix is $\begin{pmatrix} (3\lambda,2) & (0,5) \\ (2\lambda,1) & (\lambda,0) \end{pmatrix}$. We see

$$\sigma(\lambda) = \begin{cases} 5 & \text{if}\,\lambda \le 1 \\ 3\lambda+2 & \text{if}\,\lambda \ge 1 \end{cases}.$$

The difference matrix is

$$\lambda\mathbf{A} - \mathbf{B} = \begin{pmatrix} 3\lambda-2 & -5 \\ 2\lambda-1 & \lambda \end{pmatrix}.$$

This matrix has a saddle point no matter what be the value of $\lambda > 0$. If $0 < \lambda \le 1$, there is a saddle at $\langle 2,1\rangle$. If $\lambda \ge 1$, there is a saddle at $\langle 2,2\rangle$. Thus,

$$\delta(\lambda) = \begin{cases} 2\lambda-1 & \text{if}\;0 < \lambda \le 1 \\ \lambda & \text{if}\;\lambda \ge 1. \end{cases}$$

From (10),

$$\varphi(\lambda) = \left(\frac{\sigma(\lambda)+\delta(\lambda)}{2\lambda}, \frac{\sigma(\lambda)-\delta(\lambda)}{2}\right).$$

For $\lambda = 1$, we find $\varphi(\lambda) = (3,2)$. This is obviously feasible since it is the upper left entry of the original bimatrix. So the NTU-solution is (3,2), and $\lambda^* = 1$ is the equilibrium exchange rate.

7. (a) If Player II uses column 2, Player I is indifferent as to what he plays. If I uses $(1-p,p)$, Player II prefers column 2 to column 1 if $89(1-p)+98p \le 90(1-p)$, that is if $99p \le 1$. I's safety level is $\mathrm{Val}\begin{pmatrix} 11 & 10 \\ 2 & 10 \end{pmatrix} = 10$, and II's safety level is $\mathrm{Val}\begin{pmatrix} 89 & 98 \\ 90 & 0 \end{pmatrix} = 89 + \frac{1}{11}$. At the equilibrium with $p = 1/99$, the payoff vector is $(10, 90(98/99)) = (10, 89+\frac{1}{11})$. Thus both players only get their safety levels.

(b) Working together, Player I and II can achieve $\sigma = 100$. The difference matrix, $\mathbf{D} = \mathbf{A} - \mathbf{B}$, has value

$$\delta = \mathrm{Val}\begin{pmatrix} -78 & -80 \\ -96 & 10 \end{pmatrix} = \frac{-78\cdot 10 - 80\cdot 98}{10-78+80+96} = -\frac{235}{3} = -78\frac{1}{3}.$$

Therefore, the TU solution is $\varphi = ((\sigma+\delta)/2, (\sigma-\delta)/2) = (10\frac{5}{6}, 89\frac{1}{6})$. This is on the line segment joining the top two payoff vectors of the game matrix, and so is in the NTU feasible set. Player I's threat strategy is $(106/108, 2/108) = (53/54, 1/54)$. Player II's threat strategy is $(90/108, 18/108) = (5/6, 1/6)$.

Part of the reason Player I is so strong in this game is that even if Player I carries out his threat strategy, $(53/54, 1/54)$, the best Player II can do against it is to play column 1, when the payoff to the players is $((10\frac{5}{6}, 89\frac{1}{6}, 0)$, the same as given by the NTU solution.

Solutions to Exercises of Part IV

Solutions to Chap. 18

1. We find $v(\{1,2\}) = -v(\{3\}) = 4.4$, $v(\{1,3\}) = -v(\{2\}) = 4$, $v(\{2.3\}) = -v(\{1\}) = 1.5$, and $v(\emptyset) = v(N) = 0$.

(I,II):

	III: 1	2
1,1	−1	−3
1,2	4	5
2,1	−3	−2
2,2	6	2

(I,III):

	II: 1	2
1,1	−1	−3
1,2	4	5
2,1	2	6
2,2	−2	−3

(II,III):

	I: 1	2
1,1	2	1
1,2	−1	−12
2,1	−1	4
2,2	−10	1

2. $v(\emptyset) = 0$, $v(1) = 6/10$, $v(2) = 2$, $v(3) = 1$, $v(12) = 5$, $v(13) = 4$, $v(23) = 3$ and $v(123) = 16$.

I \ (II,III)	1,1	1,2	2,1	2,2
1	1	3	−1	3
2	−1	1	7	3

(I,III)

		1,1	1,2	2,1	2,2
II:	1	2	0	2	0
	2	6	2	5	2

(I,II)

		1,1	1,2	2,1	2,2
III:	1	1	−3	4	4
	2	1	1	3	1

III

		1	2
(I,II):	1,1	3	3
	1,2	5	5
	2,1	1	1
	2,2	12	5

II

		1	2
(I,III):	1,1	2	−4
	1,2	4	4
	2,1	3	11
	2,2	4	4

I

		1	2
(II,III):	1,1	3	6
	1,2	1	3
	2,1	3	9
	2,2	3	3

3. (a) $v(\emptyset) = 0$, $v(1) = 2$, $v(2) = 2$ and $v(12) = 9$.

(b) Player 2's threat strategy and MM (safety level) strategy are the same, column 2. Player 1's threat strategy is row 1, while his MM strategy is row 2. His threat is not believable, whereas 2's threat is very believable. In addition, (row 2, col 1) is a PSE.

(c) $\sigma = 9$ and $\Delta = \begin{pmatrix} -2 & 3 \\ -2 & 1 \end{pmatrix}$, so $\delta = -2$. The TU-value is $\phi = ((9 - 2)/2, (7 - 2)/2) = (7/2, 11/2)$. Player 1 gets 5, but has to make a side payment of 3/2 to Player 2.

(d) The strategy spaces are taken to be $X_1 = \{\{1\}, \{1,2\}\}$ and $X_2 = \{\{2\}, \{1,2\}\}$. The bimatrix therefore is 2 by 2:

$$\begin{array}{c} \\ \{1\} \\ \{1,2\} \end{array} \begin{array}{c} \begin{array}{cc} \{2\} & \{1,2\} \end{array} \\ \begin{pmatrix} (2,2) & (2,2) \\ (2,2) & (9/2,9/2) \end{pmatrix} \end{array}$$

4. $N = \{1, 2, 3\}$. $v(\emptyset) = 0$. $v(\{1\}) = 1$, because Player 2 can always choose j within 1 of i. Similarly, $v(\{2\}) = 1$ because Player 1 can choose $i = 5$ say, and then Player 3 can choose k within 1 of j. $v(\{3\}) = 4$ is achieved by choosing $i = 5$ and $j = 10$, say. Similarly, $v(\{1, 2\}) = 10$, $v(\{1, 3\}) = 10$, $v(\{2, 3\}) = 14$, and $v(N) = 18$.

Solutions to Chap. 19

1. A constant-sum game has $v(S) + v(N - S) = v(N)$ for all coalitions S. Therefore, a two-person constant-sum game has $v(\{1\}) + v(\{2\}) = v(\{1,2\})$ and so is inessential.

2. We have $v(\{1\}) = \mathrm{Val}\begin{pmatrix} 4 & 1 \\ 0 & 2 \end{pmatrix} = 8/5$, $v(\{2\}) = \mathrm{Val}\begin{pmatrix} 2 & 0 \\ 1 & 4 \end{pmatrix} = 8/5$, and $v(\{1,2\}) = 6$. The set of imputations is $\{(x_1, x_2) : x_1 + x_2 = 6, x_1 \geq 8/5, x_2 \geq 8/5\}$, the line segment from $(8/5, 22/5)$ to $(22/5, 8/5)$. The core is the same set. In fact, for all 2-person games, the core is always the whole set of imputations.

3. On the plane $x_1 + x_2 + x_3 = 5$:

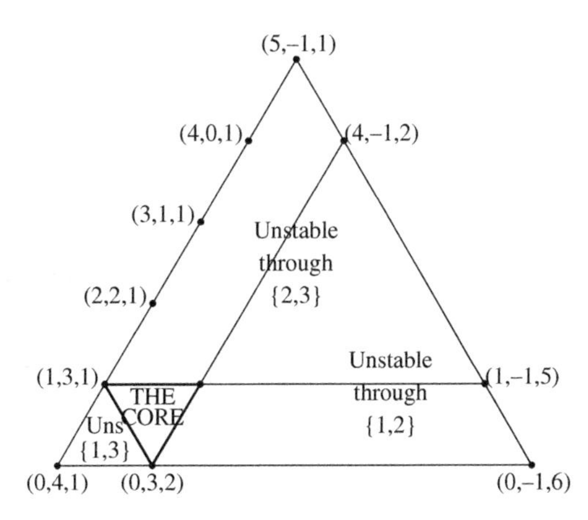

4. (a) The set of imputations is $\{(x_1, x_2, x_3) : x_1+x_2+x_3 = 3, x_1 \geq 0, x_2 \geq 0, x_3 \geq 0\}$. The core is

$$\{(x_1, x_2, x_3) : x_1 + x_2 + x_3 = 3,\ x_1 + x_2 \geq a,\ x_1 + x_3 \geq a,$$
$$x_2 + x_3 \geq a,\ \ x_1 \geq 0,\ x_2 \geq 0,\ x_3 \geq 0\}.$$

For values of $a \leq 2$, the point $(1, 1, 1)$ is in the core. For values of $a > 2$, the core is empty, since summing the corresponding sides of the inequalities $x_1 + x_2 \geq a$, $x_1 + x_3 \geq a$, and $x_2 + x_3 \geq a$, gives $6 = 2(x_1 + x_2 + x_3) \geq 3a > 6$, a contradiction.

(b) The core consists of points (x_1, x_2, x_3, x_4) such that

$$\begin{array}{llll} x_1 \geq 0 & x_1 + x_2 \geq a & x_2 + x_3 \geq a & x_1 + x_2 + x_3 \geq b \\ & & & x_1 + x_2 + x_3 + x_4 = 4. \\ x_2 \geq 0 & x_1 + x_3 \geq a & x_2 + x_4 \geq a & x_1 + x_2 + x_4 \geq b \\ x_3 \geq 0 & x_1 + x_4 \geq a & x_3 + x_4 \geq a & x_1 + x_3 + x_4 \geq b \\ x_4 \geq 0 & & & x_2 + x_3 + x_4 \geq b \end{array}$$

If $a \leq 2$ and $b \leq 3$, then $(1, 1, 1, 1)$ is in the core.
If $a > 2$, summing the inequalities involving a gives $12 = 3(x_1 + x_2 + x_3 + x_4) \geq 6a > 12$, so the core is empty.
If $b > 3$, summing the inequalities involving b gives $12 = 3(x_1 + x_2 + x_3 + x_4) \geq 4b > 12$, so the core is empty.
Thus the core is non-empty if and only if $a \leq 2$ and $b \leq 3$. If v is superadditive, then a cannot be greater than two, so the condition reduces to $b \leq 3$.

(c) The core is nonempty if and only if $f(k)/k \leq f(n)/n$ for all $k = 1, \ldots, n$. To see this, note that if $f(k)/k \leq f(n)/n$ for all $k = 1, \ldots, n$, then $\mathbf{x} = (f(n)/n, \ldots, f(n)/n)$ is in the core since for any coalition S of size $|S| = k$, we have $\sum_{i \in S} x_i = kf(n)/n \leq f(k) = v(S)$. On the other hand, suppose that $f(k)/k > f(n)/n$ for some k. Then for any imputation $\mathbf{x}$, the coalition S consisting of those players with the k smallest x_i satisfies $(1/k)\sum_{i \in S} x_i \leq (1/n)\sum_{i \in N} x_i = f(n)/n < f(k)/k$. This means that $\mathbf{x}$ is unstable through S, and so cannot be in the core.

5. Suppose $\mathbf{x} = (x_1, \ldots, x_n)$ is in the core. For $\mathbf{x}$ to be stable, we must have $\sum_{j \neq i} x_j \geq v(N - \{i\})$ for all i. This is equivalent to $x_i \leq \delta_i$ for all i. However, summing over i gives $v(N) = \sum_1^n x_i \leq \sum_1^n \delta_i < v(N)$, a contradiction. Thus, the core must be empty.

6. Suppose Player 1 is a dummy. If $\mathbf{x} = (x_1, \ldots, x_n)$ is in the core, it is an imputation, so $x_1 \geq v(\{1\}) = 0$. Suppose $x_1 > 0$. Then, $v(N - \{x_1\}) = v(N) = \sum_1^n x_i > \sum_2^n x_i$. So $\mathbf{x}$ is unstable through $N - \{x_1\}$ and cannot be in the core.

7. (a) If Players 1 and 2 are in P, and Players 3 and 4 are in Q, then $v(i) = 0$ for all i, $v(ij) = 1$ for all ij except $ij = 12$ and $ij = 34$, $v(ijk) = 1$ and $v(1234) = 2$. The core, where all coalitions are satisfied, is

$$\begin{aligned} C = \{(x_1, x_2, x_3, x_4) &: x_1 \geq 0, x_2 \geq 0, x_3 \geq 0, \\ &x_4 \geq 0, x_1 + x_2 + x_3 + x_4 = 2, \\ &x_1 + x_3 \geq 1, x_1 + x_4 \geq 1, x_2 + x_3 \geq 1, x_2 + x_4 \geq 1\}. \end{aligned}$$

Since $x_1 + x_3 \geq 1$, $x_2 + x_4 \geq 1$ and $x_1 + x_2 + x_3 + x_4 = 2$, we must have $x_1 + x_3 = 1$, $x_2 + x_4 = 1$. Similarly, we have $x_1 + x_4 = 1$, $x_2 + x_3 = 1$. The core is therefore $C = \{(x_1, x_1, 1 - x_1, 1 - x_1) : 0 \leq x_1 \leq 1$. This is the line segment joining the points $(0, 0, 1, 1)$ and $(1, 1, 0, 0)$.

(b) If Players 1 and 2 are in P, and Players 3, 4 and 5 are in Q, then the core is

$$\begin{aligned} C = \{(x_1, x_2, x_3, x_4, x_5) &: x_1 \geq 0, x_2 \geq 0, x_3 \geq 0, x_4 \geq 0, x_5 \geq 0, \\ &x_1 + x_3 \geq 1, x_1 + x_4 \geq 1, x_1 + x_5 \geq 1, x_2 + x_3 \geq 1, x_2 + x_4 \geq 1, \\ &x_2 + x_5 \geq 1, x_1 + x_2 + x_3 + x_4 \geq 2, x_1 + x_2 + x_3 + x_5 \geq 2, \\ &x_1 + x_2 + x_4 + x_5 \geq 2, x_1 + x_2 + x_3 + x_4 + x_5 = 2\}. \end{aligned}$$

These inequalities imply that $x_3 = x_4 = x_5 = 0$, and therefore that $x_1 = x_2 = 1$. The core consists of the single point, $C = \{(1, 1, 0, 0, 0)\}$.

(c) If $|P| < |Q|$, the core is the imputation, $\mathbf{x}_P$, with $x_i = 1$ for $i \in P$ and $x_i = 0$ for $i \in Q$. If $|P| > |Q|$, the core is the imputation, $\mathbf{x}_Q$, with $x_i = 1$ for $i \in Q$ and $x_i = 0$ for $i \in P$. If $|P| = |Q|$, the core is the line segment joining $\mathbf{x}_P$ and $\mathbf{x}_Q$.

8. In the core, we have $x_i + x_k \geq 1$ for $i = 1, 2$ and $k = 3, 4, 5$. Also we have $x_i + x_j + x_k \geq 2$ for $i = 1, 2$ and $j, k = 3, 4, 5$. Finally, we have $x_1 + x_2 + x_3 + x_4 + x_5 = 3$.

Since, $x_1 + x_3 \geq 1$ and $x_2 + x_4 + x_5 \geq 2$, and $x_1 + x_2 + x_3 + x_4 + x_5 = 3$, we must have equality: $x_1 + x_3 = 1$ and $x_2 + x_4 + x_5 = 2$. Similarly, $x_1 + x_3 + x_4 = 2$, which with $x_1 + x_3 = 1$ implies $x_4 = 1$, etc. The core consists of the single point, $(0, 0, 1, 1, 1)$.

Solutions to Chap. 20

1. No player can get anything acting alone, so $v(\{1\}) = v(\{2\}) = v(\{3\}) = 0$. Players 2 and 3 can do nothing together, $v(\{2,3\}) = 0$, but $v(\{1,2\}) = 30$ and $v(\{1,3\}) = 40$. The object is also worth 40 to the grand coalition, $v(\{1,2,3\}) = 40$. (Player 3 will take the object and replace it by \$40, and the players must now decide how to split this money.) To find the Shapley value using Equation (4), we first find $c_{\emptyset} = c_{\{1\}} = c_{\{2\}} = c_{\{3\}} = c_{\{2,3\}} = 0$, and $c_{\{1,2\}} = 30$, $c_{\{1,3\}} = 40$ and $c_{\{1,2,3\}} = v(\{1,2,3\})-$ the sum of the above, so $c_{\{1,2,3\}} = 40 - 40 - 30 = -30$. Thus we have $v = 30w_{\{1,2\}} + 40w_{\{1,3\}} - 30w_{\{1,2,3\}}$, and consequently, $\phi(v) = 30\phi(w_{\{1,2\}}) + 40\phi(w_{\{1,3\}}) - 30\phi(w_{\{1,2,3\}})$.

From this we may compute

$$\phi_1(v) = 30/2 + 40/2 - 30/3 = 25$$

$$\phi_2(v) = 30/2 + 0 - 30/3 = 5$$

$$\phi_3(v) = 0 + 40/2 - 30/3 = 10.$$

So 3 gets the painting for \$30, of which \$25 goes to 1 and \$5 to 2. The core is

$$C = \{(x_1, x_2, x_3) : x_1 \geq 30, x_2 \geq 0, x_3 \geq 0, x_1 + x_2 \geq 30,$$
$$x_1 + x_3 \geq 40, x_2 + x_3 \geq 0, x_1 + x_2 + x_3 = 40\}$$
$$= \{(x_1, x_2, x_3) : x_2 = 0, 30 \leq x_1 \leq 40, x_3 = 40 - x_1\}.$$

The core gives player 2 nothing, while the Shapley value gives him 5. Without 2 present, 1 and 3 would probably agree on a price of 20. With 2 present, 1 is in a better bargaining position as he can play 3 off against 2.

2. We compute the Shapley value using Theorem 2. $\phi_1(v) = (1/3)\cdot 1 + (1/6)\cdot 2 + (1/6)\cdot 3 + (1/3)\cdot 3 = 13/6$. $\phi_2(v) = (1/3)\cdot 0 + (1/6)\cdot 1 + (1/6)\cdot 7 + (1/3)\cdot 7 = 22/6 = 11/3$. $\phi_3(v) = (1/3)\cdot(-4) + (1/6)\cdot(-2) + (1/6)\cdot 3 + (1/3)\cdot 4 = 1/6$.

3. $\sum_{j\in N}\phi_j(v) = v(N)$ from Axiom 1. We must show $\phi_i(v) \geq v(\{i\})$ for all $i \in N$. Since v is superadditive, $v(\{i\}) + v(S - \{i\}) \leq v(S)$ for all S containing i. But since $\phi_i(v)$ is an average of numbers, $v(S) - v(S - \{i\})$, each of which is at least $v(\{i\})$, $\phi_i(v)$ itself must be at least $v(\{i\})$.

4. (a) By the symmetry axiom, $\phi_2(v) = \phi_3(v) = \cdots = \phi_n(v)$. If $1 \in S$, then $v(S) - v(S - \{i\}) = \left\{\begin{smallmatrix} 1 & \text{if } i \in S \\ 0 & \text{if } i \notin S \end{smallmatrix}\right.$. Therefore $\phi_2(v)$ is just the probability that 1 comes before 2 in the random ordering of the players. This probability is 1/2 by symmetry. So $\phi_2(v) = \phi_3(v) = \cdots = \phi_n(v) = 1/2$, and $\phi_1(v) = n - \phi_2(v) - \phi_3(v) - \cdots - \phi_n(v) = (n+1)/2$.

One may also use Equation (4) to compute the Shapley value. First show $c_{\{1\}} = c_{\{1,j\}} = 1$ for all $j = 2, \ldots, n$, and all other $c_S = 0$. Hence, $v = w_{\{1\}} + \sum_{j=2}^{n} w_{\{1,j\}}$, from which follows $\phi_1(v) = 1 + (n-1)/2 = (n+1)/2$ and $\phi_j(v) = 1/2$ for $j = 2, \ldots, n$.

(b) By the symmetry axiom, $\phi_1(v) = \phi_2(v)$ and $\phi_3(v) = \cdots = \phi_n(v)$. This time $\phi_3(v)$ is the probability that 1 or 2 is chosen before 3 in the random ordering. This is 1 minus the probability that 3 is chosen before 1 and 2, namely, $1 - (1/3) = 2/3$. So $\phi_3(v) = \cdots = \phi_n(v) = 2/3$ and $\phi_1(v) = \phi_2(v) = (1/2)(n - (n-2)(2/3)) = (n+4)/6$.

(c) By the symmetry axiom, $\phi_1 = \phi_2$ and $\phi_3 = \cdots = \phi_n$. If $3 \in S$, then $v(S) - v(S - \{3\})$ is 1 if 1 and 2 are in S and 0 otherwise. This implies that ϕ_3 is just the probability that 3 enters after both 1 and 2. Since each of 1, 2 and 3 have the same probability of entering after the other two, $\phi_3 = 1/3$. Then since $2\phi_1 + (n-2)\phi_3 = n$, we have $\phi_1 = \phi_2 = (n+1)/3$, and $\phi_3 = \cdots = \phi_n = 1/3$.

5. The answer is no. Here is a counterexample with $n = 4$ players. The minimal winning coalitions are $\{1,2\}$ and $\{3,4\}$. To be a weighted voting game, with weights w_i and quota q, we must have $w_1 + w_2 > q$ and $w_3 + w_4 > q$, so that $w_1 + w_2 + w_3 + w_4 > 2q$. On the other hand, $\{1,3\}$ and $\{2,4\}$ are losing coalitions so that $w_1 + w_3 \leq q$ and $w_2 + w_4 \leq q$. This gives $q_1 + q_2 + q_3 + q_4 \leq 2q$, a contradiction.

6. The winning coalitions are $\{2,3\}$, $\{2,4\}$, $\{3,4\}$ and all supersets. It appears that 1 is a dummy, so $\phi_1(v) = 0$. Also, $\phi_2(v) = \phi_3(v) = \phi_4(v)$

from symmetry. Since the sum of the $\phi_i(v)$ must be 1, we have $\phi(v) = (0, 1/3, 1/3, 1/3)$.

7. Player 2 can be in only two winning coalitions that would be losing without him, namely, $S = \{1,2\}$ and $S = \{2,3,\ldots,n\}$. Hence,

$$\phi_2(v) = \frac{(2-1)!\,(n-2)!}{n!} + \frac{(n-2)!\,(n-(n-1))!}{n!}$$
$$= \frac{2(n-2)!}{n!} = \frac{2}{n(n-1)}.$$

By symmetry,

$$\phi_3(v) = \cdots = \phi_n(v) = \frac{2}{n(n-1)}$$

and

$$\phi_1(v) = 1 - \phi_2(v) - \cdots - \phi_n(v) = 1 - (n-1)\frac{2}{n(n-1)} = \frac{n-2}{n}.$$

8. Let 1,2,3,4 denote the stockholders and let c denote the chairman of the board. The coalitions winning with 1 but losing without 1 are $\{1,4,c\}$ and $\{1,2,3\}$. So

$$\phi_1(v) = \frac{2!\,2!}{5!} + \frac{2!\,2!}{5!} = \frac{1}{30} + \frac{1}{30} = \frac{2}{30}.$$

The corresponding coalitions for 2 are $\{2,3,c\}$, $\{2,4\}$, $\{1,2,3\}$, $\{1,2,3,c\}$, $\{2,4,c\}$, and $\{1,2,4\}$. So $\phi_2(v) = 4(1/30) + 2(1/20) = 7/30$. Similarly, $\phi_3(v) = 7/30$, $\phi_4(v) = 12/30$, and $\phi_c(v) = 2/30$. The Shapley value is $(2/30, 7/30, 7/30, 12/30, 2/30)$, or, in terms of percentage power $(6.7\%, 23.3\%, 23.3\%, 40\%, 6.7\%)$.

9. (a) Let 1 denote the large party and 2, 3, 4 denote the smaller parties. Then the only winning coalitions that become losing without 2 are $S = \{1,2\}$ and $S = \{2,3,4\}$. Hence,

$$\phi_2(v) = \frac{1!\,2!}{4!} + \frac{2!\,1!}{4!} = \frac{1}{6}.$$

By symmetry, $\phi_3(v) = \phi_4(v) = 1/6$ also, and hence $\phi_1(v) = 1/2$. The large party has half the power.

(b) Let 1, 2 denote the large parties and 3, 4, 5 denote the smaller ones. The only coalitions winning with 3 and losing without 3 are $\{1,3,4\}$,

$\{1,3,5\}$, $\{2,3,4\}$, and $\{2,3,5\}$. Hence,

$$\phi_3(v) = 4\frac{2!\,2!}{5!} = \frac{2}{15}.$$

By symmetry, $\phi_4(v) = \phi_5(v) = 2/15$, and hence $\phi_1(v) = \phi_2(v) = 3/10$. The two large coalitions are less powerful than their size indicates.

10. There are three coalitions that are winning with 6 but losing without 6: $\{1,2,6\}$, $\{1,3,5,6\}$, $\{2,3,5,6\}$. Hence,

$$\phi_6(v) = \frac{2!\,3!}{6!} + 2\frac{3!\,2!}{6!} = \frac{3}{60}.$$

By symmetry, $\phi_5(v) = 3/60$. There are 7 coalitions winning with 4 but losing without 4: $\{1,2,4\}$, $\{1,3,4\}$, $\{2,3,4\}$, $\{1,3,4,5\}$, $\{1,3,4,6\}$, $\{2,3,4,5\}$, $\{2,3,4,6\}$. Hence, $\phi_4(v) = 7/60$. There are 11 coalitions winning with 3 but losing without 3: $\{1,2,3\}$, $\{1,3,4\}$, $\{2,3,4\}$, $\{1,3,4,5\}$, $\{1,3,4,6\}$, $\{1,3,5,6\}$, $\{2,3,4,5\}$, $\{2,3,4,6\}$, $\{2,3,5,6\}$, $\{1,3,4,5,6\}$, $\{2,3,4,5,6\}$. Hence,

$$\phi_3(v) = 3\frac{2!\,3!}{6!} + 6\frac{3!\,2!}{6!} + 2\frac{4!\,1!}{6!} = \frac{13}{60}.$$

Since $\phi_1(v) = \phi_2(v)$ by symmetry, we find that $\phi_1(v) = \phi_2(v) = 17/60$. The Shapley value is $(17/60, 17/60, 13/60, 7/60, 3/60, 3/60)$. In terms of percentage power, this is $(28.3\%, 28.3\%, 21.7\%, 11.7\%, 5\%, 5\%)$, which is much closer to the original intention found in Table 1.

11. Let $\phi_A(v)$ denote the Shapley value for A, of one of the big five, and let $\phi_a(v)$ denote the Shapley value of a, one of the smaller members. We must have $5\phi_A(v) + 10\phi_a(v) = 1$. Let us find $\phi_a(v)$. The only losing coalitions that become winning, when a is added to it, are the coalitions consisting of the big five and three of the other smaller nations. Thus in the random ordering, a must come in ninth and find all members of the big five there already. The number of such coalitions is the number of ways of choosing the three smaller nation members out of the remaining nine, namely $\binom{9}{3} = 9!/3!\,6!$. Thus

$$\phi_a(v) = \frac{8!\,6!}{15!} \cdot \frac{9!}{3!\,6!} = \frac{4}{15 \cdot 13 \cdot 11} = 0.001865\cdots.$$

Thus the 10 smaller nations have only 1.865% of the power, and each of the big five nations has 19.627% of the power.

12. The trip to A and return costs 14, so the value of A acting alone is $v(A) = 20 - 14 = 6$. Similarly, $v(B) = 20 - 16 = 4$, and $v(C) = 20 - 12 = 8$.

If A and B combine forces, the travel cost is 17, the cost of the trip to A then to B and return. So, $v(AB) = 40 - 17 = 23$. Similarly, $v(AC) = 40 - 17 = 23$ and $v(BC) = 40 - 18 = 22$. If all three cities combine, the least travel cost is obtained using the route from H to A to B to C and return (or the reverse), for a total cost of 19. So $v(ABC) = 60 - 19 = 41$. From these, we may compute the Shapley value as

$$\phi_A(v) = \frac{1}{3}\cdot 6 + \frac{1}{6}\cdot 19 + \frac{1}{6}\cdot 15 + \frac{1}{3}\cdot 19 = 14$$

$$\phi_B(v) = \frac{1}{3}\cdot 4 + \frac{1}{6}\cdot 17 + \frac{1}{6}\cdot 14 + \frac{1}{3}\cdot 18 = 12.5$$

$$\phi_C(v) = \frac{1}{3}\cdot 8 + \frac{1}{6}\cdot 17 + \frac{1}{6}\cdot 18 + \frac{1}{3}\cdot 18 = 14.5$$

Thus, we require A to pay $20 - 14 = 6$, B to pay $20 - 12.5 = 7.5$, and C to pay $20 - 14.5 = 5.5$ for a total of 19.

13. Consider a permutation of the n players, $\pi = (\pi_1, \pi_2, \ldots, \pi_n)$, and let π' denote the reverse permutation, $\pi' = (\pi_n, \ldots, \pi_2, \pi_1)$. Consider player i and let z_i denote the sum of the contributions player i makes to the coalitions when he enters for these two permutations. Below, it is shown that $z_i = x_i$ if $i \in B$ and $z_i = y_i$ if $i \in C$. If so, then the Shapley value for player i in this game is $x_i/2$ if $i \in B$ and $y_i/2$ if $i \in C$.

Let S_0 denote the coalition player i finds upon entering, when the players enter in the order given by π, and let $S = S_0 \cup \{i\}$. Then if the players enter in the reverse order π', player i will find coalition $\bar{S}$ (the complement of S) there when he enters and he increases it to $\bar{S}_0$. The amount player i contributes to the grand coalition is $v(S) - v(S_0)$ when entering in the order given by π and $v(\bar{S}_0) - v(\bar{S})$ when entering in the order given by π. The sum is therefore

$$z_i = v(S) - v(S_0) + v(\bar{S}_0) - v(\bar{S}) = [v(S) - v(\bar{S})] - [v(\bar{S}_0) - v(S_0)]$$

Letting T denote the common value $T = \sum_B x_j = \sum_C y_j = v(N)$, we find

$$v(S) - v(\bar{S}) = \min\left\{\sum_{j\in S\cap B} x_j, \sum_{k\in S\cap C} y_k\right\} - \min\left\{\sum_{j\in \bar{S}\cap B} x_j, \sum_{k\in \bar{S}\cap C} y_k\right\}$$

$$= \min\left\{\sum_{j\in S\cap B} x_j, \sum_{k\in S\cap C} y_k\right\} - \min\left\{T - \sum_{j\in S\cap B} x_j, T - \sum_{k\in S\cap C} y_k\right\}$$

$$= \min\left\{\sum_{j\in S\cap B} x_j, \sum_{k\in S\cap C} y_k\right\} - T + \max\left\{\sum_{j\in S\cap B} x_j, \sum_{k\in S\cap C} y_k\right\}$$

$$= \sum_{j\in S\cap B} x_j + \sum_{k\in S\cap C} y_k - T.$$

Similarly,

$$v(\bar{S}_0) - v(S_0) = \sum_{j\in \bar{S}_0\cap B} x_j + \sum_{k\in \bar{S}_0\cap C} y_k - T.$$

Since S plus $\bar{S}_0$ is the set of all players but including player i twice, the sum of the two previous displays is equal to $T + x_i + T - 2T = x_i$ if $i \in B$, and $T + T + y_i - 2T = y_i$ if $i \in C$.

14. (a) We compute the Shapley value, ϕ, by the method of Theorem 1. We have $c_\emptyset = 0$. For singleton coalitions, we have $c_{\{1\}} = 1$ and $c_{\{j\}} = 0$ for $j \neq 1$. For coalitions of two players, we have $c_{\{1,2\}} = 2 - 1 = 1$ and $c_{\{i,j\}} = 0$ for all other $i < j$. Continuing in the same way we find

$$c_{\{12\cdots k\}} = 1 \quad \text{for } k = 1, \ldots, n, \quad \text{and} \quad c_S = 0$$
$$\text{for all other coalitions, } S.$$

We can check this by checking that $v(T) = \sum_S c_S w_S(T) = \sum_{k=1}^n w_{\{1,2,\ldots,k\}}(T)$. From this, we may conclude that $\phi_i(v) = \sum_{S \text{ containing } i} c_S/|S|$, or

$$\begin{aligned}
\phi_1(v) &= 1 + (1/2) + (1/3) + \cdots + (1/n)\\
\phi_2(v) &= \quad (1/2) + (1/3) + \cdots + (1/n)\\
\phi_3(v) &= \quad (1/3) + \cdots + (1/n)\\
&\vdots\\
\phi_n(v) &= \quad (1/n).
\end{aligned}$$

(b) Similarly, $c_{\{12\cdots k\}} = a_k - a_{k-1}$ for $k = 1, \ldots, n$, and $c_S = 0$ for all other coalitions, S, where $a_0 = 0$. Then, $\phi_i(v) = \sum_i^n (a_j - a_{j-1})/j$ for $i = 1, \ldots, n$.

15. (a) Let S be an arbitrary coalition and let $m = \max\{i : i \in S\}$. Then,

$$v_k(S) = \begin{cases} -(c_k - c_{k-1}) & \text{if}\, k \le m \\ 0 & \text{if}\, k > m. \end{cases}$$

So $\sum_{k=1}^{n} v_k(S) = \sum_{k=1}^{m} -(c_k - c_{k-1}) = -c_m = v(S)$.

(b) Since the Shapley value is additive, $\phi_i(v) = \sum_{k=1}^{n} \phi_i(v_k)$. To compute $\phi_i(v_k)$, note that $\phi_i(v_k) = 0$ if $i < k$ and $\phi_i(v_k) = -(c_k - c_{k-1})P(i,k)$ for $i \ge k$, where $P(i,k)$ represents the probability that in a random ordering of the players into the grand coalition, player i is the first member of S_k to appear. $P(i,k)$ is just the probability that i is first in a random ordering of memberw of S_k, and so $P(i,k) = 1/(n-k+1)$, since there are $n-k+1$ players in S_k. Therefore,

$$\phi_i(v) = \sum_{k=1}^{i} \frac{-(c_k - c_{k-1})}{n-k+1}.$$

Thus, Player 1 pays c_1/n, Player 2 pays $c_1/n + (c_2 - c_1)/(n-1)$, etc. Since all n Players use the first part of the airfield, each player pays c_1/n for this. Since Players 2 through n use the second part of the airfield, they each pay $(c_2 - c_1)/(n-1)$, and so on.

16. The characteristic function is

$$v(S) = \begin{cases} 0 & \text{if}\, 0 \notin S \,\text{or if}\, S = \{0\}, \\ a_{k(S)} & \text{otherwise} \end{cases}$$

where $k(S) = \min\{i : i \in S - \{0\}\}$. For $i \ne 0$,

$$\phi_i(v) = \sum_{S \in \mathcal{S}_i} \frac{|S| - 1)!(m+1-|S|)!}{(m+1)!}(a_i - a_{k(S-\{i\})}),$$

where $\mathcal{S}_i = \{S \subset N : 0 \in S, i \in S, k(S) = i\}$. This is because $v(S) - v(S - \{i\}) = 0$ unless $S \in \mathcal{S}$. Then,

$$\phi_i(v) = a_i \left[\sum_{S \in \mathcal{S}_i} \frac{|S| - 1)!(m+1-|S|)!}{(m+1)!} \right]$$
$$+ \sum_{j=i+1}^{m} a_j \left[\sum_{S \in \mathcal{S}_{i,j}} \frac{|S| - 1)!(m+1-|S|)!}{(m+1)!} \right],$$

where $\mathcal{S}_{i,j} = \{S \subset N : 0 \in S, i \in S, k(S - \{i\}) = j\}$.

The coefficient of a_i is just the probability that in a random ordering of all $m+1$ players, i enters after 0 but before $1,\ldots,i-1$. This is the same as the probability that in a random ordering of $0,1,\ldots,i$, 0 enters first and i second, namely $1/((i+1)i)$.

The coefficient of a_j is just the probability that in a random ordering of all $m+1$ players, i enters after 0 and j but before $1,\ldots,j-1$. This is the same as the probability that in a random ordering of $0,1,\ldots,j$, i enters third after 0 and j in some order, namely $2/((j+1)j(j-1))$. This gives

$$\phi_i(v) = \frac{a_i}{(i+1)i} - \sum_{j=i+1}^{m} \frac{2a_j}{(j+1)j(j-1)}.$$

Similarly,

$$\begin{aligned}\phi_0(v) &= \sum_{j=1}^{m} a_j P(0 \text{ enters after } j \text{ but before } 1,\ldots,j-1) \\ &= \sum_{j=1}^{m} \frac{a_j}{(j+1)j}.\end{aligned}$$

17. (a) If $v(N-\{i\})=1$ and if $\mathbf{x}$ is in the core, then $1 = v(N-\{i\}) \le \sum_1^n x_j - x_i = 1 - x_i$, so $x_i = 0$. Thus, any player without veto power gets zero at every core point. If there are no veto players, then there can be no core points since we must have $\sum_1^n x_i = 1$.

(b) If i is a veto player, then $\mathbf{x} = \mathbf{e}_i$, the ith unit vector, is a core point, since if S is a winning coalition, then $i \in S$ and $\sum_{i\in S} x_i = 1 \ge v(S) = 1$, and if S is losing, then certainly $\sum_{i\in S} \ge v(T) = 0$.

(c) The core is the set of all vectors, $\mathbf{x}$, such that $\sum_1^n x_i = 1$, $x_i \ge 0$ for all $i \in N$, and $x_i = 0$ if i is not a veto player.

Solutions to Chap. 21

1. The core is the set of imputations, $\mathbf{x}$, such that the excesses, $e(\mathbf{x}, S)$, are negative or zero for all coalitions, S. The nucleolus is an imputation that minimizes the largest of the excesses. If the core is not empty, there is an imputation, $\mathbf{x}$, with $e(\mathbf{x}, S) \le 0$ for all S. Therefore the nucleolus also satisfies $e(\mathbf{x}, S) \le 0$ for all S and so is in the core.

2. A constant-sum game satisfies $v(S)+v(S^c) = v(N)$ for all coalitions, S. The Shapley value for Player 1 in a three person game is

$$\begin{aligned}
\phi_1 &= \frac{1}{3}v(\{1\}) + \frac{1}{6}[v(\{1,2\}) - v(\{2\})] + \frac{1}{6}[v(\{1,3\}) - v(\{3\})] \\
&\quad + \frac{1}{3}[v(N) - v(\{2,3\})] \\
&= \frac{1}{3}v(\{1\}) + \frac{1}{6}[v(\{N\}) - v(\{3\}) - v(\{2\})] \\
&\quad + \frac{1}{6}[v(\{N\}) - v(\{2\}) - v(\{3\})] + \frac{1}{3}v(\{1\}) \\
&= \frac{1}{3}[v(N) + 2v(\{1\}) - v(\{2\}) - v(\{3\})]
\end{aligned}$$

and similarly for the other two players. The excess, $e(\mathbf{x}, \{1\}) = v(\{1\}) - x_1$, is the negative of the excess, $e(\mathbf{x}, \{2,3\}) = v(\{2,3\}) - x_2 - x_3 = v(N) - v(\{1\}) - x_1 - x_2 - x_3 + x_1 = -v(\{1\}) + x_1$, since $x_1 + x_2 + x_3 = v(N)$. Since $x_i \ge v(\{i\})$ for any imputation, the maximum excess is $\max\{x_1 - v(\{1\}), x_2 - v(\{2\}), x_3 - v(\{3\})\}$. This can be made a minimum by making all three terms equal: $x_1 - v(\{1\}) = x_2 - v(\{2\}) = x_3 - v(\{3\})$ which, together with $x_1 + x_2 + x_3 = v(N)$, determines the x_i to be the same as for the Shapley value.

3. (a) The core is the set of vectors (x_1, x_2, x_3) of nonnegative numbers satisfying $x_1 + x_2 + x_3 = 1200$, $x_1 + x_2 \geq 1200$, $x_1 + x_3 \geq 1200$, and $x_2 + x_3 \geq 0$. If nonnegative numbers satisfy $x_1 + x_2 + x_3 = 1200$ and $x_1 + x_2 \geq 1200$, we must have $x_3 = 0$. Similarly, we must have $x_2 = 0$. Therefore $x_1 = 1200$. The core consists of the single point $(1200, 0, 0)$. Since the nucleolus is in the core and the core consists of one point, that point must be the nucleolus.

(b) If the players enter the grand coalition in a random order, Player B can win only if Player A enters first and B second. This happens with probability $1/6$. The amount won is 1200. So $\phi_B = (1/6)1200 = 200$. Similarly, $\phi_C = 200$, and then $\phi_A = 1200 - 200 - 200 = 800$.

(c) The Shapley value seems more reasonable to me. There is a danger that B and C will combine to demand say 1000, (500 each), so some payment to one or the other or both seems reasonable. It does not seem reasonable that A can play B and C against each other to be able to pay practically nothing.

4. The core consists of points $(x, 0, 40 - x)$ for $30 \leq x \leq 40$. We might try $(30, 0, 10)$ as a guess at the nucleolus. In the table below, we see the maximum excess is zero. The excess for either of the coalitions $\{2\}$ and $\{2, 3\}$ cannot be made smaller without making the other larger, so $x_2 = 0$. The excess for $\{1, 3\}$ can be made smaller by increasing x_1. This increases the excess for $\{3\}$. These are equal for $x_1 = 35$. This gives $(35, 0, 5)$ as the nucleolus.

Coalition	Excess	$(30, 0, 10)$	$(35, 0, 5)$
$\{1\}$	$-x_1$	-30	-35
$\{2\}$	$-x_2$	0	0
$\{3\}$	$-x_3$	-10	-5
$\{1, 2\}$	$30 - x_1 - x_2$	0	-5
$\{1, 3\}$	$40 - x_1 - x_3$	0	0
$\{2, 3\}$	$-x_2 - x_3$	-10	-5

The Shapley value is $(25, 5, 10)$. Player 2 receives 5 for just being there (to help Player 1). Since the Shapley value is not in the core and the core is not empty, we know that the nucleolus cannot be equal to the Shapley value. The nucleolus is always in the core when the core is not empty.

5. The Shapley value was found to be $(13/6, 22/6, 1/6)$ so we might try $(2, 3, 1)$ as an initial guess at the nucleolus. The largest excess occurs at either of the coalitions $\{1\}$ and $\{2,3\}$. One cannot be made larger without making the other smaller. So $x_1 = 2$ in the nucleolus. The next largest excess occurs at $\{2\}$ and $\{1,2\}$. These can be made smaller by making x_2 larger. This increases the excess of $\{1,3\}$. These are equal at $x_2 = 3.5$ and $x_3 = .5$. The nucleolus is $(2, 3.5, .5)$.

Coalition	Excess	$(2,3,1)$	$(2,3.5,.5)$
$\{1\}$	$1 - x_1$	-1	-1
$\{2\}$	$-x_2$	-3	-3.5
$\{3\}$	$-4 - x_3$	-5	-4.5
$\{1,2\}$	$2 - x_1 - x_2$	-3	-3.5
$\{1,3\}$	$-1 - x_1 - x_3$	-4	-3.5
$\{2,3\}$	$2 - x_2 - x_3$	-1	-1

6. Since the characteristic function is symmetric in players 2 through n, we may assume the nucleolus is of the form $(x_1, x, x, \ldots, x)$ for some x_1 and x. To be an imputation we must have $x_1 + (n-1)x = v(N) = n$, so $x_1 = n - (n-1)x$. The excess for S not containing 1 is $e(\mathbf{x}, S) = -|S|x$. The excess for S containing 1 is $|S| - x_1 - (|S|-1)x = -(n-|S|) + (n-|S|)x$. The smallest maximum excess is certainly less than 0 (since it is for $x = 0$) so we can see that $0 < x < 1$. The largest excess for S not containing 1 is $-x$ (when $|S| = 1$). The largest excess for S containing 1 is $-(1-x)$ (when $|S| = n-1$). The largest of these two is smallest when x is chosen to make then equal. This gives $x = 1/2$. Hence $x_1 = (n+1)/2$ and the nucleolus is the same as the Shapley value, $((n+1)/2, 1/2, ..., 1/2)$.

7. Player 1 is a dummy, so he gets zero. Players 2, 3 and 4 are symmetric, so they get the same amount, say x. Since the sum must be $v(N) = 1$, we have $3x = 1$ or $x = 1/3$. This gives $(0, 1/3, 1/3, 1/3)$ as the nucleolus.

8. (a) No player can profit without the others so $v(A) = v(B) = v(C) = 0$. Players A and B can build a road for 18 and receive 19 in return so $v(AB) = 19 - 18 = 1$. Similarly, $v(AC) = 0$, $v(BC) = 6$, and $v(ABC) = 8$, the latter requiring a road of cost 19.

(b) The Shapley values are: $\phi_A = (1/3)0 + (1/6)1 + (1/6)0 + (1/3)2 = 5/6$, $\phi_B = (1/3)0 + (1/6)1 + (1/6)6 + (1/3)8 = 23/6$, and $\phi_C = (1/3)0 + (1/6)0 + (1/6)6 + (1/3)7 = 20/6$. To build the road Player A pays $10 - 5/6 = 9 + (1/6)$, Player B pays $9 - (23/6) = 5 + (1/6)$, and Player C pays $8 - (20/6) = 4 + (2/3)$, for a total of 19.

(c) Based on the Shapley value we might try a first guess of $(1, 4, 3)$. We must always have $x_1 + x_2 + x_3 = 8$. The maximum excess occurs for A and BC. One cannot be made smaller without making the other larger, so $x_1 = 1$. The next largest excess occurs for C, so we must make x_3 larger. But as C is made smaller, B and AC get larger. We choose $x_3 = 3.5$ because then all three will be equal. Thus $(1, 3.5, 3.5)$ is the nucleolus. Player A pays 9, Player B pays 5.5 and Player C pays 4.5 for a total of 19.

Coalition	Excess	$(1,4,3)$	$(1,3.5,3.5)$
$\{A\}$	$-x_1$	-1	-1
$\{B\}$	$-x_2$	-4	-3.5
$\{C\}$	$-x_3$	-3	-3.5
$\{A,B\}$	$1 - x_1 - x_2$	-4	-3.5
$\{A,C\}$	$-x_1 - x_3$	-4	-4.5
$\{B,C\}$	$6 - x_2 - x_3$	-1	-1

9. (a) $\phi_1 = 3/2$, $\phi_2 = \phi_3 = \phi_4 = 1/2$.

(b) $\nu_1 = 3/2$, $\nu_2 = \nu_3 = \nu_4 = 1/2$.

(c) If 1 enters the coalition and finds k peasants there, he wins $f(k)$. He is equally likely to enter at any of the positions 1 through $m+1$, so his expected payoff is $\phi_1 = (f(1) + \cdots + f(m))/(m+1)$. The other players are symmetric and so receive equal amounts, $\phi_2 = \cdots = \phi_m = (f(m) - \phi_1)/m$.

(d) Players 2 through $m+1$ are symmetric, so the nucleolus must be of the form $\nu = (f(m) - my, y, y, \ldots, y)$. The largest excess, $e(\nu, S)$ for S not containing 1 occurs at $|S| = 1$ with value $-y$, decreasing in y. The largest excess for S not containing 1 is $\max_{0 \le k < m}[f(k) - f(m) + (m-k)y]$, increasing in y. The maximum excess is minimized when these are equal:

$$\max_{0 \le k < m} [-(f(m) - f(k)) + (m - k - 1)y] = 0$$

These are lines with positive slope starting at a negative value. Therefore this equation is satisfied at the first root, $y = \min_{0 \le k < m}[(f(m) - f(k))/(m - k + 1)]$. Thus, $\nu_2 = \cdots = \nu_{m+1} = \min\{(f(m) - f(k))/(m + 1 - k) : 0 \le k < m\}$, and $\nu_1 = f(m) - m * \nu_2$.

10. (a) Let us write the value in terms of profit. This normalizes the game so that $v(A) = v(B) = v(C) = v(D) = 0$. In addition, we have $v(AB) = v(CD) = 0$, $v(AC) = 4$, $v(AD) = 8$, $v(BC) = 3$ and $v(BD) = 5$.

Finally, $v(ABC) = 4$, $v(ABD) = 8$, $v(ACD) = 8$, $v(BCD) = 5$ and $v(ABCD) = 11$.

(b) $c_a = c_b = c_C = c_D = c_{AB} = c_{CD} = 0$, $c_{AC} = 4$, $c_{AD} = 8$, $c_{BC} = 3$, $c_{BD} = 5$, $c_{ABC} = -3$, $c_{ABD} = -5$, $c_{ACD} = -4$, $c_{BCD} = -3$, $c_{ABCD} = 6$. Therefore, the Shapley value is $\phi_A = 3.5$, $\phi_B = 1.833$, $\phi_C = 1.667$, $\phi_D = 4.0$.

(c) The nucleolus is $(3.5, 1.5, 1.5, 3.5)$. Under the nucleolus, A receives 13.5 for his house, B receives 21.5 for his house, C gets B's house for 21.5 and D gets A's house for 13.5. Under the Shapley value, A receives 13.5 for his house, B receives 21.833 for his house, C pays 21.333 and gets B's house and D pays 14 and gets A's house.

Coalition	Excess	$(3.5, 1.5, 1.5, 4.5)$
$\{A\}$	$-x_1$	-3.5
$\{B\}$	$-x_2$	-1.5
$\{C\}$	$-x_3$	-1.5
$\{D\}$	$-x_4$	-4.5
$\{A, B\}$	$-x_1 - x_2$	-5
$\{A, C\}$	$4 - x_1 - x_3$	-1
$\{A, D\}$	$8 - x_1 - x_4$	0
$\{B, C\}$	$3 - x_2 - x_3$	0
$\{B, D\}$	$5 - x_2 - x_4$	-1
$\{C, D\}$	$-x_3 - x_4$	-6
$\{A, B, C\}$	$x_4 - 7$	-2.5
$\{A, B, D\}$	$x_3 - 3$	-1.5
$\{A, C, D\}$	$x_2 - 3$	-1.5
$\{B, C, D\}$	$x_1 - 6$	-2.5

11. The nucleolus will certainly have (*) $x_1 \geq x_2 \geq \cdots \geq x_n$. Therefore, the maximum excess for a coalition S such that $v(S) = k$ occurs when $S = \{1, \ldots, k\}$, except for $k = 0$, when it occurs at $S = \{n\}$. The problem therefore reduces to minimizing

$$\max\{1 - x_1, 2 - x_1 - x_2, \ldots, (n-1) - x_1 - \cdots - x_{n-1}, -x_n\}$$

subject to (*) and $\sum_{i=1}^{n} x_i = n$. The last two are minimized when they are equal, giving $x_n = 0.5$. Then the previous one is minimized when $x_{n-1} = 0.5$, and so on down to $x_3 = 0.5$. Then the first two are minimized when they are equal, giving $x_2 = 1$, and therefore $x_1 = n/2$.

12. Since the game is symmetric in players 2 through n, we have $\nu_2 = \nu_3 = \cdots = \nu_n$. If $1 \in S$, the biggest excess, $e(\nu, \boldsymbol{S})$, occurs when $S = \{1, 2\}$, in which case, $e(\nu, S) = 1 - \nu_1 - \nu_2$. If $1 \notin S$, the biggest excess (and only positive excess) occurs at $S = \{2, \ldots, n\}$, in which case $e(\nu, S) = 1 - \nu_2 - \cdots - \nu_n = 1 - (n-1)\nu_2$. The largest excess is minimized if these are equal: $1 - \nu_1 - \nu_2 = 1 - (n-1)\nu_2$. Since $\nu_1 + (n-1)\nu_2 = 1$, we may solve to find $\nu_2 = 1/(2n-1)$ and $\nu_1 = n/(2n-1)$.

Solutions to Exercises of Appendix 1

1. Yes. This follows because the utility defined by (1) is linear on $\mathcal{P}^*$ in the sense that $u(\lambda p_1 + (1-\lambda)p_2) = \lambda u(p_1) + (1-\lambda)u(p_2)$. A1 is satisfied because $\lambda p_1 + (1-\lambda)q \preceq \lambda p_2 + (1-\lambda)q$ if and only if $u(\lambda p_1 + (1-\lambda)q) \leq u(\lambda p_2 + (1-\lambda)q)$, if and only if $\lambda u(p_1) + (1-\lambda)u(q) \leq \lambda u(p_2) + (1-\lambda)u(q)$, if and only if $\lambda u(p_1) \leq \lambda u(p_2)$, if and only if $p_1 \preceq p_2$ (since $\lambda > 0$). Similarly for A2, if $u(p_1) < u(p_2)$ and $u(q)$ is any given number, we can find a $\lambda > 0$ sufficiently small so that $u(p_1) < \lambda u(q) + (1-\lambda)u(p_2)$.

2. For $\mathcal{P} = \{P_1, P_2\}$, define a preference on $\mathcal{P}^*$ to be

$$(1-\theta)P_1 + \theta P_2 \prec (1-\theta')P_1 + \theta' P_2 \quad \text{if and only if}$$
$$\theta < \theta' \text{ and } \theta' \geq 1/2.$$

This is the preference of a person who prefers P_2 to P_1 but has no preference between lotteries that give probability less than $1/2$ to P_2. A1 is not satisfied since, taking $q = P_1$, $p_1 = P_1$ and $p_2 = P_2$, we have $p_1 \prec p_2$ but $\lambda p_1 + (1-\lambda)P_1 \simeq \lambda p_2 + (1-\lambda)P_1$ if $\lambda < 1/2$. A2 is still satisfied since if $p_1 \prec p_2$ and q is any other element of $\mathcal{P}^*$, we can take λ sufficiently small so that $p_1 \prec \lambda q + (1-\lambda)p_2$ and $\lambda q + (1-\lambda)p_1 \prec p_2$.

3. For $\mathcal{P} = \{P_1, P_2, P_3\}$, we may use $(\theta_1, \theta_2, \theta_3)$ to represent the element $\theta_1 P_1 + \theta_2 P_2 + \theta_3 P_3$, where $\theta_1 \geq 0$, $\theta_2 \geq 0$, $\theta_3 \geq 0$, and $\theta_1 + \theta_2 + \theta_3 = 1$. Define a preference on $\mathcal{P}^*$ to be

$$(\theta_1, \theta_2, \theta_3) \prec (\theta'_1, \theta'_2, \theta'_3) \quad \text{if and only if} \quad \theta_1 > \theta'_1 \quad \text{or}$$
$$(\theta_1 = \theta'_1 \quad \text{and} \quad \theta_3 < \theta'_3).$$

This is the preference of the person for whom it is the overriding consideration to avoid P_1 (death), but if two lotteries give the same probability to P_1, then the one that gives higher probability to P_3 is preferred. Then clearly A2 is not satisfied for $q = P_1$, $p_1 = P_2$ and $p_2 = P_3$. Checking A1 for $p_1 = (\theta_1, \theta_2, \theta_3)$ and $p_2 = (\theta'_1, \theta'_2, \theta'_3)$ reduces to showing that both sides of (4) are equivalent to $\theta_1 > \theta'_1$ or ($\theta_1 = \theta'_1$ and $\theta_3 < \theta'_3$), for all q and $\lambda > 0$. This is checked by straightforward analysis.

References

Steve Alphern and Shmuel Gal (2003), *The Theory of Search Games and Rendezvous*, International Series in Operations Research and Management Science, Springer.

Robert J. Aumann (1989), *Lectures on Game Theory*, Westview Press, Inc., Boulder, Colorado.

R. J. Aumann and L. S. Shapley (1974), *Values of Non-atomic Games*, Princeton University Press.

Robert J. Aumann and Michael B. Maschler (1995), *Repeated Games of Incomplete Information*, The MIT Press, Cambridge, Mass.

A. Baños (1968), On Pseudo-Games. *Ann. Math. Statist.* **39**, 1932–1945.

V. J. Baston, F. A. Bostock and T. S. Ferguson (1989), The Number Hides Game, *Proc. Amer. Math. Soc.* **107**, 437–447.

John D. Beasley (1990), *The Mathematics of Games*, Oxford University Press.

E. R. Berlekamp, J. H. Conway and R. K. Guy (1982), *Winning Ways for Your Mathematical Plays* (two volumes), Academic Press, London.

Eric Berne (1964), *Games People Play*, Grove Press Inc., New York.

H. Scott Bierman and L. Fernandez (1993), *Game Theory with Economic Applications*, 2nd ed. (1998), Addison-Wesley Publishing Co.

Ken Binmore (1992), *Fun and Games — A Text on Game Theory*, D.C. Heath, Lexington, Mass.

D. Blackwell and M. A. Girshick (1954), *Theory of Games and Statistical Decisions*, John Wiley & Sons, New York.

Émile Borel (1938), Traité du Calcul des Probabilités et ses Applications Volume IV, Fascicule 2, Applications aux jeux des hazard, Gautier-Villars, Paris.

C. L. Bouton (1902), Nim, a game with a complete mathematical theory, *Ann. Math.* **3**, 35–39.

G. W. Brown (1951), "Iterative Solutions of Games by Fictitious Play", in *Activity Analysis of Production and Allocation*, T.C. Koopmans (Ed.), New York: Wiley.

G. S. Call and D. J. Velleman (1993), Pascal's Matrices, *Amer. Math. Monthly* **100**, 372–376.

M. T. Carroll, M. A. Jones and E. K. Rykken (2001), The Wallet Paradox Revisited, *Math. Mag.* **74**, 378–383.

V. Chvátal (1983), *Linear Programming*, W. H. Freeman, New York.

J. H. Conway (1976), *On Numbers and Games*, Academic Press, New York.

W. H. Cutler (1975), An Optimal Strategy for Pot-Limit Poker, *Amer. Math. Monthly* **82**, 368–376.

W. H. Cutler (1976), End-Game Poker, preprint.

Melvin Dresher (1961), *Games of Strategy: Theory and Applications*, Prentice Hall, Inc. N.J.

Melvin Dresher (1962), A Sampling Inspection Problem in Arms Control Agreements: a Game-Theoretic Analysis, *Memorandum RM-2972-ARPA*, The RAND Corporation, Santa Monica, California.

Lester E. Dubins amd Leonard J. Savage (1965), *How to Gamble If You Must: Inequalities for Stochastic Processes*, McGraw-Hill, New York. 2nd ed. (1976) Dover Publications Inc., New York.

J. Eatwell, M. Milgate and P. Newman, Eds. (1987), *The New Palmgrave: Game Theory*, W. W. Norton, New York.

R. J. Evans (1979), Silverman's Game on Intervals, *Amer. Math. Monthly* **86**, 277–281.

H. Everett (1957), Recursive Games, *Contrib. Theor. Games III, Ann. Math. Studies* **39**, Princeton Univ. Press, 47–78.

C. Ferguson and T. Ferguson (2007), The Endgame in Poker, in *Optimal Play: Mathematical Studies of Games and Gambling*, Stuart Ethier and William Eadington (Eds.), Institute for the Study of Gambling and Commercial Gaming, 79–106.

T. S. Ferguson (1967), *Mathematical Statistics — A Decision-Theoretic Approach*, Academic Press, New York.

T. S. Ferguson (1998), Some chip transfer games, *Theoretical Computer Science* **191**, 157–171.

T. S. Ferguson and C. Melolidakis (1997), Last Round Betting, *J. Applied Probability* **34** 974–987.

J. Filar and K. Vrieze (1997), *Competitive Markov Decision Processes*, Springer-Verlag, New York.

Peter C. Fishburn (1988), *Nonlinear Preference and Utility Theory*, John Hopkins University Press, Baltimore.

A. S. Fraenkel (2002), Two-player games on cellular automata, in *More Games of No Chance*, MSRI Publications **42**, R. Nowakowski (Ed.), Cambridge University Press, 279–306.

L. Friedman (1971), Optimal Bluffing Strategies in Poker, *Man. Sci.* **17**, B764–B771.

S. Gal (1974), A Discrete Search Game, *SIAM J. Appl. Math.* **27**, 641–648.

D. Gale (1974), A Curious Nim-type Game, *Amer. Math. Monthly* **81**, 876–879.

M. Gardner (1978), *Mathematical Magic Show*, Vintage Books, Random House, New York.

Andrey Garnaev (2000), *Search Games and Other Applications of Game Theory*, Lecture Notes in Economics and Mathematical Systems **485**, Springer.

Robert Gibbons (1992), *Game Theory for Applied Economists*, Princeton University Press.

P. M. Grundy (1939), Mathematics and games, *Eureka* **2**, 6–8 (reprinted (1964), *Eureka* **27**, 9–11).

Richard K. Guy (1989), *Fair Game*, COMAP Mathematical Exploration Series.

R. K. Guy (1996), Impartial Games, in *Games of No Chance*, MSRI Publications **29**, R. Nowakowski (Ed.), Cambridge University Press, 61–78.

Richard K. Guy and Cedric A. B. Smith (1956), The G-values of Various Games, *Proc. Cambridge Philos. Soc.* **52**, 514–526.

G. A. Heuer and U. Leopold-Wildburger (1991), Balanced Silverman Games on General Discrete Sets, *Lecture Notes in Econ. & Math. Syst.*, No. 365, Springer-Verlag.

R. Isaacs (1955), A Card Game with Bluffing, *Amer. Math. Monthly* **62**, 99–108.

S. M. Johnson (1964), A Search Game, *Advances in Game Theory, Ann. Math. Studies* **52**, Princeton University Press, 39–48.

Samuel Karlin (1959), *Mathematical Methods and Theory in Games, Programming and Economics* (two volumes), Reprinted 1992, Dover Publications Inc., New York.

David M. Kreps (1990), *Game Theory and Economic Modeling*, Oxford University Press.

H. W. Kuhn (1997), *Classics in Game Theory*, Princeton University Press.

Emanuel Lasker (1931), *Brettspiele der Völker*, Berlin, 183–196.

H. W. Lenstra (1977–78), Nim Multiplication, in *Seminaire de Theorie des Nombres, exposé No. 11*, Univ. Bordeaux.

S. C. Little Child and G. Owen (1973), A Simple Expression for the Shapley Value in a Special Case, *Management Sci.* **20** 370–372.

R. D. Luce and H. Raiffa (1957), *Games and Decisions — Introduction and Critical Survey*, reprinted 1989, Dover Publications Inc., New York.

A. P. Maitra and W. D. Sudderth (1996), *Discrete Gambling and Stochastic Games*, Applications of Mathematics **32**, Springer.

J. J. C. McKinsey (1952), *Introduction to the Theory of Games*, McGraw-Hill, New York.

N. Megiddo (1980), On Repeated Games with Incomplete Information Played by Non-Bayesian Players, *Int. J. Game Theory* **9**, 157–167.

N. S. Mendelsohn (1946), A Psychological Game, *Amer. Math. Monthly* **53**, 86–88.

J. Milnor and L. S. Shapley (1957), On Games of Survival, *Contrib. Theor. Games III, Ann. Math. Studies* **39**, Princeton Univ. Press, 15–45.

E. H. Moore (1910), A generalization of a game called nim, *Ann. Math.* **11**, 93–94.

Peter Morris (1994), *Introduction to Game Theory*, Springer-Verlag, New York.

Kent E. Morrison (2010), The Multiplication Game, *Math. Mag.* **83**, 100–110.

Geoffrey Mott-Smith (1954), *Mathematical Puzzles for Beginners and Enthusiasts*, 2nd ed., Dover Publ. Inc.

Roger B. Myerson (1991), *Game Theory — Analysis of Conflict*, Harvard University Press.

Peter C. Ordeshook (1986), *Game Theory and Political Theory*, Cambridge University Press.

B. O'Neill (1982), A Problem of Rights Arbitration from the Talmud, *Math. Social Sci.* **2**, 345–371.

Guillermo Owen (1967), An Elementary Proof of the Minimax Theorem, *Management Sci.* **13**, 765.

Guillermo Owen (1982), *Game Theory*, 2nd ed., Academic Press.

Christos H. Papadimitriou and Kenneth Steiglitz (1982), *Combinatorial Optimization*, reprinted (1998), Dover Publications Inc., New York.

T. Parthasarathy and T. E. S. Raghavan (1971), *Some Topics in Two-Person Games*, Elsevier Publishing Co.

T. E. S. Raghavan, T. S. Ferguson, T. Parthasarathy and O. J. Vrieze, Eds. (1991) *Stochastic Games and Related Topics*, Kluwer Academic Publishers.

J. Robinson (1951), An Iterative Method of Solving a Game, *Ann. Math.* **54**, 296–301.

Robert Rosenthal (1981), Games of Perfect Information, Predatory Pricing and the Chain Store, *J. Economic Theory* **25**, 92–100.

Alvin E. Roth and Marilda A. Oliveira Sotomayor (1990), *Two-Sided Matching — A Study in Game-Theoretic Modeling and Analysis*, Cambridge University Press.

W. H. Ruckle (1983), *Geometric Games and their Applications*, Research Notes in Mathematics 82, Pitman Publishing Inc.

L. J. Savage (1954), *The Foundations of Statistics*, John Wiley & Sons, New York.

D. Schmeidler (1969), The Nucleolus of a Characteristic Function Game, *SIAM Appl. Math.* **17**, 1163–1170.

Fred. Schuh (1952), The Game of Divisions, *Nieuw Tijdschrift voor Wiskunde* **39**, 299–304.

Fred. Schuh (1968), *The Master Book of Mathematical Recreations*, translated from the 1934 edition, Dover Publ. Inc.

A. J. Schwenk (1970), Take-away Games, *Fibonacci Quart.* **8**, 225–234.

L. S. Shapley (1953a), Stochastic Games, *Proc. Nat. Acad. Sci.* **39**, 1095–1100.

L. S. Shapley (1953b), A Value for n-person Games, *Contrib. Ther. Games II, Ann. Math. Studies* **27**, Princeton University Press, 307–317.

L. S. Shapley and R. N. Snow (1950), Basic Solutions of Discrete Games, *Contrib. Theor. Games I, Ann. Math. Studies* **24**, Princeton Univ. Press, 27–35.

Martin Shubik (1982), *Game Theory in the Social Sciences*, The MIT Press.

Martin Shubik (1984), *Game Theory in the Social Sciences — Concepts and Solutions*, The MIT Press, Cambridge Mass.

David L. Silverman (1971), *Your Move*, McGraw-Hill, New York.

John Maynard Smith (1982), *Evolution and the Theory of Games*, Cambridge University Press.

S. Sorin and J. P. Ponssard (1980), The LP Formulation of Finite Zero-Sum Games with Incomplete Information, *Int. J. Game Theory* **9**, 99–105.

Sylvain Sorin (2002), *A First Course on Zero-Sum Repeated Games*, Mathématiques & Applications **37**, Springer.

R. Sprague (1936), Über mathematische Kampfspiele, *Tohoku Math. J.* **41**, 438–444.

R. Sprague (1937), Über zwei Abarten von Nim, *Tohoku Math. J.* **43**, 351–354.

Alan D. Taylor (1995), *Mathematics and Politics — Strategy, Voting, Power and Proof*, Springer-Verlag, New York.

Stef Tijs (2003), *Introduction to Game Theory*, Hindustan Book Agency, India.

John Tukey (1949), A Problem in Strategy, *Econometrica* **17**, 73.

J. von Neumann and O. Morgenstern (1944), *The Theory of Games and Economic Behavior*, Princeton University Press.

M. J. Whinihan (1963), Fibonacci Nim, *Fibonacci Quart.* **1**, 9–13.

John D. Williams (1966), *The Compleat Strategyst*, 2nd ed., McGraw-Hill, New York.

W. A. Wythoff (1907), A Modification of the Game of Nim, *Nieuw Archief voor Wiskunde* **7**, 199–202.

Index

Airport Game, 253
Aliquot Game, 24
Assignment Game, 262

Backward induction, 4
Bankruptcy Game, 256
Barycentric Coordinates, 240
Basic Endgame in Poker, 120, 126, 131, 222
Battle of the Sexes, 190
Behavioral Strategies, 129
Bertrand Duopoly, 200
Best Response, 188
Betting Tree, 167
Brouwer Fixed Point Theorem, 275

Cattle Drive, 260
Centipied Game, 193
Chance Moves, 118
Characteristic Function, 232
Chomp!, 8
Coalitions, 231
Colon Principle, 55
Colonel Blotto Games, 92, 97
Common Knowledge, 193
Complete Information, 124
Concave and Convex Games, 160
Constant Sum Game, 234
Contraction Map, 272
Cooperative and Noncooperative Games, 182
Core, 239
Cost Allocation, 252, 261
Cournot Duopoly, 197

Dawson's Chess, 33
Dim+, 24
Directed Graph, 18, 118
Disjunctive Sum of Games, 27
Dominated Strategies, 71, 72

Ending Condition, 5
Epsilon-Optimal Strategies, 140
Equilibrium Exchange Rate, 221
Essential Games, 238
Evil and Odious Numbers, 39
Excesses of an Imputation, 255
Extensive Form Games, 118, 180

Fermat 2-power, 45
Fibonacci Nim, 9
Fictitious Play, 112
Fixed Threat Point, 221
Fusion Principle, 57

Game Tree, 118
Game of Chicken, 194
Game without a Value, 116
Games, Impartial, 3
Games, Partizan, 3
General-Sum Games, 179
Glove Market, 242
Grundy's Game, 34, 36
Guess It, 136

Imputations, 237
Indifference, 80
Indifference Equations, 170
Individually Rational Equilibria, 194

Information Sets, 121
Inspection Game, 136, 150

Kayles, 32, 34, 35
Kuhn Tree, 121

Lambda-Transfer, 220
Lasker's Nim, 31
Latin Square, 74
Lexographic Order, 257
Linear Program, 105
Lottery, 264, 267
Lower Envelope, 73

Mendelsohn Games, 88
Metric Space, 271
Minimal Excludant or Mex, 19
Minimax Theorem, 65, 153, 158, 159
Misre Play Rule, 5
Mixed Strategy, 64, 153, 188
Mock Turtles, 39, 47
Moore's Nim, 17

N-position, 5
NTU = Nontransferable Utility, 182
Nash Axioms, 216
Nash Bargaining Model, 215
Near-Optimal Strategies, 154
Nim Multiplicatiion, 44
Nim-Sum, 12
Normal Play Rule, 5
Nucleolus, 255
Number Hides Game, 99

One-product Balanced Market, 253
P-position, 5
PSE = Pure Strategic Equilibrium, 186, 192
Pareto Optimality, 210
Payoff Matrix, 68
Perfect Recall, 123
Pivot Method, 108
Poker Models, 120, 167, 169, 222
Preference Relation, 264
Prisoner's Dilemma, 191, 193
Progressively Bounded Graph, 19
Progressively Finite Graph, 22
Pure Strategy, 64

Recursive Games, 140, 151
Rooted Graph, 52
Rugs = Turnablock, 46, 47, 51
Ruler, 40, 46, 51
Rulerette, 50

S-Games, 156
SE = Strategic Equalibrium, 186, 188
SG-values, 20
Safety Levels, 183
Semicontinuous Functions, 159
Shapley Axioms, 244
Shapley Iteration, 149
Silverman Games, 96
Simple Games, 249
Sprague-Grundy Function, 19
Stable Coalitions, 239
Stackelberg Duopoly, 202
Stochastic Games, 146
Subgame Perfect Equilibria, 195
Subtraction Games, 6, 38, 51
Superadditivity, 232
Symmetric Game, 86, 241

TU = Transferable Utility, 182
Tartan Theorem, 46
Terminal position, 5
Turning Corners, 42
Turning Turtles, 15, 46
Twins, 39, 45, 46

Utility Function, 265
Utility Theory, 65

Wallet Game, 173
Weighted Voting Games, 249
Wythoff's Game, 24

Zeckendorf's Theorem, 9
Zero-sum Game, 61